Endicott College Library
BEVERLY, MASSACHUSETTS

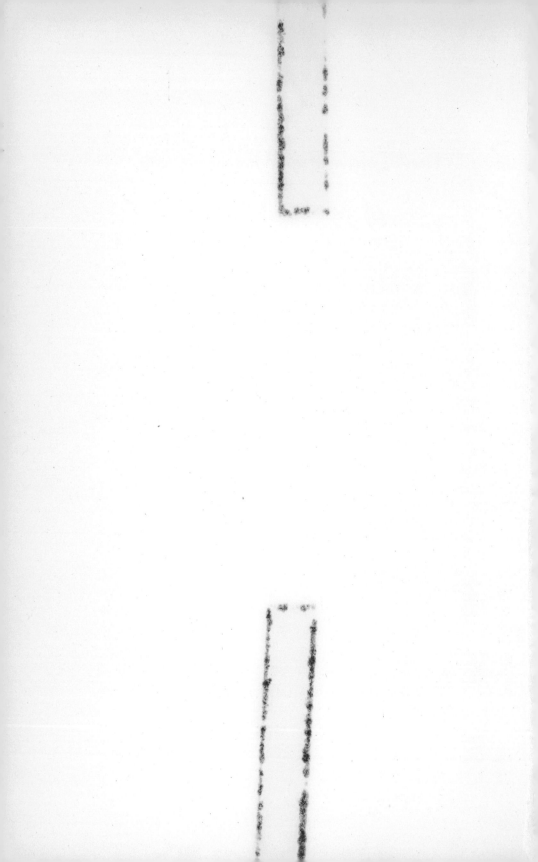

The Chemical Foundations
of Molecular Biology

The Chemical Foundations of Molecular Biology

ROBERT F. STEINER, Ph. D

*Head, Biological Macromolecules Branch
Physical Biochemistry Division
Naval Medical Research Institute
Bethesda, Maryland*

D. VAN NOSTRAND COMPANY, INC.

Princeton, New Jersey

Toronto   New York   London

D. VAN NOSTRAND COMPANY, INC.
120 Alexander St., Princeton, New Jersey (*Principal office*)
24 West 40 St., New York 18, New York

D. VAN NOSTRAND COMPANY, LTD.
358, Kensington High Street, London, W.14, England

D. VAN NOSTRAND COMPANY (Canada), LTD.
25 Hollinger Road, Toronto 16, Canada

COPYRIGHT © 1965, BY
D. VAN NOSTRAND COMPANY, INC.

Published simultaneously in Canada by
D. VAN NOSTRAND COMPANY (Canada), LTD.

No reproduction in any form of this book, in whole or in part (except for brief quotation in critical articles or reviews), may be made without written authorization from the publishers.

PRINTED IN THE UNITED STATES OF AMERICA

This book is dedicated
to my wife and daughters and
to the memory of my parents

Preface

The author of a book on molecular biology is confronted at the onset by a problem in terminology. The term "molecular biology" is of relatively recent origin and appears to be used in two quite different senses. Many authors have chosen to restrict its area of application to the chemistry of genetic processes and the genetic control of metabolism. By this definition the proper domain of molecular biology would be confined largely to the replication of nucleic acids and the directed synthesis of proteins.

I regard this usage as unfortunate and prefer to give the term its most obvious and literal meaning, namely, those aspects of biology which can be described at the molecular level. This definition is broad enough to include the topics of replication and protein synthesis, which retain their central importance.

Various practical considerations, including the desirability of avoiding a book of excessive length and price, have made it necessary to limit the coverage by excluding many subjects which could, by the above definition, be regarded as falling within the discipline of molecular biology. The additional criteria for inclusion have been, first, the *timeliness* of the subject, or the intensity of current scientific interest; and second, the extent to which it is already covered by existing texts.

The book has been written at the advanced undergraduate level. It is hoped that this will not preclude its being of use to less advanced students or to fully qualified scientists from other disciplines. The reader should be acquainted with elementary general chemistry and biology, as well as with organic chemistry, and preferably should have some knowledge of physical chemistry. For parts of Chapter 4 a knowledge of basic calculus is essential.

The author has no illusions that the choice of material will please everyone or satisfy all needs of the aspiring molecular biologist. The intended function of the book is to serve in a manner somewhat analogous to the first stage of a multistage rocket, in that it leaves the later phases of the journey to be completed in more specialized craft.

R. F. S.

Contents

1 INTRODUCTION 1
1-1 The Meaning of Molecular Biology 1
1-2 The Major Biopolymers 2
1-3 Cellular Organization 3
1-4 A Rudimentary Account of Genetic Principles 5
1-5 Substructure of the Gene 21

2 THE AMINO ACIDS 26
2-1 Preliminary Remarks 26
2-2 Optical Activity of Amino Acids 27
2-3 The Neutral Amino Acids 33
2-4 Acidic and Basic Amino Acids 39
2-5 Amino Acid Analogs 41

3 THE CHEMICAL STRUCTURE OF PROTEINS 43
3-1 Preliminary Remarks 43
3-2 The Manner of Linkage of Amino Acids in Proteins 44
3-3 The Sequential Arrangement of Amino Acids in Proteins 47
3-4 Techniques of Sequence Determination 51
3-5 The Primary Structures of Individual Proteins 62
3-6 Variations in Primary Structure 69

4 THE SIZE, SHAPE, AND ELECTRIC CHARGE OF PROTEIN MOLECULES 76
4-1 Introduction 76
4-2 Electron Microscopy 78
4-3 Methods Yielding Average Values of Molecular Weight Directly: Osmotic Pressure 81
4-4 Methods Yielding Average Values of Molecular Weight Directly: Light Scattering 83
4-5 General Aspects of the Hydrodynamic Methods 90
4-6 Methods Yielding a Measure of the Size and Shape, Subject to the Assumption of a Geometrical Model: Viscosity 99

CONTENTS

- 4-7 Methods Yielding a Measure of the Size and Shape, Subject to the Assumption of a Geometrical Model: Fluorescence Polarization — 102
- 4-8 Methods Yielding a Measure of the Size and Shape, Subject to the Assumption of a Geometrical Model: Diffusion — 105
- 4-9 Methods Yielding Both the Molecular Weight and a Measure of the Homogeneity—Ultracentrifugation — 109
- 4-10 Methods Yielding a Measure of the Homogeneity: Electrophoresis — 120

5 THE SPATIAL ORGANIZATION OF PROTEINS — 131
- 5-1 General Remarks — 131
- 5-2 Forces Involved in the Stabilization of Protein Structure — 133
- 5-3 Polypeptide Conformations — 138
- 5-4 The Determination of the Conformation of Polypeptides in the Solid State — 143
- 5-5 The Estimation of Helical Content for Proteins and Polypeptides in Solution — 147
- 5-6 The Helix-Coil Transition for Polypeptides — 152
- 5-7 The Tertiary Structure of Proteins — 157

6 STRUCTURE AND FUNCTION OF CERTAIN IMPORTANT PROTEINS — 162
- 6-1 General Remarks — 162
- 6-2 Structural Proteins: Collagen — 163
- 6-3 Contractile Proteins: Myosin and Actin — 171
- 6-4 Antibodies — 178
- 6-5 Fibrinogen — 188
- 6-6 Plasma Proteins: Serum Albumin — 194
- 6-7 Hemoglobin — 197
- 6-8 Myoglobin — 204

7 CATALYTIC PROTEINS: THE ENZYMES — 208
- 7-1 Introduction — 208
- 7-2 Enzyme Kinetics — 215
- 7-3 Structure and Activity of Trypsin — 229
- 7-4 Structure and Activity of the Chymotrypsins — 232
- 7-5 Acetylcholinesterase — 238
- 7-6 Pancreatic Ribonuclease — 242

8 THE NUCLEOTIDES — 247
- 8-1 General Remarks — 247
- 8-2 The Nucleotide Bases — 250

	8-3	The Nucleosides	256
	8-4	The Nucleotides	259
	8-5	Properties of the Nucleotides	261

9 THE DEOXYRIBONUCLEIC ACIDS 266

9-1	General Remarks	266
9-2	The Primary Structure of DNA	267
9-3	The Secondary Structure of DNA	272
9-4	Solution Properties of DNA	277
9-5	The Atypical DNA of $\phi \times 174$ Virus	279
9-6	The Denaturation of DNA	280
9-7	The Renaturation of Denatured DNA and the Formation of Molecular Hybrids	286
9-8	Biosynthetic DNA	292
9-9	The Biological Replication of DNA	295
9-10	Transforming Principle	299

10 THE RIBONUCLEIC ACIDS AND THE BIOSYNTHESIS OF PROTEINS 307

10-1	General Remarks	307
10-2	The Primary Structure of RNA	310
10-3	The Secondary Structure of RNA	314
10-4	Biosynthetic Polyribonucleotides	322
10-5	The Guided Synthesis of RNA	326
10-6	Protein Biosynthesis	329
10-7	Enzyme Induction	338
10-8	The Linear Assembly of the Polypeptide Chains of Hemoglobin	341

11 THE NUCLEIC ACIDS OF VIRUSES AS CARRIERS OF BIOLOGICAL INFORMATION 347

11-1	General Remarks	347
11-2	The Infectious RNA of Tobacco Mosaic Virus	352
11-3	Infectious RNA's of Other Viruses	360
11-4	The T-Even Bacterial Viruses	361
11-5	The Temperate Bacteriophages	379

12 THE CARBOHYDRATES AND THEIR BIOSYNTHESIS 383

12-1	The Monosaccharides	383
12-2	The Polysaccharides	389
12-3	The Biosynthesis of Carbohydrates	396
12-4	The Biosynthesis of the Polysaccharides	408

CONTENTS

13 ENERGY TRANSFORMATIONS BY BIOLOGICAL SYSTEMS   412
    13-1   Energy Storage and Utilization   412
    13-2   The Coenzymes of Oxidative Metabolism   418
    13-3   The Initial Phase of the Oxidative Metabolism of Carbohydrates   426
    13-4   The Mitochondria   432
    13-5   The TCA Cycle   433
    13-6   Oxidative Phosphorylation   438

APPENDIX A   The Primary Structure of the B-Chain of Insulin   443
APPENDIX B   Basic Thermodynamic Concepts   447
APPENDIX C   Synthesis of Polypeptides   453
APPENDIX D   Biological Oxidation and Reduction   457

INDEX   463

1
Introduction

1-1 THE MEANING OF MOLECULAR BIOLOGY

General remarks. It is desirable to begin this introductory chapter with a definition of terms. The term "molecular biology" would have been meaningless 50 years ago and would have represented little more than a tantalizing aspiration as recently as 25 years ago. Biological systems can be described from many viewpoints which do not begin to approach the molecular level. Indeed, classical biology developed in a manner which paralleled, rather than stemmed from, the contemporaneous evolution of the physical sciences. For example, the science of genetics attained a high degree of success in rendering coherent an enormous body of superficially chaotic data long before more than the dimmest notion was held of the chemical nature of genetic determinants.

Molecular biology represents an effort to account for, in detail, biological events in terms of the established principles of physics and chemistry. Since molecules are the most complex physical entities which can be described in a sharply definitive manner, such an endeavor must necessarily be concerned with the properties of molecules which occur in biological systems. Its hopes for success depend on the assumption, originally largely a matter of faith, that biological systems differ only in complexity from the simple systems which furnished the earlier triumphs of chemistry and that they obey the same laws.

In the broadest sense, molecular biology may be regarded as the study of those aspects of biological systems, or components thereof, which can be described at the molecular level. Since the area for which this is the case tends to enlarge progressively, it can be foreseen that all of biology will ultimately fall into this category. The term is also often used in a more restrictive sense to denote the molecular aspects of genetic mechanisms and the control of metabolic processes by genetic material.

This book will adhere to the more general definition. For didactic purposes, the discussion will be largely centered about the complicated giant molecules, or *biopolymers,* which occur only in living systems.

The material to be presented may be regarded as an effort to answer the following questions:

(1) What is the nature of the principal biopolymers? A complete answer to this question in any particular case would include an account of its gross size and shape, its purely chemical structure, and its geometrical organization.

(2) What is the biological function of each biopolymer and how is function related to structure? A primary objective of molecular biology is the correlation of definite biological processes with chemical events at the molecular level.

(3) How are the biopolymers positioned and oriented in the cells of living systems? This question is of course related to (2).

(4) How are the biopolymers formed from their constituents? In particular, how are the formidable energetic and entropic obstacles to their synthesis overcome by living systems?

The overwhelming volume of the available information naturally makes necessary some degree of selectivity in the choice of topics to be covered. It is inevitable that the selection should be somewhat biased, both by the particular interests of the author and by the fortuitous availability of much more information in some areas than in others.

The balance of this introductory chapter will be devoted primarily to background information which, while not in the domain of molecular biology proper, provides a needed frame of reference for much of the discussion to follow. The presentation here will be elementary, condensed, and oversimplified. It is hoped that the reader will supplement it with the more complete presentations to be found in the references at the end of this chapter.

1-2 THE MAJOR BIOPOLYMERS

Types of biopolymers. *Polysaccharides* are perhaps the simplest of the biopolymers in structure and function. They are polymers of simple sugars, or *monosaccharides,* or derivatives thereof. The biological function of the polysaccharides is relatively restricted and passive. *Starch* and *glycogen* serve as nutritional reservoirs. *Cellulose* has a function which is primarily structural. Other polysaccharides have functions which are relatively limited and specialized.

The *nucleic acids* are linear polymers of *nucleotides,* which consist of a sugar, a nitrogenous base, and phosphoric acid. They occur in two forms called *ribonucleic* and *deoxyribonucleic* acid and universally referred to as RNA and DNA, respectively. These occupy a central position in molecular biology as the directive agents for control of genetic processes and protein synthesis. A major fraction of this book will be centered about their structure and properties.

The *proteins* account for the largest portion of living systems and have the greatest diversity of structure and function. Their basic chemical structure corresponds to linear polymers of *amino acids*, which may be cross-linked. Various schemes for the classification of proteins have been proposed, but their diversity is such that no system is completely satisfactory. With regard to *function*, the proteins may be roughly grouped as follows:

(1) Structural: Proteins of this class are generally rather inert with respect to biochemical processes. They serve to maintain the form and position of organs, as components of container walls for biological fluids, as means of attachment of tissues to the skeleton, and so on.

(2) Contractile: These, by virtue of their property of contractibility, supply to the living organism the capacity of motion or of external work. The muscle proteins are of course the most familiar examples. Their contractile properties appear to reflect their ability to undergo an internal configurational change which results in a change in extension.

(3) Catalytic: The properties of living cells depend upon the continual occurrence of an extraordinary variety of chemical processes. The regulation of the rates of these reactions is the province of a class of proteins with catalytic properties. These are called *enzymes*.

(4) Transport: The transfer of an essential biological factor, from a point in the organism where it becomes available from an external source to a point where it is biochemically utilized, may be accomplished by reversible combination with a carrier protein. In this manner oxygen is transported by hemoglobin.

1-3 CELLULAR ORGANIZATION

Structural elements. The reader should be warned that the description of cellular elements to be presented here will be held to an extreme minimum and should, if possible, be supplemented by the fuller descriptions to be found in the references cited at the end of this chapter (1-5).

The cell is the basic unit of which all living organisms are constructed. The recognition of its existence and of its fundamental importance dates from the nineteenth century and the classical investigations of Schwann and Virchow.

There are limits to the extent with which one can generalize about living cells. Their size and morphology naturally vary considerably according to their function. Nevertheless, there are many features which recur for the cells of almost all higher species. Examination of cells with a microscope, using appropriate stains, reveals the presence of many differentiated bodies, which respond in different ways to stains as a consequence of their varying chemical composition. The cells of living organisms are endowed with their characteristic functions by the presence of hierarchies of biological polymers. The distribution of these within the cell is governed by the localization of particular functions in specific regions.

A basic division of the cell exists between the *nucleus* and the surrounding *cytoplasm*. Both are surrounded and confined by membranes. The cell membrane, which encloses the cytoplasm (Fig. 1-1a), possesses a complex structure and is capable of considerable specificity in regulating the passage of ions and molecules.

Within the cytoplasm, examination with the electron microscope has revealed the presence of numerous preformed bodies. These may be isolated from the nucleus and from each other by differential centrifugation of disintegrated cells. Among the most prominent of these granules are the *chloroplasts* of green plant cells and the *mitochondria*, which occur in both animal and plant cells. The former (Chapter 12) utilize the radiant energy of sunlight to form compounds essential for the synthesis of carbohydrates from water and atmospheric carbon dioxide. The mitochondria fill the role of a "powerhouse" for the cell and supply in usable form the energy required for such processes as the synthesis of proteins and nucleic acids, the transport of essential materials, and the performance of mechanical work. The source of this energy is the oxidative metabolism of foodstuffs, which is mediated by an array of enzymes localized in the mitochondria.

Another organized structural element of the cytoplasm is the *lysosome*. This contains the digestive enzymes which break down biopolymers into their smaller constituents, which can be oxidized by the mitochondrial enzyme system. Rupture of the lysosomal membrane and release of its contents result in a rapid lysis of the cell.

The *centrosomes* (or *centrioles*) become readily visible by light microscopy only when the cell is beginning to divide. They are replicated during mitosis and appear to have a directive function in this process (section 1-4).

The cytoplasm, in addition to the particulate bodies mentioned above, appears to possess a system of internal membranes which are not visible in the ordinary light microscope. These vary greatly in complexity from cell to cell. This membrane system has been called the *endoplasmic reticulum*. A large number of small granules, called *ribosomes*, occur on its surface. These are roughly spherical in shape and contain a high proportion of one form of nucleic acid, ribonucleic acid (RNA). The ribosomes are believed to be the primary sites of protein synthesis within the cell. The membrane may be synthesized by certain mysterious structures called *Golgi bodies*.

A considerable fraction of the contents (other than water) of the cytoplasm fails to sediment in a centrifugal field strong enough to deposit the ribosomes, mitochondria, and other particulate forms. This material is called the *cell sap*, or *soluble fraction*, as opposed to the particulate entities described above. It consists mainly of proteins, fats, RNA, and small molecules which are not integrated into a preformed structural element.

The cytoplasm is separated from the nucleus by the *nuclear membrane*. The nucleus contains the vitally important filaments of *chromatin*, to which the deoxyribonucleic acid (DNA) of the cell is confined. These are rather indistinct in the interval between cell divisions, but become much more compact and visible during division. During this process they appear as distinct bodies called *chromosomes*.

Another prominent preformed granule is the *nucleolus*. This undergoes cyclic changes in appearance, disappearing during cell division and reappearing at the end of division. It is rich in RNA and may be the site of protein and RNA synthesis.

1-4 A RUDIMENTARY ACCOUNT OF GENETIC PRINCIPLES

General remarks. A complete account of the discipline of genetics would require many volumes, and no attempt will be made to do justice to the subject in this book, except in the as yet relatively restricted areas where a discussion on a molecular level is feasible. However, since certain very basic principles provide a necessary background for much of the discussion to follow and since many readers may be without formal training in genetics, a brief account of these will be given here.

Mendelian genetics. The original experiments of Mendel are still the best point of departure for a discussion of classical genetics (6-9). Mendel crossed strains of garden peas (*Pisum sativum*), which differed with respect to certain easily recognizable external characteristics, such as the color of pigmentation of the flowers and color and surface form of the seeds. After protracted trial and error, 22 varieties were selected for further experimentation. All of these were "pure" strains which bred true; that is, each generation of progeny derived from a single strain contained only plants which uniformly resembled their parents. These supplied seven distinct differences of character (or *phenotype*), which was adequate for his purpose (6-9).

It was found that, if two different strains were crossed, that in each case *all* the seedlings of the first filial (F_1) generation uniformly resembled one of the parents with respect to each of the specific characteristics.

For example, a cross of a strain of peas with purple flowers with a strain producing white flowers yielded an F_1 generation whose members produced, not light purple or mottled flowers, but flowers of the same purple color as one of the parental strains.

This held for each of the seven phenotypes studied by Mendel. The type which was manifest in the F_1 generation following the cross, as purple pigmentation in the above instance, was called *dominant* and its latent alternative, which was in this case white flower color, was termed *recessive*. No intermediates appeared in the F_1 generation, whose members uniformly resembled one or other parent.

If a second (F_2) generation was raised from the F_1 generation by self pollination, the recessive types reappeared. Intermediates were again absent. The ratio of dominants to recessives was the same in the case of each of the seven characteristics tested for and was equal to 3. In the above-mentioned case of flower pigmentation, three-fourths of the F_2 progeny were purple and one-fourth white.

The results of Mendel suggested that a *segregation* occurred of the factors responsible for each of the characteristics so that they existed as discrete genetic units which were transmitted in intact form. Each genetic determinant existed in two stable *allelic* forms which retain their identity in the F_1 gencration, to separate and form new combinations in the F_2 generation.

These genetic determinants, or units of heredity, were subsequently termed *genes*. Each individual of the F_1 generation is the product of a fusion of the germ cells of the parents and receives half of its determinants from each parent. If one characteristic is singled out, as flower pigmentation, then each individual of every generation possesses two corresponding genes which control pigmentation.

An organism possessing a double quantity of a single allelic form of a gene is said to be a *homozygote*. This was the case for the original pure strains of Mendel. One which contains two different allelic forms, like the members of the F_1 generation, is called a *heterozygote*.

Thus if the dominant allelic form (for example, purple) is designated as A and the alternative recessive form (for example, white) as a, then the original strains are AA and aa. The members of the F_1 generation are all heterozygotes of the Aa type.

The F_2 generation contains AA, Aa, and aa forms. The ratio of individuals with dominant characteristics to those with recessive characteristics is exactly what would be expected for a separation and *random* recombination of allelic forms. Thus a random reshuffling would produce individuals of types AA, Aa, and aa in a ratio of 1:2:1. Since Aa forms would have dominant characteristics, this distribution would predict a 3:1 ratio of individuals having dominant characteristics (AA or Aa) to those displaying recessive traits (aa).

This pattern was followed for successive generations. For example, if the purple-flowered plants of the F_2 generation, which contained AA and Aa types in the ratio 1:2, were allowed to produce an F_3 generation, it was found that one-third of the F_2 plants (the AA forms) produced only purple progeny, while the remaining two-thirds (the Aa form) produced both purple (Aa or AA) and white (aa) progeny in a 3:1 ratio.

Mendel also made crosses between strains differing in two of his seven selected hereditary traits. The distribution of progeny was in each case that expected for an *independent* assortment of genetic traits. For example, two strains were crossed which differed in the *color* (yellow versus green) and *form* (smooth versus wrinkled) of their seeds. The dominant allelic forms were yellow (A) and smooth (B), while the green (a) and wrinkled (b) types were recessive (Fig. 1-1).

If a pure strain with yellow-smooth seeds ($AABB$) was crossed with a pure strain with green-wrinkled seeds ($aabb$), all members of the F_1 generation had yellow-smooth seeds, since all of the heterozygotes were of the $AaBb$ type and contained both dominant genes.

The F_2 generation showed all four combinations of phenotype in the ratio expected for *random* and *independent* reassortment of genes (Fig. 1-1b). That is, the probability that a given individual would have a smooth or wrinkled seed was entirely independent of whether the seed was yellow or green.

The relative frequencies of occurrence of the four possible combinations were as follows:

Properties	Gene Composition	Relative Frequency
Yellow-smooth	$AaBB$; $AABB$; $AaBb$; $AABb$	9
Green-smooth	$aaBB$; $aaBb$	3
Yellow-wrinkled	$AAbb$; $Aabb$	3
Green-wrinkled	$aabb$	1

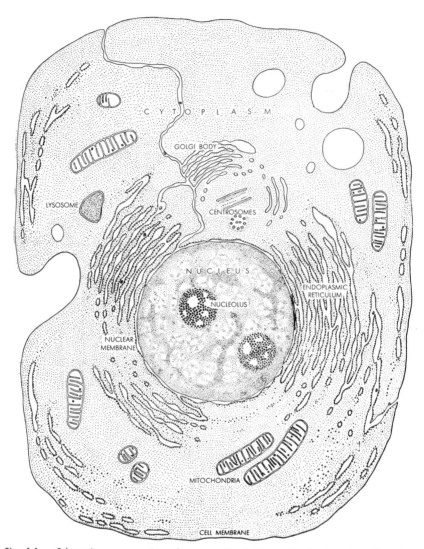

Fig. 1-1a. Schematic representation of a generalized living cell, showing the differentiated structures of the nucleus and cytoplasm. (This illustration is redrawn from J. Brachet, *Sci. Amer.*, 205, 50 [1961].)

The over-all picture of the gene which emerged from these experiments was that of a *stable* and *indivisible* unit which could exist in two or more sharply differentiated forms, to which no intermediate forms existed. Genes controlling different properties were thought to be transmitted entirely independently.

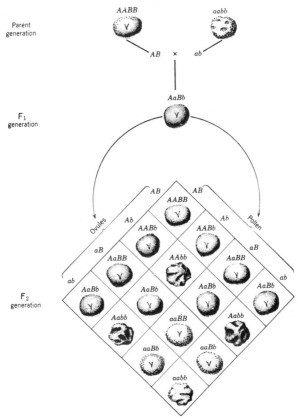

Fig. 1-1b. Illustration of the independent assortment of genes (7). A and a represent the genes for yellow and green colors, respectively, while B and b represent those for smooth and wrinkled seed surfaces.

Part of the success of this simple model in explaining Mendel's results was due to his fortunate choice of inheritable traits. In actuality there are many instances of sets of genes which do not migrate independently and which instead behave as if they were linked together in a common genetic bundle. However, a discussion of linked genes should be prefaced by an account of cell division and chromosomal movements.

Chromosomes and cell division. The existence of visible elongated structures which appear in the nucleus during cell division has been recognized since the last century. In the intervals between cell division, the chromosomes become so indistinct that their persistence was long questioned. However, it is now believed that the chromosomes retain their identity throughout the cycle of the cell, although their form and appearance change.

The number of chromosomes is a well-defined constant for the cells of each species of higher organism. In general, the ordinary *somatic* cells of higher organisms, which are not directly concerned with reproduction of the organism as a whole, possess a double set of chromosomes and are said to be *diploid*. One-half of the double set of chromosomes is derived from each parent. For example, the somatic cells of the human species contain 46 chromosomes consisting of two homologous sets of parental origin, with 23 chromosomes each. The latter are termed *haploid* sets. Each somatic cell contains two haploid sets.

The germ cells, or *gametes*, contain only a *single* haploid set. During the fertilization process a fusion of gametes from each parent occurs to form a diploid, consisting of two homologous haploid sets of paternal and maternal origin.

Replication of the somatic cells occurs by a process called *mitosis* (Fig. 1-2). The central event of mitosis is the exact duplication of each chromosome, followed by a division into two new cells, each of which receives a normal complement of chromosomes.

In terms of Fig. 1-2, the replication of chromosomes occurs at some time between stages 11 (telophase) and (b) to (d) (prophase). By the time the chromosomes become clearly visible they are already differentiated into two distinct strands called *chromatids*. As mitosis continues these become progressively shorter and thicker (stages [e] to [h]). The clear circles of Fig. 1-2 represent the *centromeres* which appear to have an essential function in the mitotic process. The two chromatids, into which each chromosome is split during the earlier phases of mitosis, continue to share a single centromere until the actual separation of chromatids occurs.

During the later stages of mitosis the chromatids separate, each with its own centromere, to produce two diploid sets of chromosomes (stages [i] and [j]). This is followed by the segregation of the two sets and the final division into two cells, each of which has a complete diploid set of chromosomes (stages [j] to [11]).

The gametes (ova and spermatozoa) of the higher organisms are formed by the diploid cells of the reproductive tract. They arise by a special kind of cell division called *meiosis*, in which the normal diploid set of chromosomes is reduced to a single haploid set.

Essentially, meiosis consists of two nuclear divisions occurring in rapid succession, *while the chromosomes divide only once* (Fig. 1-3). The result of this process is the formation of four cells each of which has a haploid number of chromosomes (Fig. 1-3).

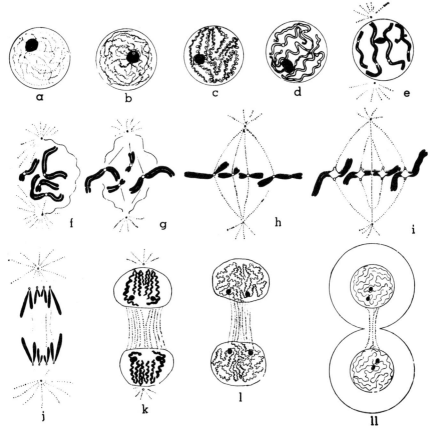

Fig. 1-2. Schematic representation of the consecutive stages of mitosis (2). Stage (a) (*interphase*) represents the state between divisions. In (b), (c), (d), (e) (*prophase*), the chromosomes become visible, each being composed of two chromatids which share a common centromere (clear circle). In (f), (g) (*prometaphase*), the *spindle* (dotted lines) forms, and the nuclear membrane disappears. In (h), (i) (*metaphase*), separation of the chromatids commences. In (j) (*anaphase*), separation is complete. In (k), (l), (II) (*telophase*), the process culminates in the formation of two complete cells.

In Fig. 1-3 the maternal and paternal chromosomes are represented as black and grey, respectively. During the initial stages of the first meiotic division (stages 2 to 4), each paternal chromosome locates the corresponding maternal chromosome and becomes very closely paired with it so that the corresponding structural elements of the two are in complete register. The original diploid set now appears as a set of *double* chromosomes (stage 3). A longitudinal splitting of each member now

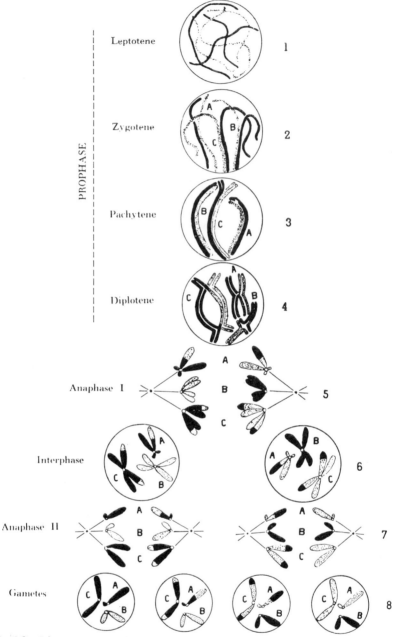

Fig. 1-3. Schematic version of meiosis (2). The maternal chromosomes are shown as grey and the paternal as black. Stages 1-4 represent the initial prophase, in which the paternal and maternal chromosomes pair closely. In the later stages of prophase (3, 4), interchange of chromosomal segments may occur. Note that each paternal and maternal chromosome appears as a pair of chromatids. In stage 5, the paired homologous chromosomes separate, but the chromatids remain joined. In stage 7, the chromatids separate to give rise to a total of four haploid cells.

occurs (stage 4) so that each double chromosome appears as two pairs which may remain linked at one or more points (stage 4). Such linkages are called *chiasmata*. In stage 4 of Fig. 1-3, chromosomes A and C have one and two chiasmata, respectively. A separation of the two sets of chromosome pairs next occurs (stage 5). The chromatids do not separate and continue to share a single centromere.

In the second meiotic division, the chromatids separate *but do not replicate* (stages 7 and 8). Thus the over-all result is the formation of four haploid cells from the original diploid.

In summary, the first meiotic division separates homologous chromosomes which have paired, while the second separates their longitudinal halves.

Let us now center attention upon the events occurring during the initial pairing of homologous maternal and paternal chromosomes (stages 3 to 5). Since the paternal and maternal chromosomes are structurally similar, zones of parallel structure and function will be placed in close juxtaposition during pairing. During the later stages of the first meiotic division, it frequently happens that the paired chromosomes undergo simultaneous breakage at the chiasmata, followed by a rejoining of segments *from different chromosomes* (stages 4 and 5). In this manner an interchange, or crossing-over, of chromosomal segments occurs. This process is of crucial importance to genetics.

Finally, it should be made clear that the detailed mechanism of meiosis varies from species to species and the model presented in Fig. 1-3 should not be regarded as applying literally to all cases. However, the end result—the formation of haploid germ cells—is the same for all.

Linked genes. Subsequent investigations, utilizing different systems, revealed that Mendel's conclusions required modification. In particular, numerous instances were found of genetic determinants which were not transmitted independently, but behaved as though they were linked in some way.

For example, in the case of the sweet pea (*Lathyrus odoratus*), it was found that the genes which determined flower color (purple versus red) and pollen shape (long versus round) did not show random assortment. The experiment was formally analogous to that already described for the crossing of strains of garden peas which differed in two characteristics. In this case, the purple (*A*) and long-pollen (*B*) allelic forms were dominant while the red (*a*) and round-pollen (*b*) forms were recessive.

Upon crossing a purple-long (*AABB*) with a red-round (*aabb*) strain it was found that the F_1 generation (*AaBb*) was uniformly of the purple-long type, as expected. However, the F_2 generation did not show the distribution expected for a random assortment of genes. Instead, the following biased distribution was observed:

Properties	Gene Composition	Relative Frequency
Purple-long	$AaBB$; $AABB$; $AaBb$; $AABb$	15.3
Purple-round	$AAbb$; $Aabb$	1.1
Red-long	$aaBB$; $aaBb$	1.2
Red-round	$aabb$	3.8

It was clear that the distribution was not of the original Mendelian type. The behavior was qualitatively that expected if the A and B allelic forms originally present in the same gamete tended to remain linked in subsequent crossings, as did also the a and b forms.

The linked character of many sets of genes came to be recognized largely as a consequence of the work of T. H. Morgan with the fruit fly, *Drosophila melanogaster*. The somatic cells of this organism contain only four different pairs of homologous chromosomes, which grow to enormous size in the case of the cells of the salivary glands.

By crossing strains of flies having different hereditary features, it was found that many of these were transmitted independently and behaved in accordance with the original principles of Mendelian genetics. However, it was also observed that many traits did not show independent transfer but were generally transmitted as a unit from parents to progeny, suggesting that they were somehow linked in a common genetic bundle.

By performing a large number of crossing experiments, Morgan was able to identify a total of four groups of linked genes, which is equal to the number of different chromosomes. Morgan therefore identified the chromosomes as the carriers of genetic determinants. Each chromosome was postulated to contain a collection of linked genes which tend to be transmitted as a unit.

Subsequent studies consistently confirmed and refined this model. Operationally, a chromosome may be regarded as a linear array of genes. The homologous chromosomes of diploid cells have the same linear distribution, so that corresponding segments contain genes of the same type.

It was also found by Morgan that the linked groups of genes were not completely indivisible and frequently showed independent assortment. This separation of genes occurring on the same chromosome was explained in terms of the interchange of chromosomal segments occurring during the initial stage of meiosis. Since the pairing of the maternal and paternal chromosomes is such as to maintain similar segments in register, the crossing-over which frequently occurs results in an *exchange* of a sequence of allelic genes.

If the formation of chiasmata and subsequent breakage and interchange occur with roughly equal probability at all points of the chromosome, then the probability that two genes located on the same chromosome will undergo separation, or recombination, is directly proportional to the linear distance between them. In other words, the probability of a chiasma occurring between two distant genes is greater than for two genes placed close to one another. In this manner, the frequency of recombination of two linked genes is a direct measure of their linear separation. This general conclusion has been verified by an enormous volume of data, although quantitative deviations from the rule sometimes occur.

This fact has been utilized to construct linear maps of chromosomes showing the mutual physical location of the genes which have been identified within them (Fig. 1-4). The correctness of such maps may be verified by checks for internal consistency. For example, if genes A and B are x units apart, while B and C are y units apart, then A and C should be about $x + y$ units apart. The units used here are *units of recombination* and are equal to the fraction of the progeny from any particular cross which differ from either parental genotype. It should be emphasized that the "distances" on genetic maps correspond only roughly to the actual physical parameters of the chromosome. This is a consequence of both the occurrence of multiple crossover and the varying susceptibility of the chromosome to breakage at different points along its length. Thus genetic maps are accurate only with respect to the relative *order* of genes.

Mutations. A gene can be recognized by crossing experiments only because it can be modified in a discontinuous manner to produce an abnormal

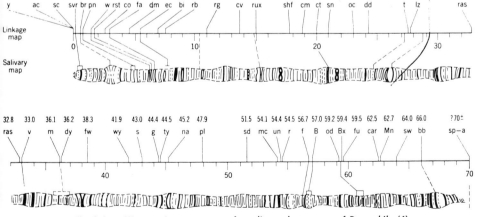

Fig. 1-4a. Microscopic appearance of a salivary chromosome of *Drosophila* (6).

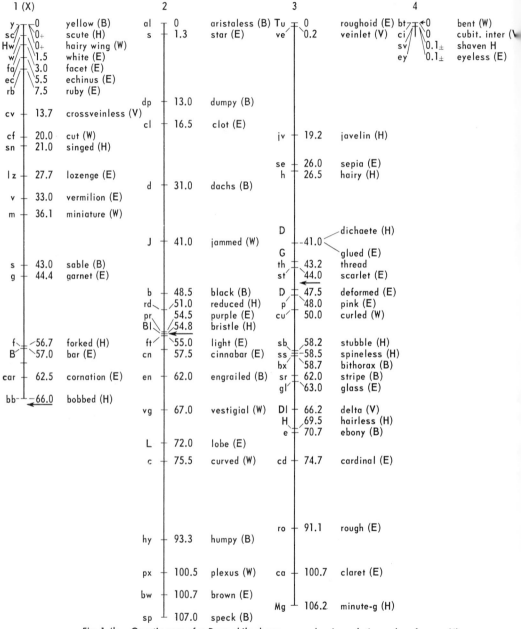

Fig. 1-4b. Genetic map of a *Drosophila* chromosome, showing relative order of genes (6).

allelic form, which determines some phenotypic characteristic. Such changes are referred to as *mutations*. They may arise naturally, as in the garden peas of Mendel, or they may be induced by exposure to ionizing radiation or chemical *mutagenic* agents. All allelic forms of a single gene arise from some mutation. In most cases this results in the loss of some function, such as the capacity to produce a particular enzyme. Such mutant genes are generally recessive.

It is only in recent times that particular mutations have come to be correlated with definite chemical events within the gene. This has provided a valuable probe of the fine structure of genes, as will be discussed in subsequent chapters.

In the case of the giant chromosomes of *Drosophila* it has been possible to correlate specific mutations induced by X-rays with visible morphological changes in the pertinent region of the chromosome. This has furnished both a confirmation of the general concept of chromosomes as gene carriers and of the validity of genetic mapping in particular instances.

Haploid organisms. The discussion of the preceding sections has been centered about higher organisms, whose somatic cells are diploid and which reproduce by a sexual mechanism. The bulk of classical genetics has been concerned with this kind of system. However, most of the more exciting current investigations have utilized bacteria and viruses, the relative simplicity of whose genetic natures has permitted a much closer approach to a description on a molecular level. Indeed, the examples to be discussed in detail in this book have been drawn from these categories.

While reproduction of a sexual nature is known to occur for some bacteria, the general mode of replication for bacteria and viruses is nonsexual.

In the bacterial cell the nucleus is not separated from the cytoplasm by a nuclear membrane. Irrespective of the stage of cell division, the nuclear region appears to consist of a system of very fine fibrils which are largely composed of DNA. There is no sign of the organization of this material into particulate entities resembling the chromosomes of higher organisms. The characteristic phases of mitosis are not observed. Instead, the entire nuclear body appears to split into two subunits without any marked alteration of form.

Nevertheless, bacteria have a genetic stability which is comparable to that of the higher organisms. The mutation rate is of similar magnitude. A bacterial mutation is generally expressed immediately, with no masking by dominance. Thus the genetic material of bacteria behaves in a manner which is basically similar to that of higher organisms. The absence of dominance indicates that these organisms are haploid; that is, only a single copy of their genetic information is present instead of a double set of chromosomes.

In the case of some strains of one bacterium, *Escherichia coli* (*E. coli*), there is relatively complete information about the genetic constitution. All the genes of *E. coli* are linearly arranged in a *single* linkage group, which is equivalent to a single chromosome. While information is less complete for other bacteria, there is reason to believe that this situation may be of general occurrence.

Viruses likewise appear to be haploid organisms. In the case of the T-even bacteriophages, whose genetic system is relatively well understood, all genes are linked in a single "chromosome" (Chapter 11). As will be described in later chapters, the relative simplicity of such haploid systems has greatly facilitated the studies which are beginning to make "molecular genetics" a reality.

Bacterial conjugation. Despite their haploid character, certain forms of bacteria can undergo a type of genetic interchange which is rather analogous to the sexual reproduction which occurs for higher organisms (12). The initial observations of this phenomenon were made by Lederberg and Tatum upon *E. coli*. Two different mutant strains were developed which differed from each other with respect to four genes. The mutations resulted in a loss of the ability to synthesize, and hence an absolute requirement for the external supply of, certain growth factors. One strain required the compounds threonine (T) and leucine (L); the other required methionine (M) and biotin (B). The two strains were designated by the symbols ($B^- M^- T^+ L^+$) and ($B^+ M^+ T^- L^-$). The minus sign indicates that the compound is not synthesized and must be present in the nutrient medium for growth to occur; the plus sign indicates that it is synthesized and need not be supplied.

Upon mixing the two strains in a common culture medium lacking all four growth factors, numerous colonies of growing bacteria appeared, despite the fact that neither strain alone could grow on this medium. Upon isolation, the new cells proved to have a genetic constitution corresponding to ($B^+ M^+ T^+ L^+$).

It was evident that some form of genetic interchange had occurred. This process was designated as *conjugation*. If cultures of the two strains were separated by a porous plug which permitted the passage of diffusible chemicals but prevented the movement of bacteria, no conjugation occurred, indicating that direct contact of cells was necessary.

It was subsequently discovered that there are two mating types in *E. coli*, one of which acts only as a genetic donor, or "male," and the other as a genetic recipient, or "female." A male form is detected by its capacity to retain its fertility even though killed by antibiotics, while the female must remain viable to retain its fertility. Thus the sole function of the male is to transfer its genetic determinant; the female must possess not only genetic material but a complete functional cell if the new cell is to develop.

The property of maleness is conferred by a *sex factor* or *F agent* (for "fertility"). This is present in the cell as an infectious, virus-like agent. During the conjugation of bacteria, many females are infected by the F agent and converted to males. The F agent is transferred only by direct contact and is never released into solution as an independent unit. Male cells, which possess the F agent, are called F^+; female cells, which lack it, are called F^-.

While the efficiency of transfer of the F agent is very high, that of the transfer of genetic information is lower by a factor of 10^4. Thus genetic interchange between F^+ and F^- forms is a rather infrequent event. However, there exist mutant male strains which transfer their genetic determinants with a much higher frequency. These forms are designated as Hfr (for "high frequency of recombination").

The transfer of genetic material from an F^+ to an F^- always results in a recombinant of the F^+ type. Thus the F agent is transferred along with the genetic determinant. The recombinants isolated from an Hfr × F^- cross are generally F^-, indicating that Hfr males usually do not transmit an F agent.

Thus Hfr strains differ from F^+ strains in two respects: (1) The sex factor is no longer present as an infectious agent. (2) The efficiency of chromosomal transfer increases by a thousandfold or more.

The transmission of genetic information from a donor to a recipient cell consists of the physical transfer of the donor chromosome, which enters the F^- cell (12). The transfer process is time-dependent and usually does not proceed to completion, being interrupted by breakage of the chromosome before conjugation is complete.

Conjugation may be artificially interrupted by agitation of the liquid culture in a Waring blender. The strong shearing stress accompanying the blending process is sufficient to rupture all chromosomes which are partially transferred.

If a single Hfr mutant containing a large number of identifiable genes is allowed to conjugate with an F^- strain and the process is interrupted by blending after varying intervals, it is found that (1) the fraction of the recognizable genes which are transferred increases with time, and that (2) the entry of particular genes always occurs in a definite sequence. This is a consequence of the mechanism of conjugation, which consists of the injection of a linear chromosomal thread into the F^- cell, so that the genes enter in linear order. The same end makes the initial penetration in each case.

The conjugation process may be monitored by transferring the interrupted culture to a series of different nutrient mediums, each of which selectively tests for one or more particular genes. For example, the presence of a T^+ gene might be tested for by transfer to a medium lacking threonine.

If genes A, B, C, D occur in that order in the donor chromosome, then if conjugation is interrupted after a given interval of time, the fraction of recombinants which contain each gene decreases from A to D. The order of genetic marker transfer may be determined by either the time of entry of each marker, or by its recombination frequency. (The recombination frequency is equal to the ratio of the number of recombinants which receive the marker to the number of Hfr cells which have mated.)

For conjugation involving a single Hfr mutant the order of gene entry is quite characteristic of the mutant. The order of recombination frequencies may be used to construct the genetic map of *E. coli*. A large number of such studies have demonstrated clearly that the genes of *E. coli* occur as a *single* linkage group or chromosome.

A comparison of the genetic maps of different Hfr mutants revealed many puzzling anomalies which could be resolved only by postulating that the F^+ strain of *E. coli* contains a single chromosome, *which consists of a closed circle of genetic markers*. The $F^+ \rightarrow$ Hfr mutation is accompanied by a rupture of the circle at one point. Entry of the chromosome into an F^- cell begins at one end of the broken chromosome. The genetic markers enter in a definite linear sequence. The sequence is different for different Hfr mutants because the break has occurred at varying points. Many different Hfr mutants have been analyzed in the above manner, and in each case the order of gene transfer is consistent with the rupture of the same circular map at a point specific for the given mutant. The complete genetic map includes hundreds of recognizable genes.

It has already been mentioned that the recombinants arising from Hfr $\times$ F^- crosses usually are in the F^- state, having failed to obtain a sex factor particle from the donor. Detailed studies have shown that in each case the sex factor is attached to one terminus of the broken chromosome of the Hfr mutant. This end is invariably opposite to the one which initially enters the F^- cell. Thus the end bearing the sex factor is the last part of the chromosome to enter the F^- cell and hence is very likely to be lost before conjugation is complete.

The basic event occurring in the $F^+ \rightarrow$ Hfr mutation is the attachment of the sex factor at a particular point of the circular *E. coli* chromosome, accompanied by the breakage of the chromosome at this point. This results in a greatly enhanced efficiency of transfer of the chromosome to an F^- cell. The sex factor is henceforth only transferable as a chromosomal attachment, being the last identifiable marker to enter the F^- cell.

1-5: SUBSTRUCTURE OF THE GENE

Functional units within genes. Classical Mendelian genetics is based upon the concept that a gene is an *indivisible* unit which retains its identity through successive replications and is only transferred by crossover in intact form. This point of view persisted for many decades and sufficed to bring an enormous volume of information into coherence.

The classical gene is basically an operational concept, which is difficult to adapt to the purposes of molecular genetics. The most intense activity of modern genetics has centered about efforts to extend the resolving power of genetic analysis to a point where the *fine structure* of the gene can be examined and correlated with the molecular nature of chromosomal material.

Early evidence as to the existence of subunits of the gene was indirect and stemmed partially from the very wide discrepancies, amounting to orders of magnitude, between the estimates of gene size obtained from different criteria. An average *upper limit* can be assigned to the length by simply dividing the known length of a chromosomal segment by the number of genes located within it. An alternative, altogether different, approach is based upon the frequency with which mutations of a particular gene are produced by ionizing radiation. By suitable analysis it is possible to estimate the effective target size.

The differences in gene size obtained by the two techniques were often outside their (ample) uncertainties, that from mutation frequency being smaller by factors as high as 10 to 100. This suggested that different aspects of gene structure were being measured. Size estimates obtained by the standard methods of genetic mapping presumably correspond to the complete gene as a genetic *functional* unit, while estimates from mutation frequency may reflect the existence of numerous sensitive loci within its structural organization.

Evidence of another kind was accumulated which was difficult to reconcile with the notion of the indivisibility of the gene. Returning to the case of two linked genes, whose dominant and recessive allelic forms are designated by A, B and a, b, respectively, a double heterozygote of the type $Aa\ Bb$ would have the characteristics of the dominant forms, since a dominant gene of each type is present.

If genes present in the same chromosome are indicated by being placed on the same side of a solidus, then such a double heterozygote could occur in the alternative forms $A\ b/a\ B$ or AB/ab, depending upon whether the dominant forms are on the same, or different, chromosomes. According to the principles of classical genetics, both arrangements should function in the same way.

Allelic forms of a single gene occur in corresponding positions of homologous chromosomes. They determine a single gross property, presumably because they influence a single primary function of the gene. When two recessive allelic forms, a and a', which arise from *different* mutations in the same gene, occur on homologous chromosomes in an a/a' arrangement, the organism is usually not normal and has mutant characteristics of the a or a' type. This is of course the basis for the classical test for allelism.

However, the test occasionally fails in one of two ways. First, instances involving *closely linked* genes have arisen for which the *cis* heterozygote Ab/aB has normal characteristics, but the *trans* heterozygote AB/ab, in which both dominant forms are present in the same chromosome, is unable to function normally and does not confer dominant characteristics. This has been explained in terms of a "position effect" whereby the mutual location of the two genes is important for their cooperative action.

Deviations of a second kind are known in which different allelic forms of a single gene *complement* each other in heterozygotes, so that a/a' functions normally (has the characteristics of the dominant A form), while a/a and a'/a' fail to carry out a function dependent on the A gene. These are almost impossible to explain in terms of the modification of a *single* functional unit which is identified with the complete gene.

Another type of evidence has arisen which is inconsistent with the view that the gene as a functional unit is identical with the gene as a unit of recombination. Thus phenomena have been observed which have been interpreted in terms of an *intragenic* recombination. This process has been designated as *transmutation*.

If a number of distinguishable mutants of a single gene are available from independent mutations, it is possible for two mutant forms to be located in different members of a pair of homologous chromosomes. It has been found that in some cases reversion to the normal form occurs with a frequency which is too high to be accounted for by reverse mutations. Thus an a/a' heterozygote may reform the normal A allele during meiosis.

This result may be explained by postulating that the a and a' mutants represent modification of different, *nonoverlapping* sites along the length of the gene and that a normal A-type gene can be reformed by the *intragenic* recombination of normal ends.

More explicitly, if A-1 and A-2 represent the two normal segments of a linear gene, and a-1 and a-2 the two corresponding altered segments, then the initial situation prior to transmutation may be represented by A-1 a-2/a-1 A-2. Intragenic crossover results in formation of A-1 A-2, which is equivalent to a normal A-type allele. Such a process is consistent with the finite probability of occurrence of a break between the normal and modified segments of a pair of genes during meiosis, followed by the "stitching together" of a set of complementary normal segments.

Thus an impressive amount of data has appeared which is difficult to reconcile with the classical picture of the gene as an irreducible unit. Indeed, the existence of a fine structure of the gene, in which separate mutational events may occur at different loci, is now universally accepted. In the case of T-2 bacteriophage, which will be considered in detail in Chapter 11, it has been possible to construct a linear map of mutation sites within a *single* gene.

The cistron concept. With the recognition that the units of function, mutation, and recombination are not identical, it became clear that the classical gene which served as the unit for all three was no longer adequate. This has led to the introduction of separate units.

A functional unit can be defined genetically, without recourse to biochemical information, by means of the cis-trans test. If two mutant alleles in the trans configuration (present on different chromosomes) produce a normal phenotype, the mutations must occur in different functional units of the gene, which are mutually complementary. As a control the cis configuration is also examined. It should of course produce a normal phenotype (Fig. 1-5).

A group of noncomplementary mutants is always found to occupy a limited segment of the genetic map. Such a segment, which corresponds to a function which is unitary by the cis-trans test, has been designated as a *cistron*. In order for a normal phenotype to be produced, a complete set of normal cistrons must be present. These need not be in the same chromosome.

Thus the gene, regarded as a *functional* unit, often behaves as if it consisted of two or more *dissociable* subunits which need not be physically attached to produce their complementary action.

Specific units of recombination and mutation have also been proposed. A unit of recombination has been defined as the smallest element in the linear map which is interchangeable, but not divisible, by genetic

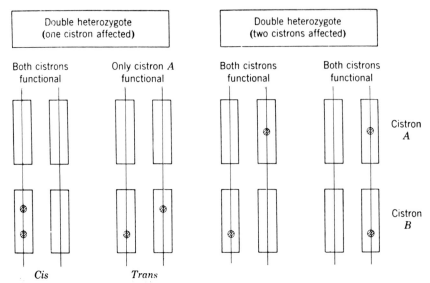

Fig. 1-5. Schematic version of cis and trans double mutations occurring in the same and in different cistrons (6).

recombination. One such element is referred to as a *recon*. The unit of mutation, the *muton*, has been defined as the smallest element which, when modified, produces a mutant form of the organism.

GENERAL REFERENCES

Structural elements of the cell:

1. *The Cell*, J. Bracket and A. Mirsky, Academic Press, New York (1959).
2. *General Cytology*, E. de Robertis, W. Nowinski, and F. Saez, Saunders, Philadelphia (1948).
3. *The Bacterial Cell*, R. Dubos, Harvard University Press, Cambridge, Mass. (1949).
4. *Frontiers in Cytology*, S. Palay, Yale University Press, New Haven (1958).
5. *Biochemical Cytology*, J. Brachet, Academic Press, New York (1957).

Genetics:

6. *The Molecular Basis of Evolution*. C. Anfinsen, Wiley, New York (1959).
7. *Evolution, Genetics, and Man*, T. Dobzhansky, Wiley, New York (1955).
8. *The Chemical Basis of Heredity*, W. McElroy and B. Glass, Johns Hopkins, Baltimore (1957).
9. *The Elements of Genetics*, C. Darlington and R. Mather, Allen and Unwin, London (1949).

Cell division:

10. D. Mazia in reference 1, vol. 3, p. 77.
11. "Second Conference on the Mechanism of Cell Division," *Annals of the N.Y. Academy of Science*, **90,** article 2, p. 345 (1960).

Bacterial conjugation:

12. *Sexuality and the Genetics of Bacteria*, Academic Press, New York (1961).

2

The Amino Acids

2-1 PRELIMINARY REMARKS

All natural proteins are built up from small subunits called *amino acids* and can be converted quantitatively into their constituent amino acids by exhaustive hydrolysis in the presence of high concentrations of acid or alkali (1-5). The number of different amino acids which occur in proteins is not large. Only about 20 occur with any frequency. This is a sufficient number to endow natural proteins with a fantastic diversity of properties and functions.

As the name implies, all the amino acids found in proteins possess (in the free state) an acidic carboxyl (COOH) group and either an amino (NH_2) or an imino (NH) group. The imino group replaces the amino group only in the cases of proline and hydroxyproline. In aqueous solution at neutral pH the carboxyl group is ionized and the amino protonated, so that the amino acid exists as a *dipolar ion*, for which a generalized structural formula can be written of the form $H_3{}^+N$—CH—$CO\overline{O}$. The symbol R represents the *side-chain*, to which the
$\quad\quad\quad\quad\quad |$
$\quad\quad\quad\quad\quad R$
different amino acids owe their chemical individuality. The carbon atom to which the amino and carboxyl groups are attached is known as the α-carbon. The attached groups are referred to as the α-*amino* and α-*carboxyl* groups. This class of amino acids is therefore called the α-*amino acids*.

The side-chains of the natural amino acids are shown in Fig. 2-1. On the basis of the nature of their side-chains, the amino acids can be grouped into three categories. Those whose side-chains contain a carboxyl group are referred to as *acidic*, while those possessing an amino or guanidinium

(—NH—C—NH$_2$) group are designated as *basic*. The balance is usually
$\parallel$
NH$_2^+$
referred to as neutral, although in several instances the side-chain contains an ionizable group.

In natural proteins the amino acids are mutually linked by *peptide bonds* to form extended chains. The peptide linkage is of the amide (CONH) type and arises by the elimination of water between an α-amino and α-carboxyl of two amino acids. Hydrolysis in the presence of high levels (1M to 10M) of acid or alkali results in the rupture of all linkages and the liberation of the free amino acids.

(2-1)
$$\begin{array}{c} \text{H H} \quad \overset{O}{\underset{\parallel}{}} \quad \text{H H} \quad \overset{O}{\underset{\parallel}{}} \quad \text{H} \\ \text{—N—C—C—N—C—C—N—} \\ | | \\ R_1 R_2 \end{array} + H_2O \rightarrow R_1 - \underset{H}{\overset{NH_3^+}{\underset{|}{C}}} - CO\bar{O} + R_2 - \underset{H}{\overset{NH_3^+}{\underset{|}{C}}} - CO\bar{O}$$

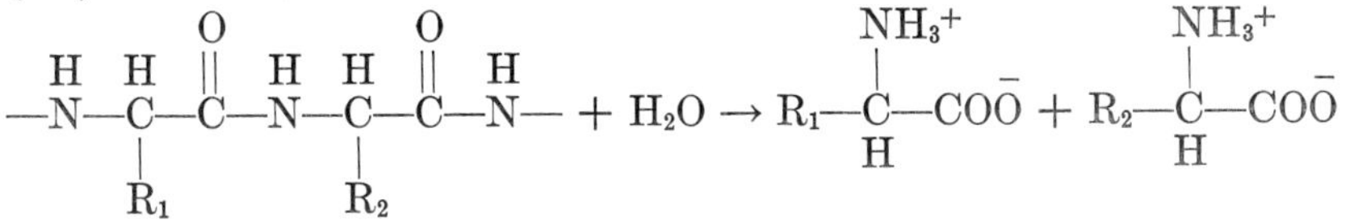

The resultant mixture of amino acids can be separated into its constituents by the chromatographic methods to be discussed in section 3-4.

2-2 OPTICAL ACTIVITY OF AMINO ACIDS (6, 7)

Many of the properties of ordinary visible light may be explained by regarding it as a collection of electromagnetic waves whose electric vectors are normal to the direction of propagation but form no particular azimuthal angle with this direction. Certain optical systems, such as the Nicol prism, have the property of transmitting only light whose direction of vibration lies in a single plane. The light emerging from such an optical element has an electric vector which oscillates in a single plane and is said to be *plane-polarized*. In contrast, unpolarized light consists of a collection of all possible orientations of the electric vector (Fig. 2-2).

One of the most important properties of many organic molecules is their ability to rotate the plane of polarization of a beam of polarized light. The parameter normally determined is the *specific rotation* [α], which is defined as the rotation in degrees of a solution containing 1 gm of material in 1 ml of solution, as examined in a 1-decimeter (10 cm) polarimeter tube (Fig. 2-2).

(2-2) $\quad [\alpha] = \dfrac{\text{observed rotation in degrees}}{\text{length of tube in dm} \times \text{concentration (gm/ml)}}$

STRUCTURE OF SIDE CHAINS (R) ON AMINO ACIDS FOUND IN PROTEINS

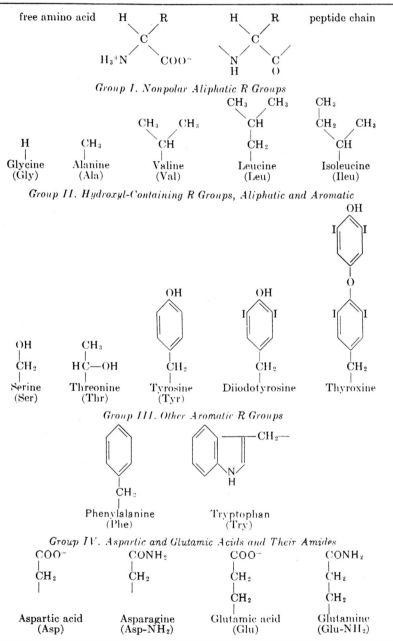

Fig. 2-1. Structures of the amino acids. In the cases of proline and hydroxyproline (Group VII) the complete structures are shown. For the other amino acids only the structures of the side chains are shown.

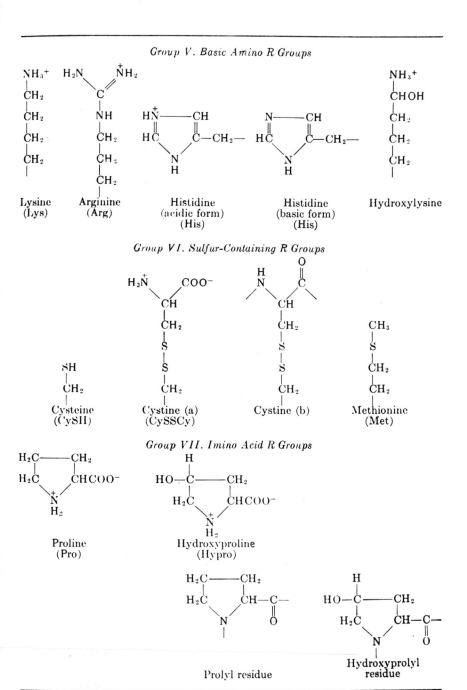

Fig. 2-1. (cont.)

In practice, determinations have often been made using the "D" line of sodium, whose wave-length is 589 mμ. In this case the specific rotation is designated as $[\alpha]_D$. Specific rotations measured at the standard temperature of 20° C, using the "D" line, are designated as $[\alpha]_D^{20}$.

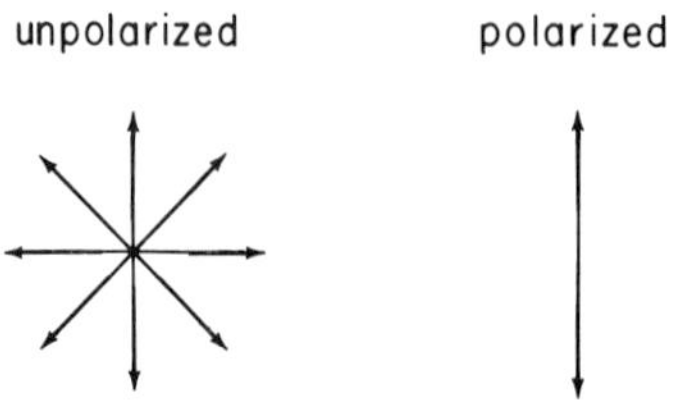

unpolarized   polarized

Fig. 2-2a. Schematic representation of unpolarized and polarized light.

The property of optical activity is conferred by the presence of molecular asymmetry. An asymmetric molecule can exist in two forms, whose three-dimensional models cannot be superimposed although the structural arrangement of atoms is exactly the same. Such optical isomers of a single asymmetric center are normally identical in all physical properties except for the direction in which they rotate the plane of polarization of light. Their mutual relationship is that of mirror images or *enantiomorphs*. The specific rotations of two optical isomers are equal in magnitude but opposite in sign. The isomer which rotates the plane of polarization to the right is referred to as *dextro-rotatory* and is said to have a positive rotation. The isomer which rotates the plane of polarization to the left is said to be *laevo-rotatory* and to have a negative rotation. The *dextro-* and *laevo-*forms are often referred to as the d- and l-forms, respectively.

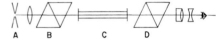

Fig. 2-2b. Diagram of a simple polarimeter. A is a slit; B is a polarizing Nicol prism; C is the cell containing the solution; D is the analyzing Nicol prism. The measurement consists essentially of determining the angle by which D must be rotated with respect to B to permit maximum transmission of light. This angle is equal to the rotation of the plane of polarization by the solution in C. In practice, the precision of the measurement is increased by supplementary optical or photoelectric devices.

In the vast majority of cases molecular asymmetry arises through the presence of an *asymmetric carbon atom* with four different substituents, as illustrated below.

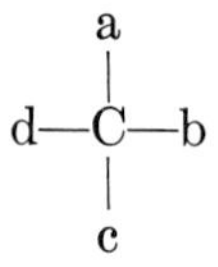

The occurrence of an asymmetric carbon suffices to confer the property of molecular asymmetry upon the molecule as a whole.

The existence of optical isomerism thus reflects the presence of *spatial* or *configurational* isomerism. The determination of the absolute configuration of an optically active organic molecule is a formidable task. It should be noted that the configuration cannot be predicted from the sign of the rotation.

The spatial configuration of the groups attached to an asymmetric carbon is often represented schematically as shown in Fig. 2-3. The illustration represents a tetrahedron, with a carbon atom in the center and its four bonds directed to the four corners. *If the corner of the tetrahedron with group c is pointed toward the observer, then the other three groups—d, a, b— will appear in that order as one proceeds in the clockwise direction.*

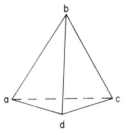

Fig. 2-3. Schematic representation of an asymmetric carbon atom.

Figure 2-4 shows three different ways of representing the two configurations of lactic acid (CH_3—$\overset{\overset{\displaystyle OH}{|}}{CH}$—COOH). It is customary to denote the two *spatial* isomers of a molecule with an asymmetric center as the D- and L-isomers. This terminology bears a confusing similarity to that used to designate the sign of rotation. However, the two are entirely distinct. Thus the D-isomer is in many cases laevo-rotatory, and vice versa.

The absolute spatial configuration of the isomers of tartaric acid (HOOC—$\overset{\overset{\displaystyle OH}{|}}{CH}$—$\overset{\overset{\displaystyle OH}{|}}{CH}$—COOH) has recently been established by X-ray diffraction. This established automatically the configuration of a large number of other organic compounds, by virtue of the known chemical transformations which interrelate them. In particular, it establishes the configuration of the α-amino acids, all of which, except for glycine, *contain an asymmetric center in the α-carbon.*

By criteria of this kind it has been shown that all the amino acids occurring in proteins are the *L-isomers*. The configuration of L-alanine, for example, is shown in Fig. 2-5. The same diagram may be used to represent the configurations of the other amino acids, if the methyl group is replaced by the appropriate side-chain.

The D-isomers of the α-amino acids have been identified in biological systems in the free state. However, their occurrence in proteins is virtually unknown.

Three amino acids have a second center of asymmetry in addition to

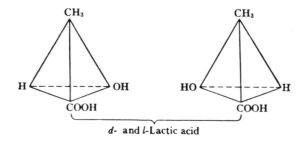
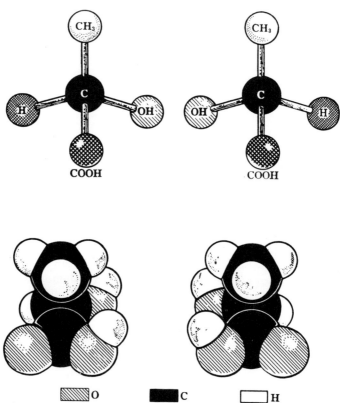

Fig. 2-4. The optical isomers of lactic acid represented in three different ways. The D-isomer is on the left and the L-isomer on the right.

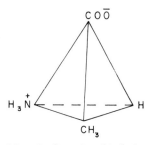
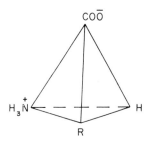

Fig. 2-5a. Configuration of L-alanine. *Fig. 2-5b.* General configuration of the L-amino acids.

the α-carbon. These are threonine, isoleucine, and hydroxyproline (6, 7).

The magnitude of the specific rotation of an amino acid is dependent upon a number of experimental parameters, including the wave-length of light, the solvent, and the state of ionization of the amino acid. When the amino acids are incorporated into a protein, the over-all rotation will, in addition, depend upon the conformation of the protein. This will be discussed in detail in Chapter 5.

The sign and magnitude of optical rotation are very sensitive to the chemical nature of the groups in the immediate vicinity of the asymmetric center. If these include an ionizable group, the rotation will depend upon its state of ionization. In addition, the nature of the solvent has a major influence.

2-3 THE NEUTRAL AMINO ACIDS

Aliphatic hydrocarbon side-chains. A glance at Fig. 2-1 shows that the class of neutral amino acids can be subdivided into several categories. The side-chains of five amino acids are noncyclic, saturated aliphatic hydrocarbons. These are glycine (R = H); alanine (R = CH_3); valine (R = $[CH_3]_2CH-$); leucine (R = $[CH_3]_2-CHCH_2-$); and isoleucine (R = CH_3-CH_2-CH-).
$|$
CH_3

Side-chains of this type contain no polar groups and are uncharged over the entire range of pH. Their purely hydrocarbon character endows them with *hydrophobic* properties; that is, a nonpolar environment is energetically preferred to water. This fact is of importance in understanding certain aspects of the stabilization of protein structure. Like all aliphatic hydrocarbons, side-chains of this class are chemically rather inert.

Aliphatic-hydroxyl side-chains. Two amino acids have noncyclic aliphatic side-chains containing a hydroxyl group. These are serine ($R = OHCH_2-$) and threonine ($R = CH_3-\underset{\underset{OH}{|}}{C}H-$). Aliphatic hydroxyls of this type have neither acidic nor basic properties and remain un-ionized at all pH's. However the hydroxyl group has the potentiality of forming hydrogen bonds. This may be of importance in stabilizing the structure of some proteins.

Sulfur-containing side-chains. The side-chains of three amino acids contain sulfur. Methionine ($R = CH_3-S-CH_2-CH_2-$) has no ionizable group in its side-chain.

The related amino acids cysteine ($R = HSCH_2-$) and cystine have the unique property of being interconvertible. Thus two molecules of cysteine may be transformed into cystine by mild oxidation. Conversely, the disulfide ($-S-S-$) bridge of cystine may be split by reduction to form two molecules of cysteine.

One frequently used method for the cleavage of cystine in proteins is oxidation by performic acid (HCOOOH) to the corresponding *cysteic acid*. Thus:

(2-3)
$$\begin{array}{c} -CHCH_2 \\ | \\ S \\ | \\ S \\ | \\ -CHCH_2 \end{array} \xrightarrow{HCOOOH} \begin{array}{c} -CHCH_2 \\ | \\ SO_3H \\ \\ SO_3H \\ | \\ -CH-CH_2 \end{array}$$

In addition to its action upon the disulfide bridge of cystine, performic acid also destroys tryptophan and oxidizes methionine to its sulfone derivative. Due allowance for these reactions must be made when these amino acids are present.

Disulfide may also be ruptured by reduction to the corresponding mercaptide ($-SH$) derivatives. Among the reagents which are effective in reducing cystine are such sulfhydryl derivatives as mercaptoethanol, mercaptoethylamine, thioglycollic acid, cysteine, and so on.

The reaction proceeds readily at neutral or alkaline pH, the sulfhydryl reagent being normally converted to its disulfide derivative.

(2-4)
$$\text{protein} \begin{array}{c} S \\ \diagup \\ \diagdown \\ S \end{array} + 2RSH \rightleftarrows \text{protein} \begin{array}{c} SH \\ \diagup \\ \diagdown \\ SH \end{array} + R-S-S-R$$

THE AMINO ACIDS

Two additional reactions of the sulfhydryl group of cysteine are of importance in protein chemistry. It may be converted to a carboxymethyl group by treatment with iodoacetate. This reaction is often used to block —SH groups.

(2-5) $R'—CH_2—SH + ICH_2\overline{COO} \rightarrow R'—CH_2—SCH_2\overline{COO} + \overline{I} + H^+$

The sulfhydryl group also combines with mercuric chloride to form a stable mercaptide.

(2-6) $R'—CH_2—SH + HgCl_2 \leftrightarrows R'—CH_2—S—HgCl + H^+ + Cl^-$

Because of its double character, cystine can be incorporated into two different chains of amino acids within a protein and serve to cross-link them. Cystine thus has an especial role in maintaining the molecular organization of proteins (section 3-2).

The sulfhydryl (SH) group of cysteine has weakly acidic properties and can ionize.

(2-7) $R'—CH_2SH \rightarrow R'—CH_2S^- + H^+$

The ionization of the sulfhydryl group of cysteine is accompanied by a major alteration in its ultraviolet absorption spectrum. An important new band arises with a maximum close to 2350 Å (Fig. 2-6).

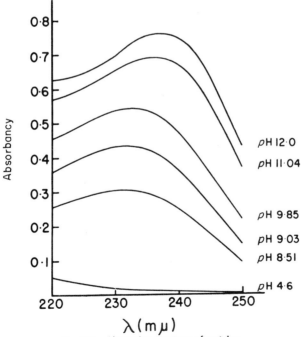

Fig. 2-6. Absorption spectrum of cysteine.

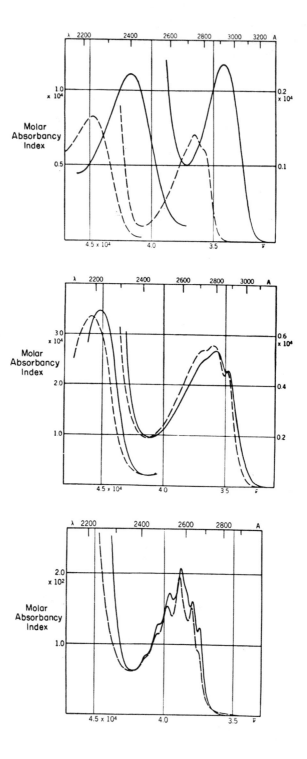

Aromatic side-chains. Three amino acids have aromatic side-chains. Two of these—tyrosine and phenylalanine—are benzene derivatives (Fig. 2-1). The side-chain of phenylalanine contains no ionizable or other polar groups and is relatively inactive chemically. However, the aromatic hydroxyl (phenolic) group of tyrosine can ionize to form a proton and a phenoxide ion (Fig. 2-1).

Both tyrosine and phenylalanine absorb ultraviolet light strongly. In both cases, the wave-length of maximum absorption is close to 2800 Å. Together with tryptophan, these residues account for most of the absorption spectra of proteins above 2300 Å.

The absorbancies of phenylalanine and tryptophan are almost invariant to pH, as neither chromophore contains an ionizable group (Figure 2-7). However, in the case of tyrosine, a major change accompanies the ionization of the phenolic hydroxyl (Fig. 2-7), the position of the maximum shifting from about 2800 Å to 2950 Å. Thus the molar absorbancy at 2950 Å furnishes an index of the fraction of tyrosine which has ionized and has frequently been utilized to monitor the ionization of tyrosine groups in proteins.

The above three aromatic amino acids all fluoresce in the ultraviolet. The wave-length of maximum intensity is close to 3000 Å for the emission bands of tyrosine and phenylalanine and near 3400 Å for that of tryptophan (Fig. 2-8). The activation spectra parallel the absorption spectra closely.

The iodinated amino acids diiodotyrosine and thyroxine (Fig. 2-1) are of very restricted occurrence in natural proteins, being essentially confined to several thyroid proteins. In the presence of I_2 and KI, a fraction of the tyrosine groups of other proteins undergo a stepwise substitution reaction to form diiodotyrosine.

α-imino acids. Proline and hydroxyproline occupy a special category in having the α-nitrogen in the form of an *imino*, rather than an amino group. This has important consequences for the molecular organization of proteins in which they occur. Hydroxyproline is of rare occurrence, having been identified only in collagen and its derivatives. The hydroxyl group of hydroxyproline appears to have properties similar to those of serine and threonine and, like the latter, does not ionize.

Fig. 2-7. Ultraviolet absorption spectra of the aromatic amino acids in 0.1 M HCL (––––) and in 0.1 M NaOH (———). The upper abscissa is the wave-length in angstroms. The lower abscissa is the frequency in wave numbers. Upper curve: tyrosine. The left ordinate refers to the left curve, for wave-lengths between 2000 and 2700 angstroms. The right ordinate refers to the curve on the right, for wave-lengths above 2600 angstroms. The two scales are necessary to represent the entire absorption curve on one graph. Middle curve: tryptophan. The two ordinates are analogous to those for the tyrosine curve. Lower curve: phenylalanine.

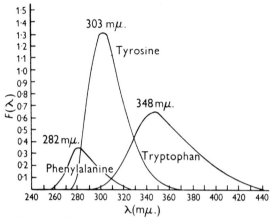

Fig. 2-8. Fluorescence spectra of the aromatic amino acids.

The ninhydrin reaction. A general reaction of great importance for analytical purposes occurs upon heating α-amino acids with *ninhydrin* (indane-1,2,3-trione-2-hydrate). The reaction is in this case confined to the α-amino site and requires the presence of the $-\underset{|}{\overset{NH_2}{C}}-CO-$ group.

(2-8) 2 [indane-1,2,3-trione] C(OH)$_2$ + H$\underset{R}{\overset{NH_3^+}{C}}$—COO$^-$ →

[bis-indanedione complex] C—N = C [indanedione] + CO$_2$ + RCHO

A bright blue color is developed with an absorption maximum at about 5700 Å. The ninhydrin reaction has been widely used for the quantitative estimation of amino acids. It is worthy of mention that the imino acids—proline and hydroxyproline—undergo a different reaction which yields a red color.

2-4 ACIDIC AND BASIC AMINO ACIDS

Acidic amino acids. Two amino acids, aspartic acid ($R = HOOCCH_2-$) and glutamic acid ($R = HOOCCH_2CH_2-$), possess side-chains containing carboxyl groups. These are strongly acidic and ionize to form a hydrogen ion and a carboxylate ion.

(2-9) $\quad\quad -CH_2-COOH \rightarrow -CH_2-C\overline{OO} + H^+$

The carbonyl group ($C = O$) is a good hydrogen bond acceptor. There has been considerable speculation as to the possible role of bonds of this type in maintaining the configuration of several proteins (section 5-2).

Aspartic and glutamic acids often occur in proteins in the form of their amide derivatives asparagine and glutamine (Fig. 2-1). The ionization of the side-chains is suppressed for these derivatives.

Basic amino acids. Lysine ($R = H_2NCH_2CH_2CH_2CH_2-$) contains a primary amino group attached to the ϵ-carbon of its side-chains. This can bind a hydrogen ion and thereby acquire a positive charge

(2-10) $\quad\quad -(CH_2)_3-NH_2 + H^+ \rightarrow -(CH_2)_3-NH_3^+$

The side-chain of arginine ($R = -(CH_2)_3-NH-\overset{\overset{\displaystyle \|}{\|}}{C}-NH_2$) contains
NH_2^+

a guanidinium group. The affinity of this group for its attached hydrogen ion is so strong that it is in ionic form over almost the entire range of pH. Only under very alkaline conditions (pH > 13) does dissociation of the proton occur.

The imidazole ring of the side-chain of histidine likewise is capable of binding a proton (Fig. 2-1).

Ionization of the amino acids. Each free α-amino acid contains at least one acidic group (the α-carboxyl) and one basic group (the α-amino). Under sufficiently alkaline conditions both groups dissociate a proton and the amino acid becomes an anion. If the pH is sufficiently acid, both groups are protonated and the amino acid exists as a cation. At intermediate pH's the α-carboxyl is ionized and the α-amino protonated so that the amino acid is a zwitterion.

(2-11) $$R-\underset{H}{\overset{NH_3^+}{C}}-COOH \underset{}{\overset{H^+}{\rightleftarrows}} R-\underset{H}{\overset{NH_3^+}{C}}-COO^- \underset{}{\overset{OH^-}{\rightleftarrows}} R-\underset{H}{\overset{NH_2}{C}}-COO^-$$

If the side-chain contains an ionizable group, an additional set of equilibria exists. Since the effective dissociation constant of each group will, in general, depend upon the state of ionization of the other two, a rigorous description of the ionization equilibria of such a system is algebraically quite complicated (8).

When amino acids are incorporated into a protein by formation of peptide linkages, the ionization of the α-amino and α-carboxyl groups is suppressed, except for the single α-amino and α-carboxyl group occurring at the ends of each polypeptide chain. Apart from these terminal groups, the only ionizable sites present in proteins are those in the side-chains.

With the exception of the guanidinium group of arginine, all of the ionizable sites present in amino acids belong to the class of weak electrolytes, whose degree of ionization depends upon the concentration of hydrogen ion.

The ionization of a collection of *isolated* and *independent* ionizable sites can be described in terms of a single dissociation constant (K).

(2-12) $$AH \rightleftarrows A^- + H^+; \quad K = \frac{[A^-][H^+]}{[AH]}$$

Dissociation constants are usually expressed as the pK, which is defined as the negative of the logarithm (to the base 10) of the dissociation constant:

(2-13) $$pK = -\log_{10} K$$

The pK is equal to the pH at which one-half the sites are ionized ($[A^-] = [AH]$).

The ionization of a system of linked groups, such as occurs in a protein, cannot be described so simply, because of the electrostatic and other factors whereby the ionizations of the different sets of groups become interdependent. It is convenient to define for each class of ionizable sites an *intrinsic dissociation constant*, which is equal to the dissociation constant that a collection of such sites would have if they ionized completely independently, without being influenced by the over-all state of ionization of the molecular unit. The approximate intrinsic constants for the various ionizable groups present in amino acids are listed in Table 2-1. They provide an approximate index of the pH zone in which ionization of a particular site may be expected to occur.

TABLE 2-1. INTRINSIC IONIZATION CONSTANTS

Group	$pK_{int.}$
Guanidinium	13
Sulfhydryl	9
Tyrosine hydroxyl	10
ε-amino	9.5
α-amino	8
Histidine imidazole	6
Carboxyl	4

2-5 AMINO ACID ANALOGS

A number of compounds structurally analogous to particular amino acids have been synthesized and examined for their utilization by biological systems in protein synthesis. In several cases such abnormal amino acids have been incorporated into proteins (9).

For example, selenomethionine, a structural analog of methionine (Fig. 2-1) in which the sulfur atom is replaced by selenium, can be utilized by a methionine-requiring mutant of *Escherichia coli* (*E. coli*) if it is grown in a methionine-free medium which contains this compound (9, 10). The cells were able to carry out the synthesis of enzymes in a normal manner and to reproduce. The progeny contained selenomethionine exclusively.

Similarly, ortho-fluoro-phenylalanine has been shown to be incorporated into the proteins of minced hen's oviduct. Amino acid analysis of such proteins indicated that phenylalanine had been partially replaced by this analog (9).

It thus appears that the powers of discrimination of the protein-synthesizing systems of living organisms are not unlimited and that mistakes in identity can occur. All instances of successful analog utilization have involved compounds closely similar in configuration and dimensions to the amino acid substituted for.

However, the protein-synthesizing apparatus has little difficulty in differentiating between normal amino acids of similar structure, such as valine and isoleucine, and no examples of interchange of such residues have appeared except by way of specific mutations leading to modified genes.

GENERAL REFERENCES

1. *Proteins, Amino Acids, and Peptides*, E. Cohn and J. Edsall, Reinhold, New York (1944).
2. *Biophysical Chemistry*, J. Edsall and J. Wyman, Academic Press, New York (1958).
3. *Physical Chemistry of Macromolecules*, C. Tanford, Wiley, New York (1961).
4. *Electrochemistry in Biology and Medicine*, T. Shedlovsky (ed.), Wiley, New York (1955).
5. *Chemistry of the Amino Acids*, J. Greenstein and M. Winitz, Wiley, New York (1961).

SPECIFIC REFERENCES

Optical activity:

6. Reference 5, p. 46.
7. Reference 3, p. 119.

Properties of amino acids:

8. Reference 1, chaps. 4, 5, 20.
9. *The Molecular Basis of Evolution*, C. Anfinsen, chap. 9, Wiley, New York (1959).
10. D. Cowie and G. Cohen, *Biochim. et Biophys. Acta*, **26**, 252 (1957).

3

The Chemical Structure of Proteins

3-1 PRELIMINARY REMARKS

Protein molecules are composed of one or more linear copolymers of α-amino acids, called *polypeptides*. The amino acids of each polypeptide chain are united by a characteristic linkage, the *peptide bond*. Because there are 20 different amino acids, the number of possible linear combinations is very great, a fact which is responsible for the extraordinary diversity of structure and function shown by natural proteins.

The range of molecular sizes encountered in natural proteins is very wide. The smallest molecules customarily regarded as true proteins have molecular weights of the order of several thousand. The largest *single* polypeptide units known have molecular weights of the order of 10^5. Proteins of higher molecular weight appear in general to consist of two or more polypeptide chains.

While the peptide bond (Fig. 3-1) accounts for the overwhelming

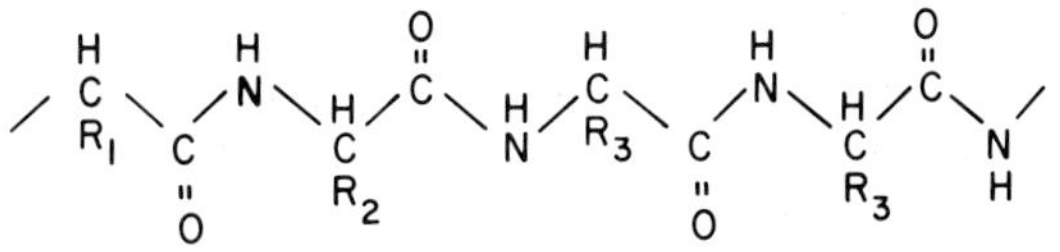

Fig. 3-1a. Schematic representation of the structure of a polypeptide chain stretched to its maximum linear extension. The R's represent side-chains.

majority of the linkages between the amino acid units of proteins, other types of linkage do occur. The most important of these involves the amino acid *cystine*, which is unique in consisting of two identical half-molecules joined by a disulfide bridge (Fig. 2-1). Because each half-molecule is bifunctional and capable of participating in two peptide bonds, a single cystine residue may be shared by two different polypeptide chains or by two different parts of the same polypeptide. In this manner the disulfide (—S—S—) bridge of cystine may serve to cross-link two different polypeptide chains or to cross-link internally a single chain.

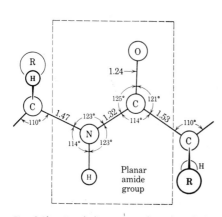

Fig. 3-1b. Bond distances and angles of the peptide linkage. Bond distances are in angstroms.

Cross-links other than the disulfide bridge are known, but they occur rather infrequently. In summary, the typical natural protein consists of one or more linear polypeptide chains, which may be connected by occasional cystine cross-links.

The polypeptide chains of natural proteins are folded, or coiled, into a highly specific three-dimensional pattern. The problem of determining the structure of any proten subdivides naturally into two phases, which require radically different experimental approaches for their solution. The present chapter will deal with the first of these. This is concerned with the number, length, and composition of the polypeptide chains; the linear arrangement of their amino acids; and the number and position of the cross-links between chains (Table 3-1). These features of the molecular organization include all those aspects which can be discussed within the traditional framework of organic chemistry, and are often referred to collectively as the *primary structure*. A discussion of the balance of the problem—the determination of the geometry of folding of the polypeptide chains, and the interactions of the sidechains—will be deferred to Chapter 5.

Many proteins contain structural elements which are not of a polypeptide nature. These are known as *conjugated* proteins. The nonpolypeptide group is often directly involved in the specific biological function of the protein. Some well-known examples of conjugated proteins are hemoglobin (section 6-7) and the cytochromes (sections 3-5 and 13-2).

3-2 THE MANNER OF LINKAGE OF AMINO ACIDS IN PROTEINS

The peptide linkage. The possibilities for the chemical linkage of the α-amino acids are severely limited by their nature. For practical purposes, the polypeptide chains of natural proteins can be regarded as consisting exclusively of linear polymers of amino acids joined by peptide bonds of the amide (CONH) type, formed by elimination of water between the α-carboxyl and α-amino groups of adjacent residues. While the side-chain carboxyls of glutamic and aspartic acids, as well as the ε-amino group of lysine, are potentially capable of participating in similar bonds, this possibility does not appear to be realized in natural polypeptides.

The bond angles and distances of the basic repeating unit of polypeptide chains (1, 2) are shown in Fig. 3-1b.

A closer look at the structure of the peptide group requires recourse to the physical technique of X-ray diffraction. To avoid disrupting the continuity of the present discussion, a detailed description of this method will be postponed to Chapter 5. For present purposes, it is sufficient to state that this kind of analysis permits determination of interatomic distances to within 0.01 Å.

The basic polypeptide repeating unit cited above contains one C—C and two N—C bonds (Fig. 3-1). The bond distance, as determined by X-ray diffraction studies with simple peptides, is 1.53 Å for the C—C linkage and about 1.47 Å for the N—C bond between the NH and CHR groups. By drawing upon the extensive data which have been accumulated for simple compounds, it can be stated that both distances are typical of those found for normal *single* bonds of these types, such as occur in ethane (CH_3—CH_3) or aminomethane (H_2N—CH_3). In particular, there is no indication of the bond shortening which would reflect the presence of some degree of double bond character (1).

The purely single bond character of these two linkages permits the prediction that essentially free rotation will be present about these bonds. This endows the polypeptide chain with the potential capacity to adopt a large number of coiled configurations.

The status of the remaining C—N bond between the C=O and NH groups (Fig. 3-1b) is very different (1). The interatomic separation, as obtained from X-ray diffraction, is in this case close to 1.32 Å. This represents a major shortening of the normal single bond distance and indicates an important degree of double bond character. It may be concluded that resonant forms of the C—N bond of the CONH group which include a C=N double bond make an important contribution to the overall resonance pattern of this group. This is by no means unexpected, in view of the likely occurrence of the resonance state
$$-\overset{\overset{O^-}{|}}{C}=\overset{}{\underset{\underset{H}{|}}{N^+}}-,$$
because of the markedly more electronegative character of oxygen, as compared with nitrogen.

The presence of an important extent of double bond character in this case has significant consequences for the configurational properties of the polypeptide chain. In contrast to the other two bonds of the peptide repeating unit, freedom of rotation will be definitely hindered, giving rise to the possibility of a configurational isomerism. This *cis-trans* isomerism is characteristic of bonds of the ethylenic type and arises from the presence of two energetically preferred angles of rotation about the double bond (1).

A further consequence of the partial double bond character of the C—N linkage is the constraint of the amide group into a planar configuration. This has proved to be the case for all the peptides thus far examined by X-ray diffraction (Chapter 5).

Finally, it should be noted that each peptide group possesses a potential hydrogen bond donor (see section 5-2) in the hydrogen atom attached to its nitrogen and a potential acceptor in its carbonyl oxygen. The resonance pattern of the peptide group is such as to place an excess of positive charge on the nitrogen and an excess of negative charge on the carbonyl oxygen. As a result, the affinity of the carbonyl for a hydrogen bond donor, and of the —NH for a hydrogen bond acceptor, is greatly enhanced. This has proved to be of vital importance with regard to the spatial configuration of proteins, as will be made clear in Chapter 5.

The cystine linkage. The second major type of primary linkage between different amino acids in proteins is the disulfide bridge of cystine (1-8). The duplex character of this amino acid renders it capable of being incorporated into two different polypeptide chains or into two different parts of the same chain, thereby serving as a cross-link.

```
                           O               O
chain 1                    ||              ||
∧∧∧∧—CH—C—NH—CH—C—NH—CH—∧∧∧∧
         R₁           |          R₂
                     CH₂
                      |
                      S
                      |
                      S                        disulfide bridge
                      |
                     CH₂   O
                O     |    ||
chain 2         ||    |
∧∧∧∧ CH—C—NH—CH—C—NH—CH ∧∧∧∧
      R₃              R₄
```

Although information as to freedom of rotation about the S—S bond is incomplete, there is considerable evidence that a definite barrier to free rotation exists, despite its single bonded character. As a consequence, it would be expected that cystine linkages could impart spatial restrictions of considerable importance and thereby take an active part in stabilizing the configuration of polypeptide chains.

The phosphate cross-linkage. A second type of cross-link has been found to occur in a few proteins. The bifunctional phosphate group is capable of forming a double ester with two different aliphatic hydroxyl groups, such as occur in the side-chains of serine.

$$-CH_2-O-\overset{\overset{\displaystyle O^-}{|}}{\underset{\underset{\displaystyle O}{\|}}{P}}-O-CH_2-$$

In this manner two different chains, or two different parts of the same chain, may be connected by a *phosphodiester* bridge between two serine residues. The proteolytic enzyme pepsin may contain such a phosphodiester cross-link.

Unusual bonds. In the case of collagen (Chapter 6) there is some evidence for the presence of ester bonds, which may involve the side-chain carboxyls of aspartic and glutamic acids and aliphatic hydroxyls of serine or threonine. It has also been suggested that imide bonds involving the side-chains may be present in this protein. There is no reason to doubt that other unusual linkages may occur in particular instances.

3-3   THE SEQUENTIAL ARRANGEMENT OF AMINO ACIDS IN PROTEINS (3, 4, 6, 7, 8)

Preliminary remarks. The general statement can be made that the linear order of amino acids within a particular protein is essentially completely specific and characteristic of the protein. In other words, proteins are not random polymers of amino acids but sharply defined chemical compounds, all members of a given class being identical.*

The complete determination of amino acid sequence, even for the smallest molecules which can be regarded as proteins, is a task demand-

* This should be qualified in view of the occasional occurrence of a *local* alteration of one or more amino acids, as a consequence of a genetic mutation, to yield a modified molecule, which may be otherwise indistinguishable in its chemical and physical properties from the normal protein. In the spirit of this discussion, this must be regarded as a distinct and different molecule.

ing enough to strain both the existing techniques and the ingenuity of the chemist to the utmost limit of their capacity. The dramatic initial successes which have been obtained in the cases of such proteins as insulin and ribonuclease represent the first fruits of a fundamental achievement which must rank with the greatest in the history of chemistry.

The difficulties inherent in the problem are intensified by the inadequacy in this instance of the standard techniques of organic chemistry. Any system for sequential analysis must depend upon sensitive techniques for the separation and identification of a multitude of small peptides, which are often of closely similar chemical and physical properties. This was achieved by the chromatographic methods to be discussed in section 3-4.

General strategy. The determination of the primary structure of a protein may be regarded as completed when the following are known: (1) the number of polypeptide chains and their respective lengths; (2) the number and position of chemical cross-links between chains; (3) the amino acid composition of each chain; and (4) the order, or linear sequence, of amino acids in each chain.

The initial phase of the experimental attack upon the problem is the procurement of accurate figures for the amino acid composition of the entire protein (9, 10). It is obvious that purity of the sample is of the very highest importance for this and for subsequent steps. Complete amino acid analysis requires rupture of all peptide bonds by exhaustive hydrolysis, usually by high concentrations of mineral acids, followed by quantitative separation of the resultant mixture of amino acids into its components by a chromatographic procedure (section 3-4). Precise determination of the relative concentration of each species permits evaluation of the mole fraction of each residue.

At this stage, the available information can be placed in the form of a table whose entries give the number of moles of each amino acid per 100,000 grams of protein (to choose a convenient round figure). Each protein molecule must contain an integral number of each kind of amino acid present. Thus, if one selects an amino acid of infrequent occurrence, there will be 100,000/M moles of residue per 100,000 gms of protein of molecular weight M, if the amino acid occurs but once, and $2 \times$ 100,000/M moles per 100,000 gms if it occurs twice. In general, if the residue occurs n times in each molecule, the number (y) of moles of that residue found in 100,000 gms of protein is given by:

(3-1) $$y = 100{,}000\, n/M$$

Or, $$M/n = 100{,}000/y$$

Thus, an accurate determination of y permits the assignment of a *minimum molecular weight*, M/n, to the protein. The true molecular weight is equal to n times this figure, where n is an unknown integer.

The choice of the correct value of n requires supplementary information from physical techniques for molecular weight determination, such as ultracentrifugation, osmotic pressure, or light-scattering. These methods, which will be discussed in detail in Chapter 4, permit evaluation of molecular weights to within 2 to 3% and hence assignment of the correct value of n. From the known molecular weight and amino acid composition an empirical formula may be written for the protein of the form $A_a, B_b, C_c, \ldots$, and so forth, where the letters stand for amino acids.

The next phase of the analysis is the determination of the number and composition of the polypeptide chains. Because of the nature of the peptide linkage, each polypeptide will have a free α-amino group at one end (called the NH_2-terminal end) and a free α-carboxyl at the other (called the COOH-terminal end). If the number of α-amino groups per molecule can be determined, this number yields directly the number of polypeptide chains in the protein. A number of reagents are known, such as the dinitrophenyl (DNP) reagent of Sanger (section 3-4), which combine selectively with α-amino groups to form conjugates stable enough to survive complete acid hydrolysis of all peptide bonds.

Conversion of all free α-amino groups of the protein to their DNP-derivatives, followed by complete hydrolysis and determination of the number of moles of DNP-labeled amino acids per mole of protein, provides the number of free NH_2-terminal amino groups and hence the number of polypeptide chains. Moreover, the identification by chromatographic means of the particular residues which are labeled locates them unequivocally at the NH_2-termini of the polypeptides.

Further progress requires a means for selective rupture of the cystine cross-links between chains, as their persistence would hopelessly confuse the sequence analysis of the individual chains. The disulfide bridges may be split by oxidation with performic acid (section 2-3) to yield two moles each of cysteic acid. Rupture of peptide bonds and alteration of amino acids are usually minimal.

The independent polypeptide chains thereby produced can be fractionated by chromatographic or other means. Purified preparations of each polypeptide can be analyzed for amino acid composition in the same manner as the intact protein.

At this stage our picture of the protein represents it as a set of one or more polypeptides, of known composition and NH_2-terminal residues, which are joined by —S—S— bridges of uncertain position.

Actual sequence determination upon the separated polypeptide chains depends upon controlled partial hydrolysis into peptides of manageable size. Fortunately, the use of end group analysis permits identification of the NH_2-terminal residue of each peptide and suffices to fix the sequence of a dipeptide. Thus the first stage of sequence determination might be partial hydrolysis to a mixture of smaller peptides, followed by conversion of all NH_2-terminal residues to their DNP derivatives. After isolation, each labeled peptide may be completely hydrolyzed and its amino acid composition and NH_2-terminal residue determined.

The fragmentary information assembled in this manner may serve to determine the order of fairly extended sequences. As an illustration of the principles involved, let us consider a very simple hypothetical example—the determination of the sequence of a pentapeptide (A,B,C,D,E), where D is NH_2-terminal.

Since the NH_2-terminal residue is known, the sequence can be written: D-(A,B,C,E). From the digest obtained by partial acid hydrolysis, the dipeptides D-A, A-C, B-E, and C-B are isolated and identified, as well as the tripeptides D-(A,C) and A-(B,C).

The presence of D-A identifies the two residues at the NH_2-terminal end. The identification of A-C and the tripeptide A-(B,C) permits enlargement of the NH_2-terminal sequence to D-A-C. The isolation of C-B shows B to be the next residue. Since only E remains, it must be the COOH-terminal amino acid, and the complete sequence of the pentapeptide can be written as D-A-C-B-E.

The above determination is simple largely because each amino acid occurs but once, so that each dipeptide can arise from one and only one position in the original peptide. In actuality, this is rarely the case, and there is considerable ambiguity in assigning the smaller peptides. The presence of any singly occurring residue is always most welcome and often provides a nucleus of known sequence which can be enlarged.

It is often most convenient to hydrolyze the separated polypeptide chains into nonoverlapping peptide fragments by enzymatic hydrolysis, thereby producing fragments of more manageable size which can be separated and analyzed further (section 3-4). A comparison of the fragments produced by different enzymes often suffices to determine their linear order (section 3-5).

There remains the problem of locating the —S—S— bridges. The basic approach is to hydrolyze the unoxidized protein into fragments containing intact disulfide bridges (Fig. 3-2). These are separated and oxidized into two cysteic acid peptides each. Determination of the amino acid composition of these usually permits their location within the known sequences of the individual polypeptides.

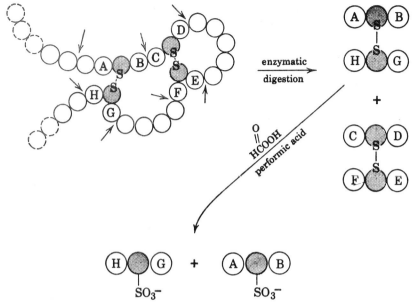

Fig. 3-2. Location of cystine residues within a protein (7).

It is clear from the preceding that no precise rules exist for sequence determination. The general method (Table 3-1) is best illustrated by considering particular examples.

3-4 TECHNIQUES OF SEQUENCE DETERMINATION

Polypeptide degradation. Both the determination of amino acid composition and the far more difficult problem of sequence analysis depend upon methods for the complete or partial degradation of polypeptides with minimal destruction of individual amino acids. While the energetics of the hydrolytic scission of peptide bonds favor the forward reaction strongly, the attainment of a reasonably rapid *rate* of hydrolysis requires either high levels of H^+ or OH^- or the addition of an enzyme catalyst (7, 8, 9).

Hydrolysis by acid or alkali is essentially nonspecific and ruptures peptide bonds between all pairs of amino acids, although quantitative differences exist in the rates of cleavage. Hydrolysis by concentrated mineral acids (usually about $6M$) is generally the method of choice for the extreme degradation of polypeptides. Tryptophan is the only amino acid which is extensively damaged. Alkaline hydrolysis is accompanied by important damage to a number of residues, including serine, threonine, and cystine, and is consequently less often used. However, tryptophan is much more stable to alkali than to acid.

The rates of splitting in acid media vary somewhat according to the nature of the amino acids linked. Linkages involving serine or threonine are especially labile, so that these residues generally occur in partial acid hydrolyzates as the NH_2-terminal residues of peptide fragments. Conversely, bonds involving isoleucine or valine are rather resistant to acid hydrolysis.

Because of the basically random character of acid-catalyzed hydrolysis, many overlapping peptides will be produced, which originate from the same linear sequence of amino acids. Thus the tetrapeptide A-B-C-D might, on partial acid hydrolysis, give rise to the overlapping tripeptides A-B-C and B-C-D.

When a polypeptide is partially hydrolyzed by acid, a complicated mixture of peptides is produced, whose complexity depends upon the extent of hydrolysis. Let us consider the hydrolysis of a linear polypeptide 100 residues long and containing all 20 amino acids. A finite amount of splitting of each peptide bond begins at once, resulting in the initial formation of over 10,000 different peptides. (Because of the unequal rates of hydrolysis of bonds between different pairs of amino acids, many of these will however be present in insignificant quantities.) As the reaction progresses the concentration of each peptide passes through a maximum value and then declines. The shorter the peptide, the later does it attain its maximum concentration. As the reaction approaches completion the number of different peptides present in measurable quantity continually decreases. In the limit, when hydrolysis is complete, only 20 components are present, corresponding to the free amino acids.

It is clear from the preceding that only in the later stages of acid hydrolysis is the mixture of peptides likely to be simple enough to be analyzed.

The action of proteolytic enzymes proceeds under such mild conditions as to make any appreciable damage to amino acids very unlikely. Moreover, the specificity of enzymes is relatively limited and exact, so that an enzymatic hydrolyzate contains nonoverlapping peptides.

Among the more frequently employed peptidases are *trypsin*, which splits bonds involving the carboxyl groups of arginine and lysine, and *α-chymotrypsin*, which hydrolyzes linkages including the carboxyl groups of tyrosine, phenylalanine, tryptophan, and methionine. The specificities of *pepsin* and *subtilisin* are relatively broad, and a large number of different linkages are attacked. Figure 3-3 illustrates the selective hydrol-

THE CHEMICAL STRUCTURE OF PROTEINS

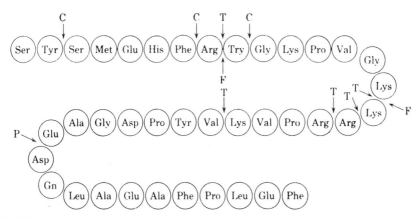

Fig. 3-3. Points of attack by the enzymes trypsin (T), chymotrypsin (C), and fibrinolysin (F) upon porcine corticotropin (7).

ysis of particular bonds of the peptide hormone *corticotropin* by various enzymes.

The enzyme *carboxypeptidase A* and *carboxypeptidase B* have an unusual specificity which renders them of particular utility in sequence studies. They selectively degrade peptide linkages at the COOH-terminal end of the chain in a stepwise manner, splitting off the terminal residue first, then the group next to it, and so forth. Residue specificity is somewhat relaxed in this case. However, the rate of hydrolysis does depend upon the character of the terminal amino acids, being most rapid for aromatic residues and becoming nil if proline is the penultimate residue.

Analysis of amino acid and peptide mixtures. A vital step in sequence determination is the resolution of the complex mixtures of peptides into their purified components. Only a highly condensed account of the various procedures will be presented here, which should be supplemented by the more detailed descriptions available in the literature (10-17).

Since the peptides are often rather similar in their chemical properties, relatively subtle techniques must be used for their fractionation. The initial mixture is often too complicated to permit separation by a single-step process. In this case, it is often convenient to make a preliminary separation into groups of peptides, using a method which depends upon some nonspecific property, such as charge.

Resolution by means of *ionophoresis* utilizes the varying rates of movement of peptides in an electrical field. This is usually carried out upon a supporting gel matrix, which serves to stabilize boundaries and minimize mixing by convective or other disturbances. Acrylamide or silica gel have been often used. Sections of gel may be cut out at the end of the separation and eluted.

Ionophoretic separation may be based upon differences in either *isoelectric point* (the pH at which no migration occurs in an electrical field) or velocity of migration (mobility). If the materials to be separated differ in isoelectric point, then at an intermediate pH the directions of migration in an electrical field will be opposite, components of isoelectric point less than the pH of the buffer migrating toward the anode and those of greater isoelectric point moving toward the cathode. Alternatively, the difference in the rates of migration of similarly charged species may provide a basis for their resolution.

In *paper ionophoresis* (or *paper electrophoresis*) the supporting matrix is a rectangular strip of filter paper. The passage of electric current through the buffer-saturated strip results in the progressive resolution of the mixture into its components, which migrate as distinct bands moving at different rates. If several strips are run simultaneously, the location of the bands on one strip by staining with a dye serves as a guide for the cutting out and quantitative elution of the bands on the other strips (15). A frequently used apparatus is shown in Fig. 3-4.

The methods of highest resolving power for the separation of complex mixtures of amino acids or peptides are chromatographic in nature. All variants of the chromatographic approach depend upon quantitative dif-

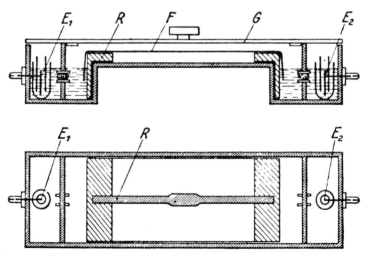

Fig. 3-4a. One type of apparatus for paper electrophoresis (9). E_1 and E_2 are electrodes; F is the filter paper strip, to which the mixture is applied as a narrow band; R is a support; and G is a cover to minimize evaporation. The electrode compartments are filled with buffer.

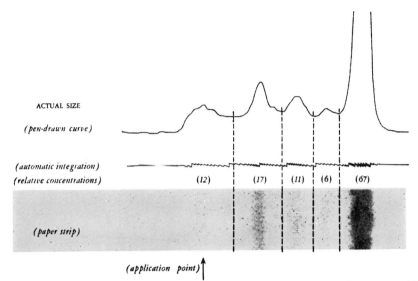

Fig. 3-4b. A strained strip and a densitometer tracing showing the distribution of components (9).

ferences in the rates of extraction of the different components from some type of supporting medium. This may be filter paper or a column of powder or gel contained in a glass cylinder.

Depending upon the nature of the procedure adopted, the supporting medium may serve as a relatively inert carrier of solvent or may take an active part in the adsorption of the materials to be separated. In the latter category, *adsorption chromatography* utilizes a column of adsorbing material, such as activated charcoal. If a solution of the mixture is poured on the column and pure solvent added continuously, the components will appear in the effluent solution in the inverse order of their affinity for the adsorbing medium (14).

Among the most widely used methods for the resolution of amino acids and peptides is column chromatography using *ion-exchange resins* (10, 11, 13). These are supplied as small beads consisting of cross-linked organic polymers containing numerous charged groups. An example is the sulfonated polystyrene employed by Moore and Stein (Fig. 3-5). The polystyrene chains are linked by occasional —CH_2— groups so as to maintain a three-dimensional gel structure which is open enough to allow extensive penetration by water, electrolytes, and amino acids.

As in the case of adsorption chromatography, the rate at which each component of the mixture is transported down the column by buffer depends upon the tenacity with which it is retained by the medium. The factors governing the rates of movement are easily understood qualita-

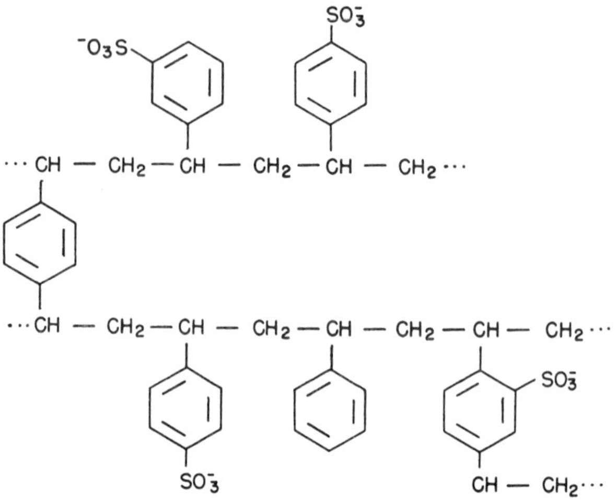

Fig. 3-5. Schematic picture of a sulfonated polystyrene resin (1).

tively (10-13). To maintain electrical neutrality in the resin interior the fixed charges attached permanently to the resin must be balanced by transient ions of opposite charge. These may be salt ions or charged amino acids. Thus electrostatic factors will tend to retard the movement of amino acids of net charge opposite to that of the resin. Because the average charge of an amino acid is governed by the pH of the buffer used, the latter is a very important parameter in separations of this kind. In the case of the sulfonated polystyrene (Dowex) resins, the amino acids with extra carboxyl groups and hence a net negative charge, such as aspartic acid, will tend to move more rapidly at neutral pH than those with additional amino groups and hence a net positive charge, such as lysine.

In addition to electrostatic factors, van der Waals interactions between the uncharged portions of the amino acid and the benzene rings of the polystyrene network can also be of major importance. At present the process is essentially an empirical one and the relative rates of migration of different species cannot be predicted quantitatively on theoretical grounds.

The rates of appearance of the components in the effluent solution, which is collected as a series of fractions, are monitored colorimetrically by the use of the ninhydrin reaction (section 2-3). The data take the form of a plot of ninhydrin color values as a function of volume of effluent (Fig. 3-6). Such a chromatogram appears as a series of discrete peaks, each of which corresponds to a single amino acid or peptide whose concentration in the original mixture is directly proportional to the area

THE CHEMICAL STRUCTURE OF PROTEINS

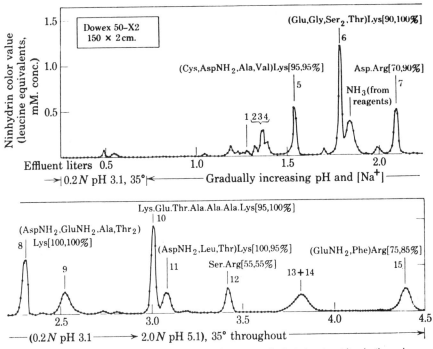

Fig. 3-6. Separation of the peptides produced by trypsin hydrolysis of oxidized ribonuclease upon Dowex 50 (7).

under the peak. If conditions are carefully controlled, the amino acids of an unknown mixture can be identified by their positions of appearance on the chromatogram, after calibration with known systems.

In addition to the sulfonated polystyrene resins mentioned above, various cellulose derivatives, such as diethylaminoethyl (DEAE) cellulose, have come into wide use in recent years. This cationic material has proved to be especially useful for separating mixtures of proteins.

Still another chromatographic technique is based primarily upon differences in molecular *size* rather than charge. This approach is often known as the "molecular sieve" method. The column material consists of granules of cross-linked dextran, a polysaccharide (Chapter 12). This substance is generally known by its commercial name, *Sephadex*. By varying the degree of cross-linking the effective porosity of the granules may be controlled.

The experimental technique is similar to that of other forms of column chromatography. For Sephadex of a given degree of porosity, the extent of penetration by a particular protein or peptide depends upon its molecular size. Molecules whose size is small in comparison with the effective pore size will enter the granules readily. Their movement down the column is thereby retarded. Molecules whose size is large in comparison with the pore size will be excluded entirely and confined to the solution external to the granules. Molecules of intermediate size will move down the column at rates which depend primarily upon their dimensions.

Thus the order of appearance of the components in the effluent solution is that of their molecular size, the largest component appearing first. The technique is particularly efficient in separating two components of widely differing size.

Paper chromatography depends upon the continuous redistribution of each solute between two liquid phases, one of which is mobile and the other stationary (14, 15, 16, 17). In this case the stationary phase, water, is immobilized by a supporting medium of filter paper. Many of the properties of filter paper are best explained if it is regarded as possessing a dual structure consisting of an amorphous phase supported by a fibrous network. The former retains the aqueous phase, which exchanges solute with a mobile organic phase.

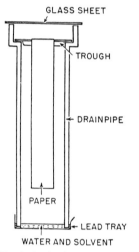

Fig. 3-7. A simple device for one-dimensional paper chromatography (descending). The hanging strip is allowed to become saturated with the vapors of the developing solution. The sample is applied in a narrow band at the top of the strip, which is placed in contact with the developing solution in the upper trough (14).

A small drop of protein hydrolyzate is placed near the end of a strip of moist filter paper. The adjacent end dips into a trough or reservoir of *developing solution* (Fig. 3-7). This contains an organic liquid, such as propanol. The developing solution slowly penetrates the filter paper strip through capillarity. The advancing liquid front moves at a definite velocity until it reaches the other end of the filter paper. The direction of flow may be either upward, in *ascending* paper chromatography, or downward, for *descending*.

As the developing solution advances, the amino acids or peptides move in the same direction but at a slower rate. The parameter R_f has been defined as the ratio of the rate of movement of a particular solute to the rate of movement of the liquid front. For a particular system it is a reproducible quantity which may be used to identify an unknown residue.

The migration of a particular component is the result of a continual exchange between the stationary water phase and the moving organic phase. The greater the solubility of the component in the organic liquid relative to that in water, the more rapidly it will tend to move. Because the effective number of extractions is very large, a small difference in solubility between two species is amplified into a large difference in rates of movement.

However, the filter paper itself is by no means entirely inert in the separatory process. There is strong evidence that amino acids may be reversibly adsorbed by filter paper and that this may influence the rates of transport. For this reason, paper chromatography, like the ion-exchange chromatography discussed earlier, is essentially an empirical procedure.

After development has been completed the paper is dried and sprayed with ninhydrin. The amino acids and peptides are thereby located as blue spots on the filter paper. By the use of known amino acids under exactly the same conditions the position characteristic of each residue may be determined and used to identify it on an unknown chromatogram.

Often a single developing solvent is inadequate to resolve completely all components of a complex mixture. In this case the paper may be dried, remoistened, and then exposed to the action of a second solvent, the direction of flow being at right angles to that of the initial treatment. By such *two-dimensional* paper chromatography, it is generally possible to resolve all the amino acids in a protein hydrolyzate into separate spots (Fig. 3-8). Less than a milligram of protein is needed.

End group determination. The determination of amino acid sequence in peptides has been immeasurably facilitated by the availability of several methods for selectively labeling the NH_2-terminal residue.

The most widely used technique for labeling NH_2-terminal residues is probably the dinitrophenyl (DNP) method (8). The reagent, 1:2:4 fluorodinitrobenzene, combines specifically with α-amino groups to form a stable DNP-conjugate.

(3-2) $NO_2\text{-}C_6H_3(F)(NO_2) + NH_2CHCO\text{—protein}$
                                              $|$
                                              R

$\rightarrow NO_2\text{-}C_6H_3(NO_2)\text{-}NHCHCO\text{—protein} + HF$
                                       $|$
                                       R

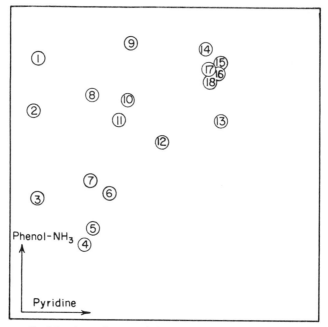

Fig. 3-8. A two-dimensional chromatogram of amino acids (7).

The DNP linkage is stable enough to survive complete acid hydrolysis of the protein or peptide, which liberates the DNP derivatives of the NH_2-terminal amino acid. This is a bright yellow compound which can be extracted, fractionated chromatographically, and estimated colorimetrically from its absorption of visible light. Quantitative determination of the number of moles of DNP-labeled amino acids per mole of intact protein gives the number of NH_2-terminal residues, and hence the number of noncyclic polypeptide chains.

An important application of the DNP technique is the identification of the NH_2-terminal residues of peptides. The labeled peptide is hydrolyzed completely and the NH_2-terminal conjugated residue separated chromatographically and identified. Identification may be made from the position of the DNP-derivative on a partition chromatogram or indirectly by a comparison of the paper chromatograms of hydrolyzates made before and after labeling.

Unfortunately, the DNP reaction is not confined entirely to α-amino groups. The ϵ-amino group of lysine also reacts to form a DNP derivative, which can be differentiated by its chromatographic behavior.

Other reagents have been used to label α-amino groups, including phenylisothiocyanate, which reacts to form a phenylthiocarbamyl (PTC) derivative. By heating in anhydrous HCl the terminal residue is selectively split off and converted to a phenylthiohydantoin derivative, which can be extracted. This can be identified by its position on a paper chromatogram. Alternatively, the balance of the peptide may be recovered and analyzed, thereby identifying the amino acid split off. The process may be repeated. The reaction is as follows:

(3-18)

$$\underset{\text{C}_6\text{H}_5}{\text{NCS}} + \underset{R}{\text{H}_2\text{N—CHCO—peptide}} \xrightarrow{\text{pyridine}} \underset{\text{C}_6\text{H}_5}{\text{NHCSNH}}\text{—}\underset{R}{\text{CHCONH}}\text{—peptide}$$

$$\text{PTC—peptide} \xrightarrow{\text{anhydrous HCl}} \underset{\substack{\\ \text{phenylthiohydantoin}}}{\text{C}_6\text{H}_5\text{—}\underset{\substack{\text{CO} \diagdown \\ \quad \text{CH} \\ \quad | \\ \quad R}}{\overset{\text{N———CS}}{|}}\underset{\diagup}{\text{NH}}} + \text{H}_2\text{N—peptide}$$

The analysis of COOH-terminal residues is less advanced than that of NH$_2$-terminals. Stepwise hydrolysis by the enzyme carboxypeptidase has been the most successful technique up to the present. The release of amino acids may be followed by paper chromatography. Either the free amino acids or their DNP-derivatives may be observed. The determination of COOH-terminal sequence requires precise quantitative estimation of the rates of release of free amino acids. The order in which the concentrations of free amino acids attain their limiting values is that of their separation from the COOH-terminal residue.

"Fingerprinting." The detection of sequence differences between proteins differing subtly in primary structure, such as homologous proteins derived from different species, has been greatly facilitated by the fingerprinting technique. Basically this involves parallel digestion of the two proteins by one or more proteolytic enzymes, followed by a comparison of their paper chromatographic or paper electrophoretic patterns. If sequence differences are confined to one or two zones of the primary structure, the peptide fragments in which these occur will in general migrate at different rates and thus will be found at different loci in the chromatograms, while fragments of identical sequence will occur at the same position.

Thus the presence of sequential isomerism will be reflected by spots which are present on one chromatogram, but not on the other. Elution and quantitative amino acid analysis of these permits the pinpointing of the particular amino acids for which substitution has occurred. Combined with sequential analysis of the fragments, this information permits specification of the exact nature of the change in primary structure.

3-5: THE PRIMARY STRUCTURES OF INDIVIDUAL PROTEINS

Insulin. The determination of the amino acid sequence of insulin by Sanger and his associates is one of the supreme achievements of protein chemistry (8, 18-23). Indeed, the dramatic success obtained in this instance has furnished the principal impetus to the subsequent accomplishments in this field, with respect to both the invaluable practical information accumulated as to technique and to the psychological stimulus of the

TABLE 3-1. STEPS INVOLVED IN PRIMARY STRUCTURE DETERMINATION

I. Amino acid composition
    (a) Complete acid hydrolysis
    (b) Chromatographic fractionation of hydrolyzate
    (c) Quantitative analysis for amino acids
II. Molecular weight determination
    (a) Minimum molecular weight from amino acid content
    (b) Determination of molecular weight by a physical method
III. Number and length of polypeptide chains
    (a) End group analysis by DNP method
    (b) Splitting of —S—S— cross-links by oxidation
    (c) Fractionation into individual polypeptide chains
IV. Amino acid sequence of polypeptides
    (a) Partial acid or enzymatic hydrolysis into small peptides
    (b) Chromatographic fractionation of acid digest
    (c) Amino acid analysis of peptide fragments
    (d) Fitting of peptides into a sequence of amino acids
V. Location of cross-links
    (a) Partial hydrolysis without prior rupture of —S—S— bridges
    (b) Analysis of cystine-containing peptides
    (c) Location of cystine peptides within amino acid sequence

TABLE 3-2. AMINO ACID COMPOSITION OF BEEF INSULIN

| | No. Groups | |
Amino Acid	A-Chain	B-Chain
Phenylalanine	0	3
Valine	2	3
Aspartic acid	2	1
Glutamic acid	4	3
Histidine	0	2
Leucine	2	4
½ Cystine	4	2
Glycine	1	3
Serine	2	1
Alanine	1	2
Tyrosine	2	2
Arginine	0	1
Isoleucine	1	0
Threonine	0	1
Proline	0	1
Lysine	0	1

TABLE 3-3. ABBREVIATIONS OF THE AMINO ACIDS

Name	Abbreviation
Glycine	Gly
Alanine	Ala
Valine	Val
Leucine	Leu
Serine	Ser
Threonine	Thr
Tyrosine	Tyr
Phenylalanine	Phe
Tryptophan	Try
Aspartic acid	Asp
Isoleucine	Ileu
Glutamic acid	Glu
Lysine	Lys
Arginine	Arg
Histidine	His
Cysteine	CySH
Cystine	CySSCy
Methionine	Met
Proline	Pro
Hydroxyproline	Hypro
Cysteic acid	$CySO_3H$
Asparagine	$AspNH_2$
Glutamine	$GluNH_2$

demonstration that the general problem was by no means insurmountable.

Insulin is a protein hormone which is important in stimulating the oxidative utilization of glucose by tissues. Its administration serves to relieve the symptoms of diabetes.

That insulin was the first protein to yield to this kind of analysis is due both to its ready availability in a purified state and to the fact that the primary structure of its ultimate molecular unit is relatively simple. In aqueous solution, insulin appears to exist as a unit of molecular weight 12,000, which polymerizes reversibly to dimers and trimers of molecular weight 24,000 and 36,000.

The minimum molecular weight of insulin, as determined from amino acid analysis, is close to 6,000. This suggested that the 12,000 unit observed in aqueous solution might be a stable dimer. This indeed proved to be the case. Subsequent physical studies showed that in organic solvents, such as glacial acetic acid or dimethylformamide, insulin exists as a unit of molecular weight 6,000, which can be identified as the basic monomer unit.

Use of the DNP labeling technique showed that the monomer contained two NH_2-terminal residues—phenylalanine and glycine. It thus consisted of two polypeptide chains joined by one or more cystine bridges. The chain with NH_2-terminal glycine was designated as the A-chain and that with NH_2-terminal phenylalanine, the B-chain.

By performic acid oxidation of the cystine cross-links to cysteic acid, it proved possible to separate the A- and B-chains and to obtain them in purified form. Their amino acid compositions, as well as that of the intact molecule, are cited in Table 3-2. The A- and B-chains are 20 and 30 residues long, respectively.

Sequential analysis. The details of the determination of the sequence of the B-chain of insulin are summarized in Appendix A. Basically, the method depended upon controlled partial acid hydrolysis to a mixture of peptides, which were fractionated chromatographically, hydrolyzed completely, and analyzed for amino acids.

As an illustration, let us consider the sequence determination of the NH_2-terminal pentapeptide. Conversion of the B-fraction to its DNP derivative prior to partial acid hydrolysis yielded a mixture of peptides containing DNP-labeled phenylalanine:

> DNP—Phe
> DNP—Phe, Val
> DNP—Phe, Asp, Val
> DNP—Phe, Asp, Glu, Val

Here and elsewhere, each amino acid will be abbreviated by its first three letters (Table 3-2). Phenylalanine was clearly the NH_2-terminal residue. The identification of the DNP-Phe-Val dipeptide indicated that valine must follow phenylalanine in the terminal sequence. The presence of a tripeptide containing these two residues and aspartic acid permitted specification of the terminal tripeptide as

Phe - Val - Asp

Similarly, the detection of the terminal tetrapeptide containing glutamic acid in addition to the above three permitted enlargement of the terminal sequence to

Phe - Val - Asp - Glu

Since aspartic acid is singly occurring in the B-chain, any peptide containing it must originate from the NH_2-terminal sequence. Thus the subsequent isolation of a fragment containing the above residues plus histidine identified the terminal pentapeptide as

Phe - Val - Asp - Glu - His

A knowledge of the terminal sequence proved to be of great assistance in solving the remainder of the problem, for it provided a fixed point in the primary structure, to which many sequences could be referred. Examination of the other small fragments produced by partial acid hydrolysis provided a series of partial sequences which collectively accounted for most of the B-chain (Fig. 3-9). However, because of the preferential hydrolysis of certain bonds, some nonoverlapping sequences were obtained, whose order was uncertain. For example, the lability of the peptide bond involving serine resulted in the invariable presence of this residue at the NH_2-terminal position of the peptides in which it occurred.

PEPTIDES OBTAINED FROM PHENYLALANYL CHAIN OF INSULIN (FRACTION B)

Dipeptides	Phe.Val Asp.Glu His.Leu CySO$_3$H.Gly Val.Asp Glu.His Leu.CySO$_3$H	Thr.Pro Lys.Ala	Gly.Glu Agr.Gly Glu.Arg	Tyr.Leu Val.CySO$_3$H Leu.Val CySO$_3$H.Gly	Ser.His Leu.Val. Glu.Ala Ala.Leu His.Leu Val.Glu Gly.Phe
Tripeptides	Phe.Val.Asp His.Leu.CySO$_3$H Val.Asp.Glu Leu.CySO$_3$H Glu.His.Leu	Pro.Lys.Ala	Gly.Glu.Orn (Glu, Arg, Gly)	Tyr.Leu.Val Leu.Val.CySO$_3$H Val.CySO$_3$H.Gly	Ser.His.Leu Val.Glu.Ala Leu.Val.Glu Ala.(Tyr, Leu)
Higher peptides	Phe.Val.Asp.Glu His.Leu.CySO$_3$H.Gly Phe.Val.Asp.Glu.His Glu.His.Leu.CySO$_3$H	Thr.Pro.Lys.Ala		Tyr.Leu.Val.CySO$_3$H Leu.Val.CySO$_3$H.Gly	Ser.His.Leu.Val Leu.Val.Glu.Ala His.Leu.Val.Glu Ser.His.Leu.Val.Glu Ser.His.Leu.Val.Glu.Ala
Sequence in fraction B	Phe.Val.Asp.Glu.His.Leu.CySO$_3$H.Gly	Thr.Pro.Lys.Ala	Gly.Glu.Arg.Gly	Try.Leu.Val.CySO$_3$H.Gly	Ser.His.Leu.Val.Glu.Ala Others

Fig. 3.9a. Peptides obtained by acid hydrolysis of B-chain of insulin (9).

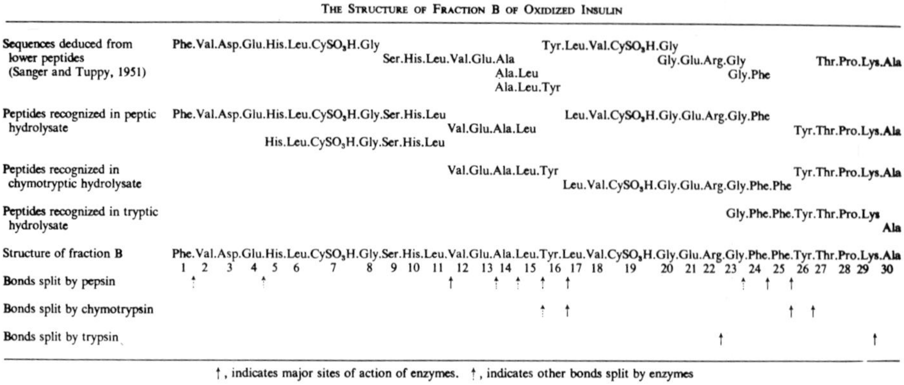

The Structure of Fraction B of Oxidized Insulin

Sequences deduced from lower peptides (Sanger and Tuppy, 1951)	Phe.Val.Asp.Glu.His.Leu.CySO₃H.Gly Ser.His.Leu.Val.Glu.Ala Ala.Leu Ala.Leu.Tyr	Tyr.Leu.Val.CySO₃H.Gly Gly.Glu.Arg.Gly Gly.Phe		Thr.Pro.Lys.Ala
Peptides recognized in peptic hydrolysate	Phe.Val.Asp.Glu.His.Leu.CySO₃H.Gly.Ser.His.Leu Val.Glu.Ala.Leu His.Leu.CySO₃H.Gly.Ser.His.Leu	Leu.Val.CySO₃H.Gly.Glu.Arg.Gly.Phe		Tyr.Thr.Pro.Lys.Ala
Peptides recognized in chymotryptic hydrolysate		Val.Glu.Ala.Leu.Tyr Leu.Val.CySO₃H.Gly.Glu.Arg.Gly.Phe.Phe		Tyr.Thr.Pro.Lys.Ala
Peptides recognized in tryptic hydrolysate			Gly.Phe.Phe.Tyr.Thr.Pro.Lys Ala	
Structure of fraction B	Phe.Val.Asp.Glu.His.Leu.CySO₃H.Gly.Ser.His.Leu.Val.Glu.Ala.Leu.Tyr.Leu.Val.CySO₃H.Gly.Glu.Arg.Gly.Phe.Phe.Tyr.Thr.Pro.Lys.Ala 1 2 3 4 5 6 7 8 9 10 11 12 13 14 15 16 17 18 19 20 21 22 23 24 25 26 27 28 29 30			
Bonds split by pepsin	↑ ↑ ↑ ↑ ↑ ↑ ↑ ↑ ↑ ↑			
Bonds split by chymotrypsin	↑ ↑ ↑ ↑			
Bonds split by trypsin	↑ ↑			

↑, indicates major sites of action of enzymes. ↑, indicates other bonds split by enzymes

Fig. 3-9b. Peptides obtained by enzymatic hydrolysis of B-chain of insulin (9).

The location of the partial sequences within the complete linear structure of the B-chain required the use of enzymatic hydrolysis (Fig. 3-9). In this manner fragments were obtained which overlapped the partial sequences obtained by acid hydrolysis and permitted their arrangement in the correct linear order to yield the complete amino acid sequence of the B-chain (Fig. 3-9).

The sequence of the A-chain was solved by entirely analogous procedures. There remained the location of the three disulfide bridges which united the A- and B-chains. It may be recalled that the separated chains were obtained by performic acid oxidation of these cross-links. The oxidized A-chain contained four and the oxidized B-chain, two, residues of cysteic acid originating from the three cystine residues.

The assignment of the correct positions of the three —S—S— bridges required the examination of fragments obtained from intact insulin under conditions where the disulfide bonds were not altered (9, 23). This proved to be difficult since the normal conditions of acid hydrolysis led to a random rearrangement of the —S—S— bonds. The proper choice of conditions required a good deal of exploratory work.

Examination of fragments containing a cystine bridge permitted their location within the known sequences of the A- and B-chains, thereby virtually completing the determination of the primary structure of insulin. Subsequent work showed that several residues initially identified as aspartic or glutamic acid really existed in the intact protein as asparagine or glutamine. The amide groups were split off under the conditions of complete acid hydrolysis and milder methods were required for their detection.

The complete covalent structure of bovine insulin is shown in Fig. 3-10.

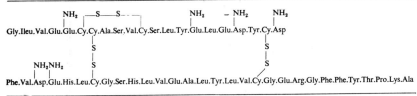

Fig. 3-10. Complete primary structure of bovine insulin. Note positions of amide groups.

The status of other proteins. At the present time, the primary structures of only a handful of true proteins are known in their entirety, although the list can be expected to expand rapidly in the next few years. A discussion of the enzyme *ribonuclease*, whose structure was the next to be solved after that of insulin, will be deferred to Chapter 7.

The primary structure of the protein subunit of *tobacco mosaic virus* (Chapter 11) was subsequently solved by Fraenkel-Conrat and associates (24). This virus, which is the infective agent of a well-known disease of tobacco plants, consists of a single molecule of ribonucleic acid (molecular weight 2×10^6) which is complexed with about 2,000 identical protein subunits of molecular weight 17,500.

Each subunit consists of a single polypeptide chain containing 158 amino acids. There is no evidence for any form of cross-linking, and no cystine has been detected within the molecule. The tentative primary structure is cited in Table 3-4.

TABLE 3-4. THE PRIMARY STRUCTURE OF THE PROTEIN SUBUNIT OF TOBACCO MOSAIC VIRUS

N-Acetyl-Ser-Tyr-Ser-Ileu-Thr-Thr-Pro-Ser-GluNH$_2$-Phe-Val-Phe-Leu-Ser-Ser-Ala-Try-Ala-Asp-Pro-Ileu-Glu-Leu-Ileu-Leu-Asp-CySH-Thr-AspNH$_2$-Ala-Leu-Gly-AspNH$_2$-GluNH$_2$-Phe-GluNH$_2$-Thr-GluNH$_2$-GluNH$_2$-Ala-Arg-Thr-Val-GluNH$_2$-Val-Arg-GluNH$_2$-Phe-Ser-GluNH$_2$-Val-Try-Lys-Pro-Ser-Pro-GluNH$_2$-Val-Thr-Val-Arg-Phe-Pro-Asp-Ser-Asp-Phe-Lys-Val-Tyr-Arg-Tyr-AspNH$_2$-Ala-Val-Leu-Asp-Pro-Leu-Val-Thr-Ala-Leu-Leu-Gly-Ala-Phe-Asp-Thr-Arg-AspNH$_2$-Arg-Ileu-Ileu-GluNH$_2$-Val-Glu-AspNH$_2$-GluNH$_2$-Ala-AspNH$_2$-Pro-Thr-Thr-Ala-Glu-Thr-Leu-Asp-Ala-Thr-Arg-Arg-Val-Asp-Asp-Ala-Thr-Val-Ala-Ileu-Arg-Ser-Ala-Asp-Ileu-AspNH$_2$-Leu-Ileu-Val-Glu-Leu-Ileu-Arg-Gly-Thr-Gly-Ser-Tyr-AspNH$_2$-Arg-Ser-Ser-Phe-Glu-Ser-Ser-Ser-Gly-Leu-Val-Try-Thr-Ser-Gly-Pro-Ala-Thr

In addition to the examples cited above, the primary structures of several biologically active peptides of intermediate size have been determined. These include the β-melanocyte stimulating hormone (β-MSH, 18 amino acid residues) and corticotropin (ACTH, 39 amino acid residues), both of which are produced by the pituitary gland. The se-

Hydrolytic Agent	Amino Acid Sequence
Trypsin	Ser.Tyr.Ser(Met.Glu.His,Phe)Arg
Chymotrypsin	Arg.Try( )
Trypsin	Try.Gly.Lys.Pro.Val.Gly.Lys
Trypsin	Lys.Arg
Trypsin	Lys.Arg.Arg
Trypsin	Arg.Pro.Val.Lys
Acid	Pro(Val,Lys,Val,Tyr)
Trypsin	Val.Tyr.Pro.Ala.Gly.Glu(Asp,Asp,Glu,Ala,Ser,Glu,Ala,Phe,Pro,Leu,Glu,Phe)
Acid	Ala(Gly,Glu,Asp)
Pepsin	Asp.Glu
Pepsin	Asp(Glu,Ala)
Pepsin	Asp(Glu,Ala)Ser
Pepsin	Glu(Ala,Ser)
Pepsin	Ser.Glu
Pepsin	Ser(Glu,Ala)
Pepsin	Ser(Glu,Ala,Phe)
Pepsin	Glu,Ala.Phe
Pepsin	Phe(Pro,Leu,Glu)
Pepsin	Pro(Leu,Glu,Phe)

Complete sequence: Ser.Tyr.Ser.Met.Glu.His.Phe.Arg.Try.Gly.Lys.Pro.Val.Gly.Lys.Lys.Arg.Arg.Pro.Val.Lys.Val.Tyr.Pro.Ala.Gly.Glu.Asp.Asp.Glu.Ala.Ser.Glu.Ala.
1 2 3 4 5 6 7 8 9 10 11 12 13 14 15 16 17 18 19 20 21 22 23 24 25 26 27 28 29 30 31 32 33 34
Phe.Pro.Leu.Glu.Phe
35 36 37 38 39

Fig. 3-11. The primary structure of porcine corticotropin and the peptides derived therefrom by the action of proteolytic enzymes (7). Determination of the sequences of the fragments permitted the piecing together of the complete primary structure.

quences of both peptides have been determined by methods basically similar to those outlined earlier (Figs. 3-11 and 3-12).

The primary structure of *lysozyme* has recently been almost com-

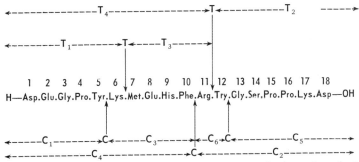

Fig. 3-12. Primary structure of β-melanocyte-stimulating hormone, showing points of attack by trypsin (T) and chymotrypsin (C). T_1-T_4 and C_1-C_6 represent peptide fragments isolated and characterized (9).

pletely determined. This protein is an enzyme of very wide distribution. It occurs in particularly high concentration in egg white. It attacks a constituent of the cell walls of many bacteria, thereby promoting lysis of the cells.

The molecular weight of lysozyme is 14,300. It consists of a single polypeptide chain cross-linked by four cystine bridges. The tentative primary structure proposed by Canfield is shown in Fig. 3-13. The positions of the cross-links remain to be determined (26).

Finally, the amino acid sequence of human *cytochrome c* is now known in its entirety. This is a *conjugated* protein; that is, it contains a group which is not a polypeptide. In the case of cytochrome c the group is a *heme* (section 13-2). Cytochrome c occurs in the mitochondria and forms part of the *respiratory chain* (Chapter 13) which figures in the final phase of the oxidative metabolism of carbohydrates and other foodstuffs.

Human heart cytochrome c consists of a single polypeptide chain of molecular weight 13,000. No cross-linking is present. The proposed primary structure is shown in Fig. 3-14.

3-6: VARIATIONS IN PRIMARY STRUCTURE

Species variation. A recurrent feature of comparative biochemistry is the mediation of parallel biological functions in widely separated species by proteins of similar, but not identical, primary structure. This is under-

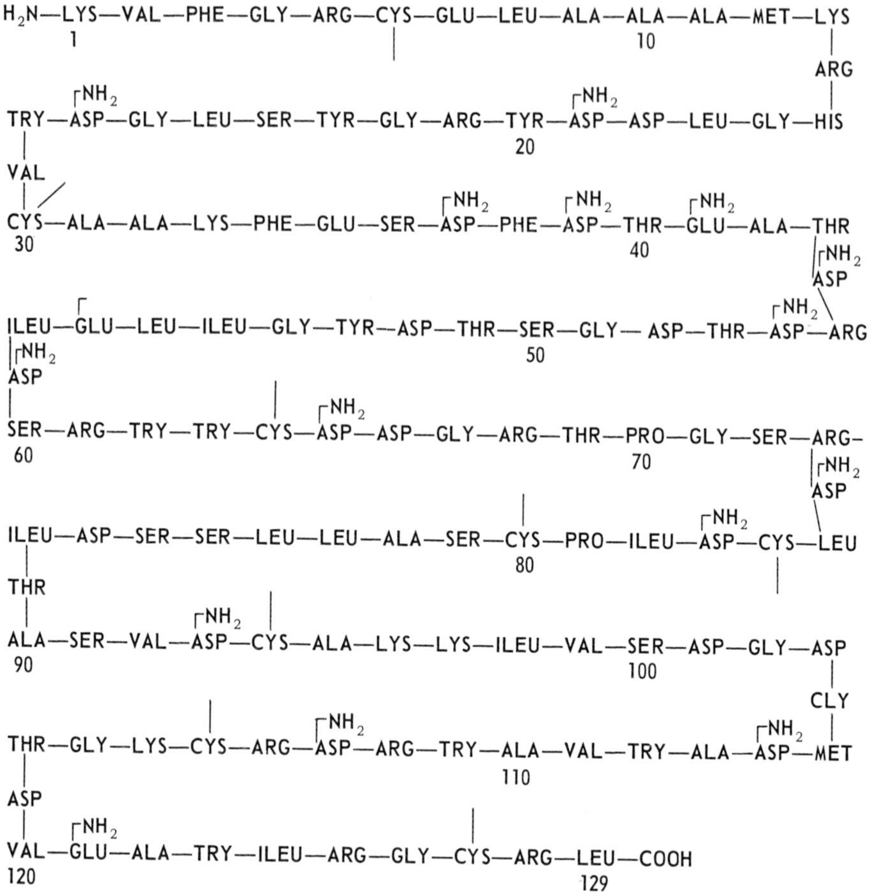

70 THE CHEMICAL FOUNDATIONS OF MOLECULAR BIOLOGY

```
H₂N—LYS—VAL—PHE—GLY—ARG—CYS—GLU—LEU—ALA—ALA—ALA—MET—LYS
    1                           |                      10              |
                                                                       ARG
         ⌈NH₂                                  ⌈NH₂                    |
TRY—ASP—GLY—LEU—SER—TYR—GLY—ARG—TYR—ASP—ASP—LEU—GLY—HIS
 |                              20
VAL
 |   /                      ⌈NH₂     ⌈NH₂         ⌈NH₂
CYS—ALA—ALA—LYS—PHE—GLU—SER—ASP—PHE—ASP—THR—GLU—ALA—THR
30                                      40                       |⌈NH₂
                                                                 ASP
                                                          ⌈NH₂  \
ILEU—⌈GLU—LEU—ILEU—GLY—TYR—ASP—THR—SER—GLY—ASP—THR—ASP—ARG
 |⌈NH₂                             50
 ASP
 |                        ⌈NH₂
SER—ARG—TRY—TRY—CYS—ASP—ASP—GLY—ARG—THR—PRO—GLY—SER—ARG-
60                                              70                |⌈NH₂
                                                                  ASP
                             |                  ⌈NH₂
ILEU—ASP—SER—SER—LEU—LEU—ALA—SER—CYS—PRO—ILEU—ASP—CYS—LEU
 |                               80                             |
THR
 |          ⌈NH₂     |
ALA—SER—VAL—ASP—CYS—ALA—LYS—LYS—ILEU—VAL—SER—ASP—GLY—ASP
90                                       100                     |
                                                                CLY
            |    ⌈NH₂                                     ⌈NH₂   |
THR—GLY—LYS—CYS—ARG—ASP—ARG—TRY—ALA—VAL—TRY—ALA—ASP—MET
 |                              110
ASP
 |    ⌈NH₂                          |
VAL—GLU—ALA—TRY—ILEU—ARG—GLY—CYS—ARG—LEU—COOH
120                                 129
```

Fig. 3.13. Tentative amino acid sequence of lysozyme (26).

standable in terms of the interrelationship of structure and function for biologically active proteins. In general, certain parts of the primary structure of such proteins are much more essential to their function than others. A protein can often tolerate an alteration in sequence in a noncritical area without impairment of its function provided that the critical arrangement of amino acids in the center of activity remains intact.

The recurrence in very dissimilar organisms of homologous proteins of similar primary structure and biological function is to be expected if such proteins arise from closely similar genes present in the two species. It is not possible to verify this concept by the standard methods of genetics, since the classical tests for allelism (Chapter 1) apply only to organisms which can be crossed.

THE CHEMICAL STRUCTURE OF PROTEINS

```
Acetyl—Gly—Asp—Val—Glu—Lys—Gly—Lys—Lys—Ileu—Phe—
       1                                          10
Val—GluNH₂        Ala
Ileu—Met—Lys—Cys—Ser—GluNH₂—CyS—His—Thr—Val—Glu—Lys—
 11   12          15                              20
                  └──── Heme ────┘
Gly—Gly—Lys—His—Lys—Thr—Gly—Pro—AspNH₂—Leu—His—Gly—
                    30
Leu—Phe—Gly—Arg—Lys—Thr—Gly—GluNH₂—Ala—Pro—Gly—
                        40
Phe—Thr             Asp
Tyr—Ser—Tyr—Thr—Ala—Ala—AspNH₂—Lys—AspNH₂—Lys—Gly—
 46   47            50
        Thr     Lys     Glu
Ileu—Ileu—Try—Gly—Glu—Asp—Thr—Leu—Met—Glu—Tyr—Leu—Glu—
       58       60    62
AspNH₂—Pro—Lys—Lys—Tyr—Illeu—Pro—Gly—Thr—Lys—Met—Illeu—
 70                                               80
       Ala                  Thr         Glu
Phe—Val—Gly—Illeu—Lys—Lys—Lys—Glu—Glu—Arg—Ala—Asp—Leu
 83                         89   90          92
Ileu—Ala—Tyr—Leu—Lys—Lys—Ala—Thr—AspNH₂—GluCOOH
                 100                    104
```

Fig. 3-14. Primary structure of human heart cytochrome c (27). The residues given in italics are those found in the corresponding positions in horse heart cytochrome c.

If this model is provisionally accepted, it is not surprising to find minor differences in amino acid sequence in homologous proteins occurring in different species. Mutations occurring in the course of evolutionary time which result in a modification of the amino acid sequence in a region of secondary importance might have no deleterious effect upon the viability of the organism, so that the altered gene could be passed on to subsequent generations. On the other hand, a mutation leading to an alteration in the "active center" might well be lethal so that the gene would not be perpetuated.

A parallel study of the primary structures of homologous proteins isolated from different species should therefore provide some idea of the rates at which viable mutations have occurred in the course of evolutionary time, as well as an index of the minimum structure which is indispensable for biological activity. Such investigations have been greatly facilitated by the use of the "fingerprinting" technique discussed in section 3-4.

As an example of sequential differences in homologous proteins, let us consider the case of insulins derived from different mammalian species.

Sanger and co-workers have determined the primary structures of insulins from five species (Table 3-5). The B-chains were identical in all

TABLE 3-5. SPECIES VARIATION IN A-CHAIN OF INSULIN

Species	Sequence
Beef	$CySO_3H$-Ala-Ser-Val
Sheep	$CySO_3H$-Ala-Gly-Val
Pig	$CySO_3H$-Thr-Ser-Ileu
Horse	$CySO_3H$-Thr-Gly-Ileu
Sperm whale	$CySO_3H$-Thr-Ser-Ileu

cases. The only differences observed were localized within the disulfide loop of the A-chain, suggesting that this zone is not very critical with regard to hormonal activity.

A second example is provided by the hormone adrenocorticotropin (ACTH). Definite differences in sequence occur in certain restricted areas between preparations obtained from sheep and from pigs (Table 3-6). The structure of the balance of the molecule is identical for the

TABLE 3-6. SPECIES VARIATION IN ACTH

Species	Sequence
Sheep	Ala-Gly-Glu-Asp-Asp-Glu-Ala-Ser-Glu-NH_2
Pig	Asp-Gly-Ala-Glu-NH_2-Asp-Glu-NH_2-Leu-Ala-Glu

two. The zones in which alterations occur have been independently shown to be unessential for activity.

It is difficult to generalize as to the nature of the variations. In general they appear to take the form of substitutions rather than deletions or inversions. The nature of the substituting residue is not always the same for a particular amino acid. Thus in the cases of sheep and pig ACTH (Table 3-6), alanine is replaced by aspartic acid in one region and by leucine in another. We shall return to this question in Chapter 11.

Mutational alterations. The existing evidence on the biochemical consequences of mutations has tended strongly to favor the concept that the biosynthesis of particular proteins is mediated by individual genes, at least in many cases. This idea finds its most forthright expression in the phrase "one gene—one enzyme." The recent progress in the mapping of genetic fine structure is beginning to make it possible to discuss gene function in chemical terms, rather than abstract formalisms (Chapter 11).

We have seen that biologically active proteins can often tolerate considerable alterations in the sequence of noncritical zones of their primary structure. This raises the possibility that nonlethal mutations might give rise to isomeric forms of the same protein in different individuals of the same species.

A number of examples of such isomerism have been uncovered, and the list is likely to expand as more information as to primary structure becomes available. The classical example is the alteration in the primary structure of human hemoglobin (Chapter 6) occurring in individuals suffering from the hereditary disease sickle-cell anemia.

The transmission of this serious disease obeys the laws of Mendelian genetics. Sickle-cell hemoglobin (hemoglobin S), which differs from normal hemoglobin (hemoglobin A) in electrophoretic mobility (Chapter 4), replaces the latter form in anemic individuals and coexists with it in individuals with *sickle-cell trait*, an asymptomatic condition.

Genetic analysis has traced this disease to a single mutant gene. In genetic terms, the anemic individual may be described as homozygous for this gene (Chapter 1), while the individual with the sickle-cell trait is heterozygous. If both allelic forms are present, both types of hemoglobin are manufactured, while if the individual has a double set of mutant genes, only hemoglobin S is produced (7).

Fingerprinting studies have localized the difference in primary structure between hemoglobin A and S at a single point in the sequence:

A . . . . Val-His-Leu-Thr-Pro-*Glu*-Glu-Lys
S . . . . Val-His-Leu-Thr-Pro-*Val*-Glu-Lys

This result provides a direct correlation between a mutational event in a specific genetic locus and a definite change in the structure of a phenotypic protein related to this locus. The allelic genes for hemoglobins A and S appear to be related to the same structural aspect, the sequence at one unique point.

Another instructive example is provided by the tryptophan synthetase enzyme system of *Escherichia coli* (*E. coli*). This has been shown by Yanofsky and co-workers (25) to contain two separable protein components, designated as A and B, whose combined action is essential for the synthesis of tryptophan. Most of the available information is concerned with the A protein, which has been obtained in purified form and shown to have a molecular weight close to 29,500.

About 150 mutants of *E. coli* incapable of forming a normal A protein have been identified after treatment with ultraviolet light and chemical mutagenic agents. The altered site in all of these has been located in a particular region of the genetic map of *E. coli*, which is referred to as the A gene. Two main groups of mutations exist. The first of these produce molecules analogous to the normal A protein, while the second class form no protein of this type.

An entire group of mutations of the first type have been detected which produce altered forms of the A protein differing only in the penultimate amino acid at the COOH-terminal end of the molecule (Table 3-7). Since

TABLE 3-7. MUTATIONAL ALTERATIONS IN THE A PROTEIN OF E. COLI

Mutation	Sequence
Normal	Asp-(Pro$_2$, Ala$_2$)-(Glu-NH$_2$, Leu)-*Gly*-Phe
A-23	-*Arg*-
A23FR2	-*Ser*-

genetic mapping studies have placed the A-23 mutations at the same site, the various substitutions presumably arise from different modifications at the point of the A gene responsible for this particular residue.

Other examples are provided by several plant and bacterial viruses. We shall pursue this subject further in Chapter 11.

GENERAL REFERENCES

1. *Biophysical Chemistry*, J. T. Edsall and J. Wyman, Academic Press, New York (1958).
2. *The Proteins*, H. Neurath and K. Bailey (ed.), vol. 1, chaps. 2 and 3, Academic Press, New York (1953).
3. *The Physical Chemistry of Macromolecules*, C. Tanford, chap. 1, Wiley, New York (1961).
4. *The Molecular Basis of Evolution*, C. Anfinsen, chaps. 5, 6, and 7, Wiley, New York (1959).
5. *Currents in Biochemical Research*, D. Green (ed.), p. 434, Interscience, New York (1956).
6. *The Chemical Structure of Proteins*, W. Stein and S. Moore, *Scientific American*, **204**, 81 (1961).
7. C. Anfinsen and R. Redfield, in *Advances in Protein Chemistry*, vol. 11, Academic Press, New York (1956).
8. F. Sanger in *Advances in Protein Chemistry*, vol. 7, Academic Press, New York (1952).
9. *Analytical Methods of Protein Chemistry*, P. Alexander and R. Block, vol. 2, chaps. 1 and 12, Pergamon Press, London (1960).

SPECIFIC REFERENCES

Column chromatography

10. S. Moore and W. Stein in *Advances in Protein Chemistry*, vol. 11, Academic Press, New York (1956).
11. S. Moore, D. Spackman, and W. Stein, *Anal. Chem.*, **30**, 1185 (1958).
12. S. Moore and W. Stein, *J. Biol. Chem.*, **192**, 663 (1951).
13. S. Moore and W. Stein, *J. Biol. Chem.*, **211**, 893 (1954).

Paper chromatography

14. H. Cassidy, in *Technique of Organic Chemistry,"* A. Weissberger (ed.), vol. 10, p. 1933, Interscience, New York (1957).
15. R. Block, E. Durrum, and G. Zweig, *Paper Chromatography and Paper Electrophoresis*, Academic Press, New York (1958).
16. *Chemistry of the Amino Acids*, J. Greenstein and M. Winitz, vol. 1, p. 750, Wiley, New York (1961).
17. A. Levy, *Nature*, **174**, 126 (1954).

Amino acid sequence of insulin

18. F. Sanger, *Biochem. J.*, **44**, 126 (1949).
19. F. Sanger, *Biochem. J.*, **45**, 563 (1949).
20. F. Sanger and E. Thompson, *Biochem. J.*, **53**, 353 (1953).
21. F. Sanger and H. Tuppy, *Biochem. J.*, **49**, 463 (1951).
22. F. Sanger, E. Thompson, and R. Kitai, *Biochem. J.*, **59**, 509 (1955).
23. A. Ryle and F. Sanger, *Biochem. J.*, **60**, 535 (1955).

Amino acid sequence of subunit of tobacco mosaic virus

24. C. Knight, in *Protein Structure and Function*, Brookhaven Symposia in Biology, no. 13 (1960).

Mutational alteration of protein structure

25. C. Yanofsky, U. Henning, D. Helinski, and B. Carlton, *Fed. Proc.*, vol. 22, part 1, p. 75 (1963).

Amino acid sequence of lysozyme

26. R. Canfield, *J. Biol. Chem.*, **238**, 2691 (1963).

Amino acid sequence of cytochrome c

27. H. Matsubara and E. Smith, *J. Biol. Chem.*, **238**, 2732 (1963).

4

The Size, Shape, and Electric Charge of Protein Molecules

4-1: INTRODUCTION

The geometry of protein molecules is, in general, as sharply defined as their primary structures. The polypeptide chains of each protein are folded into a definite three-dimensional pattern which is quite specific and characteristic of the protein.

The task of determining the spatial geometry of a protein is perhaps even more formidable than that of establishing its primary structure. The problem has generally been solved in progressive stages, which may be compared to a series of increasingly less blurred photographs of the same subject.

The physical techniques applicable to protein *solutions** (1, 2, 3, 4), which will be discussed in this chapter, may be regarded as producing an image so fuzzy that only the bare outline and general shape of the molecule are revealed, with no visible detail. A sharper image requires the use of techniques, especially X-ray diffraction, which are applicable to proteins in the *solid (crystalline) state*.

The problem of determining the detailed internal fine structure of a protein, including the geometry of folding of the polypeptide backbone and the nature of the side-chain interactions, will be discussed in detail in Chapter 5. The most powerful structural probe is undoubtedly X-ray diffraction. However, the practical difficulties associated with this technique are such that it has as yet been applied successfully to only a few proteins.

For the vast majority of proteins the only information available as to

---

* Unless otherwise stated, the solvent will be assumed to be water.

the molecular geometry has been obtained by the less powerful, but much less difficult, solution methods. These cannot, in general, yield direct information as to the *detailed* molecular architecture, although valuable inferences have been obtained in some instances. Their basic role is to provide the essential background data which must precede the application of the more sensitive solid state methods.

A latent ambiguity in all studies of this kind is the relationship of the physical parameters obtained for isolated proteins to those prevailing in their natural environment, from which they have been more or less violently wrenched. In other words, to what extent are the purified proteins artifacts of preparation?

No general answer can be given to this question. In the case of relatively stable proteins which occur naturally in free solution, such as the plasma proteins, it is often possible to compare physical measurements upon the protein in an environment not too dissimilar from its biological milieu to those obtained for the purified system. In other cases circumstantial evidence, such as the reproducibility of the system or the duplication *in vitro* of its biological function, can partially relieve doubts as to the persistence of its original structure. The problem is most pronounced for proteins which are integrated into preformed structural elements of the cell, such as collagen (Chapter 6). In these cases a combination of physical measurements upon the isolated protein with direct electron microscopic observations upon the biological system in which it occurs has often permitted a description of the morphology of the latter on a molecular level (Chapter 6).

The kinds of information obtainable by the physical methods applicable to protein solutions include the following:

(1) The molecular homogeneity, or the number and relative amounts of the discrete molecular species present in a particular preparation. A knowledge of the molecular homogeneity is of course essential if structural probes are to yield meaningful information.

The techniques which yield information of this kind all depend upon the differing rates of movement of the various components in a force field, which in the case of *sedimentation* is centrifugal and in the case of *electrophoresis* is electrical.

(2) The molecular weight. This can be obtained unambiguously by several techniques provided that the material is monodisperse. If more than one component is present, average values of the molecular weight will be obtained, the type of average depending upon the nature of the technique.

The molecular weight can be obtained without the assumption of any geometrical model for the molecule by the techniques of osmotic pressure, light scattering, sedimentation equilibrium, and sedimentation-diffusion.

(3) The size and shape parameters. The problem of determining the shape and spatial extension cannot in general be solved exactly by the solution methods. For most proteins the best which can be done is to obtain the dimensions of the smoothed geometrical model which best approximates the actual molecular shape, which is of course somewhat irregular. The relevancy of the apparent dimensions obtained in this way depends upon the closeness with which the assumed model fits the actual molecule.

4-2: ELECTRON MICROSCOPY

While electron microscopy is not a solution method, it yields information of a similar nature and is best discussed in conjunction with techniques of this kind (5, 6, 7).

The estimation of the size and shape of proteins or other biopolymers by direct observation has obvious advantages. Since biopolymers are much too small to be observed by ordinary microscopes using visible light, it is necessary to make use of the electron microscope, in which ordinary light is replaced by beams of electrons, which are focused by electromagnetic "lenses." The very short wave-length of electron beams would in theory permit an extremely high resolving power—about 0.05 Å. However, various practical difficulties prevent this limit from being approached. The *effective* resolving power is of the order of 5-10 Å. This is sufficient to make electron microscopy a tool of great utility for the viruses and larger proteins.

A schematic model of an electron microscope is shown in Fig. 4-1. The electron source is a tungsten filament, from which the beam is accelerated by a high voltage (about 100 kilovolts). The beam passes successively through the condensing lens, the specimen, an objective lens, and finally a projector lens, which produces a magnified image of the specimen on a fluorescent screen. This may be viewed directly or photographed.

The "lenses" are magnetic fields produced by the passage of electric current through coils. They have a focusing effect on the electron beam which can be controlled by varying the current.

The sample to be observed in the electron microscope must be supported by a film which is transparent to electrons. Among the few materials which have the necessary mechanical strength when in the form of films of the necessary thinness (about 200 Å) is polyvinyl formal

SIZE, SHAPE, AND ELECTRIC CHARGE OF PROTEIN MOLECULES

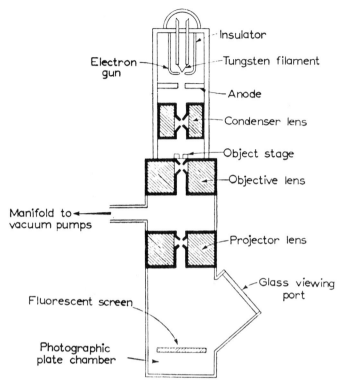

Fig. 4-1. Schematic diagram of an electron microscope (8).

(Formvar). The film, plus the specimen, is mounted upon a grid of copper wire mesh.

The protein may be applied to the supporting film by spraying from solution and drying. As the chamber of the electron microscope must necessarily be under high vacuum, the use of the technique is confined to water-free samples.

Since the protein itself often has insufficient contrast, it is frequently necessary to enhance its visibility by staining or shadow-casting. The usual staining procedures involve the addition of some material which is opaque to electrons and which combines selectively with the protein. This is followed by removal of excess stain by washing. Phosphotungstic acid has been frequently used for this purpose.

If molecules on a supporting film are stained with phosphotungstic acid, but incompletely rinsed, an image may be obtained in which the staining effect appears to be inverted. The low density of protein, in comparison with the high density of the surrounding phosphotungstic

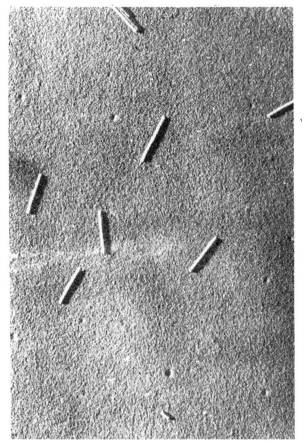

Fig. 4.2. Electron microscope photograph of tobacco mosaic virus (8). The length of the virus particles is close to 3000 Å.

acid produces the appearance of a white image on a dark background, giving a very high contrast. This technique is referred to as *negative staining*.

The alternative procedure is to increase contrast by evaporating a high density metal on to the sample. If the specimen is set at an angle to the evaporating source, the metal will pile up on one side of the particles preferentially. On the side away from the source there is an uncoated area, or "shadow." The high contrast thereby introduced extends the range of visibility down to particles about 40 Å in diameter (15 Å in the case of fibers).

Perhaps the most severe limitation of the technique arises from the requirement of a high vacuum. Thus no protein can be studied in its natural hydrated state. In addition, the technique is particularly liable to artifacts arising from aggregation or fragmentation of the molecules upon drying.

Among the most successful applications of electron microscopy to molecular biology has been the determination of the shape and dimensions of viruses (Fig. 4-2). It was by this technique that the size and shape of tobacco mosaic virus (TMV) were determined.

A second important area of application has been the examination of such fibrous systems as collagen, the primary component of connective tissue, and fibrin, whose formation is responsible for the clotting of blood.

In principle, electron microscopy would appear to offer a relatively unambiguous approach to the determination of the over-all size and shape of proteins. The method requires no arbitrary choice of a geometrical model and the problem of hydration, which beclouds the solution techniques, does not arise. Unfortunately, the effective resolving power of this technique at present is not sufficient to make it useful for the smaller and more symmetric proteins (of molecular weight less than 10^5).

4-3 METHODS YIELDING AVERAGE VALUES OF MOLECULAR WEIGHT DIRECTLY: OSMOTIC PRESSURE

General remarks. Most of the methods for molecular weight determination which have been generally adopted for small molecules, such as freezing point depression and boiling point elevation, become very insensitive for substances of molecular weight greater than about 10^3. Thus the study of biopolymers has required the development of entirely new techniques for molecular characterization.

One of the earliest methods to be applied successfully to the problem of molecular weight determination was osmotic pressure (8, 9, 10). This is useful for molecular weights between 10^3 and 10^5. The apparatus required is relatively simple, and measurements are rapid and convenient.

The protein solution is contained by a membrane which is in contact with solvent. The porosity of the membrane is such as to permit passage of the solvent molecules, as well as any electrolyte present, but not the protein. When this semipermeable membrane is in contact with pure solvent, the chemical potential of solvent on the solution side of the membrane will be less than that on the solvent side, as a consequence of the dilution of solvent by protein. The system is thus thermodynamically unstable, and transport of solvent across the membrane will occur until the chemical potentials on the two sides are equal. If the protein solution is confined, the only way this can occur is through an increase in hydrostatic pressure on the solution side.

If the solution compartment is connected to a vertical capillary open at the top, the increase in hydrostatic pressure will be reflected by a rise of liquid in the capillary (Fig. 4-3). The height of the liquid column at equilibrium is a direct measure of the osmotic pressure, which may be expressed as centimeters of water, or of Hg, or as dynes per square centimeter.

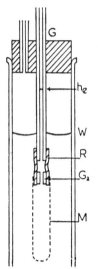

Fig. 4-3. Diagram of a simple osmometer (8). M is the membrane, in the form of a sac; G_2 is a glass connection; R is a short piece of rubber tubing; W is the surface of the solvent; and h_e represents the height of liquid in the capillary (G).

Theory. The basic equation of osmotic pressure is as follows:

$$(4\text{-}1) \qquad \frac{\pi}{RTC} = \frac{1}{M} + BC$$

where π = osmotic pressure, C = weight concentration of solute, R = gas constant, T = absolute temperature, M = molecular weight.

The parameter B is called the *second virial coefficient.* Its magnitude depends upon the nature and importance of the protein-protein interactions occurring in solution. In general, the magnitude of B decreases with decreasing electrostatic charge of the solute and with increasing concentration of electrolyte. In the limit, for an *ideal solution,* B is zero and equation 4-1 reduces to:

$$(4\text{-}2) \qquad \frac{\pi}{RTC} = \frac{1}{M}$$

If B is not negligibly small, it is necessary to make a linear extrapolation of π/RTC versus C. The reciprocal of the intercept at zero concentration yields the molecular weight (8, 9).

The case of a system which is polydisperse with respect to molecular weight deserves special mention. In this case the molecular weight obtained by osmotic pressure is a *number average* value (M_n), defined by:

$$(4\text{-}3) \qquad M_n = \frac{\Sigma n_i M_i}{\Sigma n_i}$$

where n_i is the number of moles of the ith species, of molecular weight M_i.

This kind of molecular weight average is characteristically obtained from the *colligative* methods, such as freezing point depression and boiling point elevation, which depend upon a thermodynamic equilibrium between two phases.

Techniques. For convenience, and to minimize any slow change of the protein with time, it is desirable that equilibrium be attained as rapidly as possible. Thus the choice of a suitable membrane is very important. Cellophane and cellulose nitrate membranes, whose porosity can be controlled by appropriate pretreatment, have often been used.

The capillary used (Fig. 4-3) has usually had a bore about one millimeter in diameter. The observed rise of liquid in the capillary is partly due to capillarity, for which a correction must be made. This can be minimized by introducing a layer of toluene on top of the protein solution, so as to fill partially the capillary bore. This procedure also serves to accelerate the rise due to capillarity. The height of liquid in the capillary may be read with a cathetometer.

4-4 METHODS YIELDING AVERAGE VALUES OF MOLECULAR WEIGHT DIRECTLY: LIGHT SCATTERING

General remarks. The scattering of light by small independent particles is a general property of all forms of matter and arises from the polarizability of the electrons of such systems. As a consequence of this electronic polarizability, such a particle will behave like an oscillating dipole when exposed to a periodically varying electric and magnetic field, such as occurs in a beam of light. The basic theory of electromagnetism predicts that such an oscillating dipole must emit, or scatter, radiation in all directions (11, 12, 13).

For a system of completely independent scattering particles, the intensity of scattered light is additive and directly proportional to the number of particles in unit volume (12, 13, 14). This would be the case, for example, for a gas at low pressure. At the other extreme, a completely ordered system, such as a perfect crystal at absolute zero, will scatter no light as a consequence of complete destructive interference of light dispersed from the different scattering centers. Ordinary solutions will normally represent an intermediate case.

The technique of light scattering involves the irradiation of a solution contained in a glass cell with cylindrical symmetry by a parallel and monochromatic beam of light (Fig. 4-4). The quantities of interest are the intensity of scattered light and its dependence upon the angle of

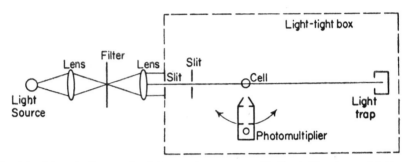

Fig. 4-4. Schematic diagram of a light-scattering apparatus (2). The incident beam of light produced by the light source (usually a Hg arc) is rendered parallel and monochromatic by the system of slits, lenses, and a filter. The cell containing the solution is usually of a cylindrical, or erlenmeyer, shape. The photomultiplier tube is mounted upon a movable arm which rotates about an axis upon which the platform holding the cell is placed. The light trap is a darkened cavity which is designed to eliminate the reflection of the incident beam by the wall of the apparatus.

observation (the angle between the scattered ray and the incident beam).

The information which can, in principle, be obtained from light scattering measurements upon solutions of biopolymers includes the molecular weight and a quantity characterizing the spatial extension of the molecule. If the molecule is known to be approximated by one of several simple geometrical shapes, as a sphere, or thin rod, then its characteristic dimension may be computed (13, 14, 15, 16).

Theory. The theory of the scattering of light by small independent particles, which was first developed by Lord Rayleigh a century ago, predicts that the intensity of light (i_θ) scattered in a direction forming an angle θ with the direction of the incident beam will be

(a) directly proportional to the intensity of the incident beam (I_o);
(b) inversely proportional to the square of the distance (r) between the illuminated volume and the point of observation
(c) inversely proportional to the fourth power of the wave-length (λ)
(d) directly proportional to $(1 + \cos^2 \theta)$

(4-4) $$i_\theta = \frac{K' \, I_o(1 + \cos^2 \theta)}{\lambda^4 \; r^2}$$

The constant K' depends upon the size and refractive index of the particle. It is convenient to eliminate factors (a), (b), and (d) by defining a *reduced intensity* R_θ which incorporates these factors.

$$(4\text{-}5) \qquad R_\theta = \frac{i_\theta r^2}{I_o(1 + \cos^2 \theta)}$$

The reduced intensity is the parameter which is determined experimentally. If the particles have no dimension comparable in magnitude to the wave-length of the incident light, then R_θ will be independent of θ. If its dimension is greater than about $\lambda/20$, destructive interference of light scattered from different parts of the same molecule results in the progressive reduction of R_θ with increasing θ.

For particles which are small in comparison with the wave-length, it is sufficient to determine R_θ at only one angle, usually 90°. In this case R_θ may be replaced by R_{90}.

Sufficiently dilute solutions of biopolymers often approximate to the above case of completely independent particles. The fact that light scattering occurs for a condensed phase is a consequence of local fluctuations in concentration. In any small region of the solution the concentration fluctuates continually from moment to moment as a result of the thermal, or *Brownian*, motion of the solute molecules, although the concentration averaged over a long period of time is equal to that in the solution as a whole. If fluctuations did not occur and the concentration were uniform everywhere, the system would be completely ordered and no light would be scattered.

The theory of fluctuations gives for the reduced intensity (in excess of that of the solvent) of a solution of independent isotropic particles of molecular weight M and concentration C, which are small in comparison with λ:

$$(4\text{-}6) \qquad \frac{KC}{R_{90}} = \frac{1}{M}$$

where $K = 2\pi^2 n_o^2 (dn/dC)^2/N_o \lambda^4$; $n_o =$ refractive index of solvent and $N_o =$ Avogadro's number. The quantity dn/dC is the *refractive increment* of the solute and is normally constant for concentrations of 2% or less. It is equal to the slope of the (linear) variation of the refractive index of the solution with the concentration of solute.

In general the particles will not be completely independent. This is the case for nonideal solutions. Thus equation (4-6) must be replaced by the more general expression:

$$(4\text{-}7) \qquad \frac{KC}{R_{90}} = \frac{1}{M} + AC$$

Equation (4-7) is analogous in form to equation (4-1) and the constant A is related to the constant B occurring in the latter by:

$$(4\text{-}8) \qquad A = 2B$$

The molecular weight of the solute is equal to the reciprocal of the intercept at zero concentration of $K(C/R_{90})$ plotted as a function of concentration. Equation (4-7) is valid only for particles small in comparison with (less than about 1/20 of) the wave-length of incident light.

If the solute molecules are not negligibly small in comparison with the wave-length of incident light, then R_θ depends upon θ, as a consequence of interference of light scattered from different parts of the same molecule (12, 13). This always has the effect of reducing the intensity of light scattered at the higher angles. For this case the reduced intensity at infinite dilution is given by:

$$(4\text{-}9) \qquad R_\theta = KCP_\theta M$$

Here P_θ is the *internal interference* factor and contains the angular dependence of R_θ. The form of P_θ depends upon the shape and dimension of the scattering particle. P_θ always has a limiting value of unity at $\theta = 0$ and decreases with increasing θ ($0 \lesssim \theta \lesssim \pi$).

At finite concentrations equation (4-9) may be rewritten as

$$(4\text{-}10) \qquad K\frac{C}{R_\theta} = \frac{1}{MP_\theta} + AC$$

It has been shown that P_θ^{-1} can be expanded into a series in $\sin^2(\theta/2)$ whose first two terms are independent of the particle shape (12, 16).

$$(4\text{-}11) \qquad P_\theta^{-1} = 1 + \frac{16}{3}\pi^2(R_G/\lambda')^2 \sin^2\frac{\theta}{2} \cdots$$

Here λ' is the wave-length in the medium and is equal to λ/n_o. The quantity R_G is the *radius of gyration* and is defined by

$$(4\text{-}12) \qquad R_G^2 = \frac{\Sigma n_i R_i^2}{\Sigma n_i}$$

where n_i is the number of small subelements of the particle at a distance R_i from the center of mass.

The radius of gyration is the only shape parameter which can be obtained from light scattering without making some assumption as to the form of the particle.

Equation (4-11) provides the basis for obtaining the radius of gyration from the angular dependence of reduced intensity. The usual practice is to evaluate M and R_G from a double extrapolation to zero angle and zero concentration. KC/R_θ is plotted as a function of $\sin^2 \theta/2 + KC$, where K

is an arbitrary constant of any convenient magnitude. The resultant two-dimensional plot is known as a *Zimm grid* (Fig. 4-5).

The intercept at zero angle and zero concentration is equal to the reciprocal of the molecular weight. The limiting slope at $\theta = 0$ and $C = 0$ of $K(C/R_\theta)$ as a function of $\sin^2(\theta/2)$ is given by:

$$(4\text{-}13) \qquad \text{slope} = \frac{16}{3M} \pi^2 \left(\frac{R_G}{\lambda'}\right)^2$$

Equations (4-11) and (4-13) provide the basis for the evaluation of the radius of gyration from the angular dependence of reduced intensity. It must be emphasized that equation (4-13) applies to the *limiting* slope at $\theta = 0$. As curvature is usually present, it is important to extend measurements to as low angles as possible.

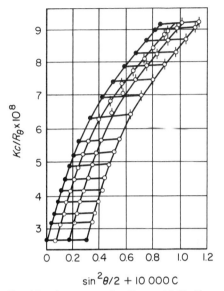

Fig. 4-5. A representative Zimm grid (2). The intercept at zero angle and zero concentration is equal to the reciprocal of the molecular weight. KC/R_θ is plotted as a function of $\sin^2 \theta/2$ for three concentrations. The term 10,000 C is added to $\sin^2\theta/2$ in the abscissa to separate the three angular plots and facilitate extrapolation to zero concentration. The plot at the left (•) represents the extrapolated limiting curve of KCR/R_θ as a function of $\sin^2\theta/2$ at $C = 0$. Each point in the extrapolated curve is obtained by linear extrapolation of the three points corresponding to the same angle for the three curves at finite concentrations.

If the shape of the particle is known independently, it is possible to compute its characteristic dimension from the radius of gyration, provided that the actual shape can be approximated by one of several models of simple geometry. Among these are the following:

(a) Thin, rigid rod. This model would apply to an elongated, cylindrical or prolate ellipsoidal particle, whose width is small in comparison to the wave-length of light. A classical example of such a particle is tobacco mosaic virus (Fig. 4-2). The length (3000 Å) computed for this virus by light scattering has been shown to agree with that obtained by electron microscopy. For this model the radius of gyration is related to the length (L) by:

$$L = \sqrt{12}\, R_G$$

(b) Sphere. The properties of a number of proteins suggest that they do not deviate greatly from spherical symmetry. In this case,

$$c = \text{radius} = \sqrt{5/3}\, R_G$$

(c) Random coil. This model will be discussed in detail in section 4-5. It corresponds to a completely unorganized polymer whose configuration is governed by purely statistical factors. For this model R_G is related to the *average squared* value of the separation of the ends ($\langle \bar{r}^2 \rangle$) by:

$$\langle \bar{r}^2 \rangle = 6 R_G^2$$

For the above simple models, it is possible to obtain precise expressions for P_θ:

(4-14) $\quad$ rod: $\quad P_\theta = \dfrac{1}{X} \displaystyle\int_0^{2X} \dfrac{\sin w}{w}\, dw - \left(\dfrac{\sin X}{X}\right)^2$

where $\quad X = 2\pi \dfrac{L}{\lambda'} \sin \dfrac{\theta}{2}$

coil: $\quad P_\theta = \dfrac{2}{Y^2}(Y - 1 + e^{-Y})$

where $\quad Y = \dfrac{8}{3}\pi^2 \dfrac{\langle \bar{r}^2 \rangle}{\lambda'^2} \sin^2 \dfrac{\theta}{2}$

sphere: $\quad P_\theta = 9\left(\dfrac{\sin Z - Z \cos Z}{Z^3}\right)^2$

where $\quad Z = 4\pi \dfrac{c}{\lambda'} \sin \dfrac{\theta}{2}$

The quantity P_θ is shown as a function of particle dimension for the above three shapes in Fig. 4-6. In all three cases the first two terms of the series expansion of $1/P_\theta$ as a function of $\sin^2(\theta/2)$ are those of equation (4-11).

If the solute consists of several species of different molecular weight, the molecular weight obtained from light scattering is the weight average value (M_w), which is defined by:

(4-15) $$M_w = \dfrac{\Sigma M_i C_i}{\Sigma C_i}$$

where C_i is the *weight* concentration of the species of molecular weight M_i.

Technique. A parallel and monochromatic beam of light is necessary if the equations cited above are to be applicable. A mercury arc is frequently used as a light source. Both the 4360 Å and the 5460 Å line, when isolated by suitable filters, have been often used. A parallel beam is produced by an appropriate combination of slits and lenses.

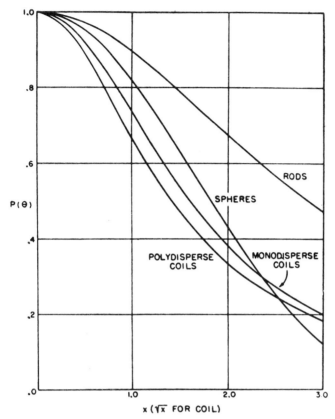

Fig. 4-6. Values of P_θ as a function of X (or Y for coils, Z for spheres) (15). The curve for polydisperse coils refers to a molecular weight distribution for which $M_w = 2M_n$.

The cylindrical scattering cell is placed on the axis of rotation of a movable arm, on which is mounted a photomultiplier tube. Scattered intensity is monitored by the photocell over an angular range usually extending from 135° to 20° or less. In order to compute reduced intensities, the intensity at each angle relative to that of the transmitted beam (I_o) must be obtained. Since the latter is usually larger than the intensity of scattered light by a factor of 10^3 or more, it is necessary to reduce I_o by a known factor in order to compare the two accurately. This is accomplished by introducing a series of darkened glass plates of known transmission into the beam when measuring I_o. The intensity of scattered or transmitted light is directly proportional to the photocurrent, which is registered by a galvanometer.

One of the practical difficulties encountered by this method is the great sensitivity of light scattering to the presence of large extraneous particles. The contribution of "dust" can overwhelm that of the solute and introduce severe errors. For this reason, extreme precautions must be taken to free the solution from such impurities. This may be done by centrifugation or filtration.

4-5 GENERAL ASPECTS OF THE HYDRODYNAMIC METHODS

General remarks. At this point a break in the continuity is necessary. The remaining methods are all of a hydrodynamic nature and depend upon the *motion* of particles in solution. As such, they share certain problems of theory and technique, a discussion of which should preface an account of the individual methods.

The first difficulty encountered in estimating the shape and dimensions of proteins in solution is that of defining and limiting the problem. As in the case of any irregular shape, the complete description of the shape of a typical protein would require a very large number of parameters. Moreover, the existing methods are insufficiently sensitive to permit the resolution of minor details of structure.

The protein chemist is thus forced to limit his objectives to determining which of a number of simplified models best describes the behavior of a particular protein. The first of these to be considered is appropriate to the case of a completely structureless, or *denatured,* protein.

Random coils. The possibility of free rotation about two of the three bonds of the repeating unit of polypeptides endows them with the potential capacity of adopting a large number of configurations. Under conditions where the intramolecular secondary bonds stabilizing the molecular organization of proteins are disrupted, they tend to revert to this random condition (4). This is often the case in the presence of certain denaturing agents, as high concentrations of urea.

A flexible polymer molecule, with no restraints upon the configurations it may assume, is said to be in the *randomly coiled* state. A formal model for such a system which provides a basis for the mathematical analysis of its properties is that of a series of short links forming a fixed angle (ϕ) with each other (Fig. 4-7). If the links possess complete freedom of rotation, then all possible orientations of each link with respect to the preceding link lie on the surface of a cone whose apex coincides with the junction of the two and whose apex angle is equal to ϕ.

The distribution of configurations for such a system is governed entirely by statistical factors. If the individual members of a system of random coils in solution could be observed simultaneously, they would be found to be a collection of all possible configurations ranging from very compact to highly extended.

The average extension will of course depend upon the value of ϕ. The smaller the value of ϕ, the greater will be the extension. In the limit, as ϕ approaches 0, the polymer will approach the completely extended rodlike state. It is not, of course, possible to define a unique dimension for a system of random coils. However, it is possible to characterize the spatial extension by an *average* quantity, the *root mean square end-to-end separation*. This is defined as the square root of the average of the squared separations of the termini of the chain.

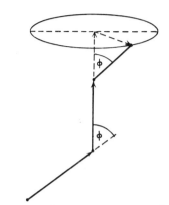

Fig. 4-7. Schematic representation of a randomly coiled polymer (4).

(4-16) $$\langle \bar{r}^2 \rangle^{1/2} = \{\Sigma n_i r_i^2 / \Sigma n_i\}^{1/2}$$

where n_i is the number of chains of end-to-end distance r_i.

Statistical analysis has yielded the following expression for the value of $\langle \bar{r}^2 \rangle$ for a chain of n links of length l and bond angle ϕ:

(4-17) $$\langle \bar{r}^2 \rangle = nl^2 \frac{1 + \cos \phi}{1 - \cos \phi}$$

The distribution of end-to-end separations has been shown to be of the Gaussian form.

(4-18) $$p(r)\, dr = \left(\frac{3}{2\pi \langle \bar{r}^2 \rangle}\right)^{3/2} 4\pi r^2 e^{-3r^2/2\langle \bar{r}^2 \rangle}\, dr$$

where $p(r)\, dr$ is the probability of an extension lying between r and $r + dr$.

As a rule, proteins are found in the randomly coiled configuration only under conditions where their internal structure is largely or entirely lost. *Native* proteins possess a high degree of molecular organization and exist in sharply defined three-dimensional patterns. In general, their hydrodynamic and other properties are those of rigid particles which are impermeable to solvent.

Hydrodynamically equivalent shapes. The hydrodynamic methods which have been developed for the study of biopolymers in solution all involve the movement of the solute molecules in a viscous medium. The theories which have been developed for the interpretation of such measurements require the assumption of a geometrically simple shape in order to obtain useful solutions. The actual shapes of natural proteins are not known in detail. Because of the presence of side chains of varying size, it is not to be expected that they will be entirely regular in form (Fig. 4-8).

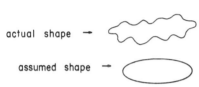

Fig. 4-8. Nature of the approximation involved in representing actual proteins by ellipsoidal models.

Thus the estimation of the dimensions of proteins by the hydrodynamic techniques requires that the actual shape be approximated by a geometrical model for which an exact theoretical treatment is possible. It has been possible to describe the properties of native proteins in solution adequately in terms of rigid bodies which are impermeable to solvent. This is a consequence of the highly organized character of native proteins, which appear to have very little *internal* space which is accessible to solvent.

The geometrical shape which has been most generally assumed is that of *ellipsoids of revolution* (Fig. 4-9). These may formally be regarded as generated by the rotation of an ellipse about either of its two axes. Rotation about the long axis (a) produces an elongated *prolate ellipsoid* (Fig. 4-9). Rotation about the short axis (b) results in a flattened *oblate ellipsoid*.

The hydrodynamic properties of ellipsoids of revolution can be described in terms of only two parameters—the volume and the ratio of the lengths of the two axes (a/b). A quantity of particular importance is the *frictional coefficient* (f), which is equal to the force which resists motion of the particle in the solvent. This arises from the viscous nature of the solvent and increases with increasing velocity of motion of the particle. When the particle is subject to a constant accelerating force, which may, for example, be gravitational, centrifugal, or electrical, the rate of migration will increase until the accelerating force is exactly balanced by the opposing frictional force. At this point the velocity attains its limiting value and is constant thereafter. A similar process occurs in the case of a parachutist who jumps from a great height. Under the (constant) gravitational field, his rate of fall increases only up to a certain *terminal velocity*, which depends upon the characteristics of his parachute.

For particles of a particular volume, the frictional coefficient is at its

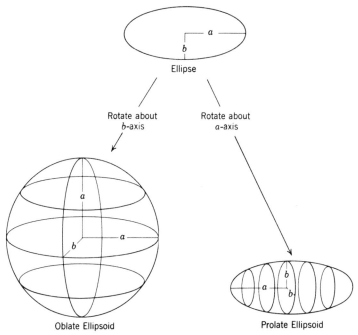

Fig. 4-9. Generation of prolate and oblate ellipsoid by rotation of corresponding ellipses (4).

minimum value for a spherical shape and increases with increasing asymmetry. For the limiting case of a sphere of radius c, moving with velocity v in a solvent of viscosity η_0, we have from *Stokes'* law:

(4-19) frictional force $= f_o v$; where $f_o =$ frictional coefficient $= 6\pi c \eta_0$

It is convenient to represent the effects of increasing asymmetry in terms of the unsolvated *frictional ratio* $(f/f_o)_u$, which is equal to the ratio of the frictional coefficient for a particle of a particular shape to that of a sphere of the same volume. For the ellipsoidal model the following relationships have been derived between the frictional ratio and the axial ratio ($J = b/a$ for prolate; $J = a/b$ for oblate):

(4-20) $J < 1$ (prolate) $\left(\dfrac{f}{f_o}\right)_u = \dfrac{(1 - J^2)^{1/2}}{J^{2/3} \ln\left[1 + (1 - J^2)^{1/2}\right]/J}$

(4-21) $J > 1$ (oblate) $\left(\dfrac{f}{f_o}\right)_u = \dfrac{(J^2 - 1)^{1/2}}{J^{2/3} \tan^{-1}(J^2 - 1)^{1/2}}$

Thus if the molecular volume is known it is, in principle, possible to estimate the axial ratio from the frictional coefficient. The reliability of the result will depend upon the adequacy of the ellipsoidal model (Fig.

4-8). The approximate nature of the calculation must however be emphasized. Science in some measure resembles politics in being an art of the possible. The chief defense of this procedure is that it is the best available at present.

Another hydrodynamic parameter of interest is the *rotational relaxation time*, which is a measure of the rotary Brownian motion of the particle. For a spherical particle the relaxation time ρ_s may be defined as follows.

Each member of an assembly of rigid spherical particles is imagined to have a particular radial direction which is differentiated in some way. If, at zero time, all such radii are oriented in the same direction, then the progress of Brownian motion will be accompanied by a transition to a more random orientation. A measure of the disorientation produced in this way is the angle θ formed by each radius with the initial direction. The time required for the average value of cosine θ to fall from unity to $1/e$ (where e is the base of natural logarithms) is defined as the relaxation time ρ_s.

For a rigid spherical particle of molar volume V, the relaxation time is given by

(4-22) $\quad \rho_s = \dfrac{3\eta_o V}{RT}$ where η_o is the solvent viscosity; R, the gas constant; and T, the absolute temperature.

In the case of an ellipsoid of revolution, two relaxation times (ρ_1 and ρ_2) are required to describe the rotational diffusion about the two axes. The quantities ρ_1 and ρ_2 are defined in the same way as ρ_s.

Solvation. In general, the deviation of proteins in aqueous solution from the behavior expected for the limiting case of a rigid sphere cannot be accounted for solely by molecular asymmetry. The presence of numerous polar groups in the side-chains results in the binding of an appreciable amount of water by the protein (Fig. 4-10). This bound and immobilized solvent migrates with the protein and may be regarded as forming part of the effective kinetic unit (1).

Fig. 4-10. Hydration of a protein molecule.

The presence of bound *water of hydration* results in an increase in the effective volume of the molecule. It is this solvated volume, rather than the anhydrous volume, which determines the frictional coefficient of the molecule. If δ grams of water (density = 1) are immobilized per gm of protein of density ρ then the molecular volume is increased by the factor $(\rho^{-1} + \delta)/\rho^{-1}$ or $1 + \rho\delta$. The radius of the equivalent sphere is enlarged by the factor $(1 + \rho\delta)^{1/3}$. The frictional coefficient (equation 4-19) is also increased by this factor.

The combined effects of solvation and molecular asymmetry are given by the product of $(f/f_o)_u$ and $(1 + \rho\delta)^{1/3}$.

$$(4\text{-}23) \qquad \frac{f}{f_o} = \left(\frac{f}{f_o}\right)_u (1 + \rho\delta)^{1/3}$$

Unfortunately, it is very difficult to determine δ directly. Various indirect estimations place the order of magnitude of δ as $0.2 - 0.3$ for most proteins.

Finally, it should be pointed out that the presence of hydration does not influence the calculation of molecular parameters by the light scattering and osmotic pressure techniques discussed earlier. The molecular weights computed in these cases correspond to the anhydrous molecules. Hydration is important only for methods in which the *rate* of motion of the molecule is the important quantity.

A parameter of importance in ultracentrifugation is the *partial specific volume* ($\overline{V}$), which is defined as dV/dG, where V is the volume of the solution and G is the total weight of solute in grams. In practice, $\overline{V}$ is independent of concentration for the usual range of values of the latter. Experimentally, it is determined from measurements of the increment in density per unit concentration of protein.

Concentration gradients. In the cases of several of the hydrodynamic techniques for the study of proteins including diffusion, electrophoresis, and ultracentrifugation, it is essential to be able to monitor the variation of concentration with distance. The most commonly employed methods utilize the refractive index of protein solutions as an index of their concentration.

For protein concentrations of several per cent or less, the increase in refractive index (n) over that of the solvent (n_o) is directly proportional to the concentration (C) of protein. Thus a curve of $n\text{-}n_o$, or Δn, versus distance (x) is equivalent to a plot of C versus x multiplied by a constant.

$$(4\text{-}24) \qquad \Delta n = KC$$

Also, a plot of the first derivative with respect to distance, or *gradient*, of refractive index versus distance may be converted to a curve of concentration gradient as a function of distance,

$$(4\text{-}25) \qquad \frac{dn}{dx} = K\frac{dC}{dx}$$

Those hydrodynamic methods, including the three mentioned earlier, which involve the transport of the protein solute, generally are concerned with the change in the shape or position, or both, of a boundary between two protein solutions or between protein and pure solvent. The most frequently used of the optical methods which have been developed for this purpose is the *schlieren* technique (Fig. 4-11). This gives a plot of the

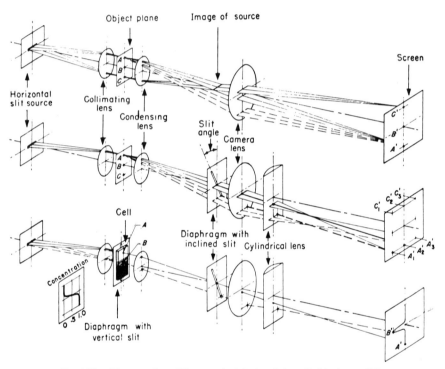

Fig. 4-11. Diagram of a schlieren optical device of the cylindrical type (17).

gradient of the refractive index (and hence of the concentration gradient) across the boundary.

Figure 4-12 compares the variation of the concentration and its gradient with distance for boundaries of various shapes. It should be noted that on either side of the boundary there is a zone where the concentration is either zero (on the solvent side) or has a constant value (or *plateau*). In both cases $dC/dx = 0$. The position of maximum height of dC/dx is that of the midpoint of the boundary, where the concentration is equal to one-half of its plateau value. This position is often referred to as the position of the *peak*.

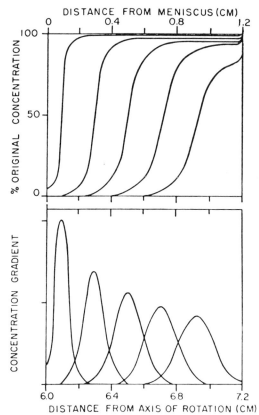

Fig. 4-12. The variation of concentration (upper) and concentration gradient (lower) with distance for the case of ultracentrifugation (4). The movement of a single boundary is shown. The lower set of curves corresponds to the first derivative with respect to distance of the upper set. Note the progressive broadening of the boundary by diffusion.

For a given plateau concentration, the peak becomes sharper and higher the more narrow the boundary. It should be noted that, irrespective of the shape of the boundary, the increase in concentration across it is given by:

$$\Delta C = \int_{x_1}^{x_2} \frac{dC}{dx}\, dx \tag{4-25}$$

where x_2 and x_1 are two (arbitrary) positions in the regions of constant concentration on either side of the peak. Or, in equivalent terms, ΔC is equal to the *area* under the peak.

If two consecutive boundaries are present, as in the case of two protein species migrating at different speeds, then two peaks of the concentration gradient are present, whose areas are equal to the concentrations of the two components.

Schlieren optics. The schlieren technique is the most commonly used method for observing the migration of a solute under the influence of an external force, as well as the change in concentration accompanying diffusion (17). This optical system permits a photographic record of the variation of refractive index *gradient* (and hence of concentration gradient) with distance in the direction of migration. The rate of movement of each boundary, and hence of the molecular species associated with the boundary, may thereby be determined (Fig. 4-11 and 4-12).

Consider first the upper arrangement of Fig. 4-11. Light rays from the horizontal slit source are rendered parallel by the collimating lens. Suppose that a thin diaphragm, containing three small holes (A,B,C) arranged in a vertical row, is placed between the collimating and condensing lenses. The optical elements are centered about a common axis and so arranged that: (1) the condensing lens forms an image of the slit before the camera lens; (2) the camera lens forms sharp images (A',B',C') of the holes upon the screen.

The positions of the optical elements shown in the upper part of Fig. 4-11 are unchanged during the remaining manipulations.

The family of rays emerging from each of the three holes of the diaphragm form a planar sheet which is always parallel to the slit source but diverges in width. It is ultimately focused again to a point on the screen. If a cylindrical lens with vertical axis is introduced into the system (Fig. 4-11, middle), then the three-point images on the screen are replaced by horizontal line images ($A'_1 A'_2 A'_3$, and so on). The *vertical* positions of the images are not affected and would be uninfluenced by a downward deviation of the sheet at A (broken lines, Fig. 4-11, middle).

If a diaphragm with a narrow inclined slit is placed in the plane of the slit image, then only the central ray (indicated by the heavy line) can pass through the system of slits and lenses to form a *point* image A'_2 on the screen, instead of the horizontal line. Similarly, the rays passing through B and C will also be represented by single points placed along a vertical line passing through A'_2 (since all light passing through A, B, and C is brought to a common slit image at the second diaphragm).

We come now to a crucial point. In the absence of the inclined slit, a downward deviation (dashed lines) of the planar sheet of rays at A, as might be caused by a prism, would not alter the position of the line image on the screen. However, if the inclined slit is present, a sufficient downward displacement will result in only the ray (heavy dashed line) at one edge of the sheet passing through the remaining optical elements to produce a point image corresponding to A at A'_1. The horizontal displacement of any light point from the central line $C'_2\ A'_2$ is directly proportional to the downward displacement encountered by the corresponding sheet of rays at the object diaphragm. Light from B or C is similarly affected.

This optical system can thus translate a *vertical* deviation of light into a *horizontal* displacement of the corresponding light point on a screen. No alteration occurs of the vertical coordinate of the light point, which corresponds to a particular level in the object diaphragm.

Replacement of the three holes A, B, and C by three narrow horizontal slits increases the intensity at the screen but has no effect upon the shape or position of the image, since the optical system has been arranged to bring all rays passing through a particular vertical level in the object diaphragm to a common focus.

Next a transparent flat-walled cell containing a refractive index gradient is substituted for the diaphragm so that the maximum gradient, which corresponds to the center of the boundary, is in the same position as B. The corresponding point B' will have the greatest horizontal displacement from the central line of the screen, as shown in the lower third of Fig. 4-11.

The displacements of light points corresponding to successive vertical positions within the boundary are proportional to their respective refractive index gradients. Since the variation of refractive index is continuous, the collection of light points forms a smooth continuous curve on the screen. Light traversing the cell in the zones of constant refractive index above and below the boundary, as at A, is not deviated and the corresponding points fall on the center line.

The schlieren optical system thus produces a curve whose horizontal displacement is directly proportional to the refractive index gradient at the corresponding vertical level of the cell.

4-6 METHODS YIELDING A MEASURE OF THE SIZE AND SHAPE, SUBJECT TO THE ASSUMPTION OF A GEOMETRICAL MODEL: VISCOSITY

General remarks. Of all the hydrodynamic parameters, viscosity (18-24) is probably the simplest to measure experimentally. The most common method of determination involves the timed flow of a volume of solution through a capillary (Fig. 4-13). The ratio of the flow time of the solution to that of the solvent is called the *relative viscosity* (η_r).

(4-27) $$\eta_r = \frac{t_{\text{solution}}}{t_{\text{solvent}}}$$

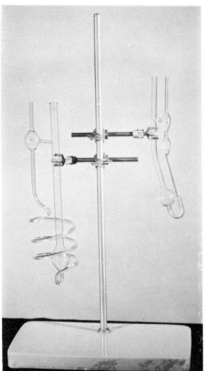

Fig. 4-13. Two types of capillary viscometer. That on the right is of the Ostwald-Fenske type. The solution is drawn by suction into the upper part of the right-hand arm of the viscometer until its meniscus is above the upper of the two small bulbs in series. Upon release of the suction, the time required for the meniscus to fall between two marks above and below the lower bulb is measured. This is the efflux time for the solution. The viscometer on the left represents a modification utilizing a long capillary of relatively wide bore. In this manner the rate of shear of the solution flowing through the capillary is kept to a low value, thereby minimizing orientation effects.

It is also convenient to define the *specific viscosity* (η_{sp}) by

(4-28) $\quad \eta_{sp} = \eta_r - 1$

For purposes of extrapolation it is helpful to convert the specific viscosity to the *reduced specific viscosity*, η_{sp}/C where C is the concentration (gms per ml).

Finally, the quantity of primary interest for the study of the shape of a protein solute is the *intrinsic viscosity*, $[\eta]$, which is defined as the limiting value of η_{sp}/C obtained by extrapolation to zero concentration. The extrapolation is often simplified by the linear variation of η_{sp}/C with C at concentrations less than about 1%.

Theory. The theories which have been developed to relate the intrinsic viscosity to the shape parameters of the dissolved particles depend upon the assumption that no orientation of the particles occurs in the capillary. This is not a problem except in the case of molecules of extreme asymmetry, such as DNA. The occurrence of orientation effects is reflected by the dependence of η_{sp} upon the rate of flow through the capillary. The existence of this factor may be checked for by examining η_{sp} as a function of the hydrostatic pressure head forcing the liquid through the capillary. If it is present, an extrapolation to zero pressure head, where orientation should be absent, should be made for each concentration.

For the case of rigid, impermeable spheres which are not solvated, the intrinsic viscosity is given by the *Einstein equation* (18, 19)

(4-29) $$[\eta] = 2.5/\rho$$

where ρ is the density of the (unsolvated) particle. Or, in terms of the volume fraction, ϕ:

(4-30) $$\frac{\eta_{sp}}{\phi} = 2.5$$

The value of η_{sp}/ϕ given by equation (4-30) represents the theoretical minimum value. Values of $[\eta]$ greater than $2.5/\rho$ may result from solvation or molecular asymmetry, or both. The effect of solvation is to increase the effective volume fraction, since for hydrodynamic purposes the bound solvent behaves like part of the dissolved particle. If δ is the weight of solvent bound by unit weight of solute, then the total volume fraction of solute is given by $(1 + \delta/\bar{V}\rho_o)\bar{V}C$, where ρ_o is the density of the solvent.

Unfortunately reliable values of δ are not yet available for biopolymers. Most estimates of its magnitude for proteins are of the order of 0.2-0.3.

For many proteins the value of $[\eta]\rho$ exceeds the Einstein value by a factor which is too large to be accounted for by any reasonable degree of solvation. This can only arise from a deviation from the impermeable spherical model.

If the assumption of impermeability to solvent is retained, the equation of Simha (20) may be utilized to compute the axial ratio of the equivalent ellipsoid of revolution whose intrinsic viscosity would correspond to that of the protein in question.

Prolate ellipsoids ($J' > 1$):

(4-31) $$\left[\frac{\eta_{sp}}{\phi}\right]_{C=0} = \frac{J'^2}{15(\ln 2J' - \frac{3}{2})} + \frac{J'^2}{5(\ln 2J' - \frac{1}{2})} + \frac{14}{15}$$

where J' = axial ratio = a/b for prolate and b/a for oblate.

Oblate ellipsoids ($J' < 1$):

(4-32) $$\left[\frac{\eta_{sp}}{\phi}\right]_{C=0} = \frac{16}{15} \frac{1/J'}{\tan^{-1}(1/J')}$$

Formally, any particular value of $[\eta]$ can be accounted for by either a prolate or oblate ellipsoid of the appropriate axial ratio. Its use requires some assumption as to the value of the solvation, δ. As a consequence of these uncertainties and of the imperfect character of the ellipsoidal model, apparent values of the axial ratio obtained from viscosity alone must be regarded as only semiquantitative.

For the other extreme of particle shape, the random coil, the concepts summarized above are not applicable (21-24). For a homologous series of randomly coiled molecules differing in molecular weight, such as a series of polymer fractions, the intrinsic viscosity has been shown to depend upon the molecular weight as follows (21):

(4-33) $$[\eta] = KM^a$$

where K and a are constants characteristic of the particular system. The constant a ranges in magnitude from 0.5 to 1.0. In practice, the obeying of equation 4-33 by a series of fractions of the same polymeric material is evidence that its configuration in solution is of the randomly coiled type.

For randomly coiled molecules, the sedimentation coefficient, S (section 4-9), and the intrinsic viscosity are related by the Flory-Mandelkern equation:

(4-34) $$S[\eta]^{1/3}M^{-2/3} = \Phi^{1/3}P^{-1}(1 - \overline{V}\rho_o)/\eta_o N_o$$

where N_o is Avogadro's number. The product $\Phi^{1/3}P^{-1}$ is a universal constant whose magnitude is close to 2.5×10^6 for a wide range of polymer systems. By the use of equation 4-34 the molecular weight may be computed from the intrinsic viscosity and sedimentation coefficient, *provided that the configuration of the system is that of randomly coiled chains.*

Technique. Considerable latitude exists in the design of capillary viscometers (23). The efflux time is directly proportional to L/R^4, where L is the capillary length and R the radius. Viscometer design normally represents a compromise between: (1) the increase in precision with longer efflux times, (2) the increasing sensitivity to dust with decreasing capillary radius, and (3) the desirability of minimizing shear rate, which increases with decreasing radius, so as to avoid orientation effects.

In practice, because of (2) and (3), it is usually preferable to increase the efflux time by increasing L. This may be done without unduly augmenting the size of the viscometer by having the capillary in the form of a spiral (Fig. 4-13). For further details, reference 23 should be consulted.

4-7 METHODS YIELDING A MEASURE OF THE SIZE AND SHAPE, SUBJECT TO THE ASSUMPTION OF A GEOMETRICAL MODEL: FLUORESCENCE POLARIZATION

General remarks. Proteins do not ordinarily fluoresce at visible wavelengths. They may however be converted to fluorescent derivatives by chemical conjugation with a fluorescent dye (25, 26, 27). The labeled protein thereby acquires the fluorescence characteristics of the dye. There does not in general appear to be any important modification in the structure of the protein, provided that the degree of labeling is low (about 1 to 2 residues per molecule).

If a solution of labeled protein is irradiated by a parallel beam of monochromatic light, whose wave-length is such as to activate the fluorescence of the label, then the emitted radiation observed at an angle of 90° to the incident beam will be partially polarized. The degree of polarization depends upon both the characteristics of the label and the molecular properties of the protein.

Measurements of fluorescence polarization can yield a value for the average rotational relaxation time of the labeled protein (section 4-5). This in turn provides information as to its size and shape and its structural rigidity (25, 26).

Theory. The degree of polarization (P) is defined by:

$$(4\text{-}35) \qquad P = \frac{I_V - I_H}{I_V + I_H}$$

where I_V and I_H are the relative intensities of the vertically and horizontally polarized components of the fluorescent light, respectively. That is, I_V and I_H are equal to the intensities of light transmitted by a polarizer with its optical axis vertical and horizontal, respectively.

The degree of polarization may be easily measured by means of a photocell set at 90° to the incident beam and in series with a polaroid or Nicol prism analyzer.

The polarization of fluorescence is a function of (1) the viscosity and temperature of the solvent; (2) the size, shape, and rigidity of the protein; and (3) the *excited lifetime* (τ) of the fluorescent label. This is equal to the interval of time which, on the average, elapses between excitation and emission.

The quantity which is measured by fluorescence polarization is the *mean rotational relaxation* time (ρ_h) of the labeled protein. This is an average of the relaxation times for rotation about the characteristic axes of the particle and, for an ellipsoid of revolution, is defined as follows:

$$(4\text{-}36) \qquad \frac{1}{\rho_h} = \frac{1}{3}\left(\frac{1}{\rho_1} + \frac{2}{\rho_2}\right)$$

where ρ_1 and ρ_2 are the relaxation times about the major and minor axes, respectively.

The quantity ρ_h is, of course, dependent upon the absolute temperature

(T) and the solvent viscosity (η_0), being proportional to η_0/T. It is customary to tabulate ρ_h for the standard conditions of water at 20° C.

The polarization at a particular temperature and viscosity is related to ρ_h by (for unpolarized incident light):

$$(4\text{-}37) \qquad \left(\frac{1}{P} + \frac{1}{3}\right) = \left(\frac{1}{P_o} + \frac{1}{3}\right)\left(1 + \frac{3\tau_o}{\rho_h}\right)$$

Here P_o, which depends primarily upon the characteristics of the label, is equal to the limiting value of P approached as T/η_o approaches zero. It may be obtained by the (linear) extrapolation of $1/P + 1/3$ versus T/η_o to $T/\eta_o = 0$ (Fig. 4-14).

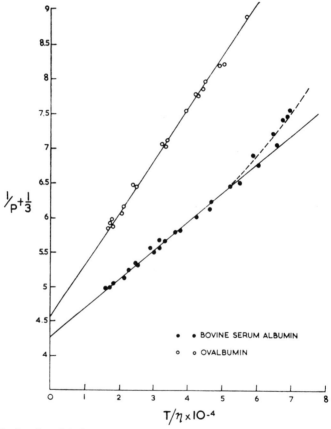

Fig. 4-14. Location of P_o by extrapolation for a conjugate of bovine serum albumin (25). The viscosity is varied by altering the temperature of the solvent. Once P_o has been obtained, the relaxation time at 20° may be computed from the value of the polarization at this temperature, using equation 4-37.

If P_o and τ are known, ρ_h may be computed from equation 4-37. It is a function of (1) the *volume* of the protein, including hydration, (2) the asymmetry, and (3) the internal rigidity.

For a perfectly rigid unhydrated sphere, ρ_s is given by:

$$\rho_s = \frac{3\eta_o V}{RT} \tag{4-38}$$

where R = gas constant, V = molar volume.

The ratio of ρ_h to the value of ρ_s computed for a molecule of the same molecular weight is an index of the deviation of the protein from the limiting case of an anhydrous sphere. Values of ρ_h/ρ_s greater than unity may reflect the presence of hydration or molecular asymmetry, or both (27). Values of ρ_h/ρ_s *less* than unity arise from the absence of complete structural rigidity and indicate the presence of internal rotations.

Technique. Among the fluorescent labels which have been extensively utilized for this purpose is 1-dimethylaminonaphthalene-5-sulfonyl chloride (DNS). This reacts with the ϵ-amino groups of the side-chains of lysine to form sulfonamido (—SONH—) derivatives.

(4-39)

DNS-SO₂Cl + H₂N—Protein → DNS-SO₂NH—Protein + HCl

Of the various devices which have been used for the measurement of the degree of polarization, perhaps the simplest and most satisfactory employs a photomultiplier tube to monitor the fluorescent radiation at 90° to the incident beam (27). A Nicol or polaroid analyzer is placed between the solution and the photocell. A comparison of the photocurrents produced with the axis of the analyzer in the vertical and horizontal positions permits computation of the polarization directly (equation 4-35).

4-8 METHODS YIELDING A MEASURE OF SIZE AND SHAPE, SUBJECT TO THE ASSUMPTION OF A GEOMETRICAL MODEL: DIFFUSION

General remarks. If the individual solute molecules of a protein solution could be observed directly, they would be seen to be in violent and erratic motion. This thermal, or *Brownian*, movement results from the continual bombardment of the macromolecules by solvent molecules. The trajectory of any individual particle would appear as a sequence of short linear tracks corresponding to the intervals between collisions.

The Brownian motion of each macromolecule is, in the absence of any external force, completely random. Thus the net distance traveled during a given time interval may be in any direction. As a consequence, an imaginary planar boundary dividing a solution of uniform concentration will, for finite time intervals, be crossed by equal numbers of solute molecules going from left to right and from right to left.

However, if the concentration of solute on the right of the dividing plane is greater than that on the left and if a solute molecule may with equal probability move in either direction, then the number crossing the plane from right to left will be greater than for the reverse direction. In other words, the net transport of solute, in the absence of an external force, is always such as to diminish a concentration gradient.

If a vessel is divided into two compartments by a removable partition and pure solvent is placed in one compartment and a protein, or other, solution in the second, then the cautious withdrawal of the partition will result in the production of an initially very sharp concentration gradient at the boundary (Fig. 4-15). As a consequence of diffusion of the solute

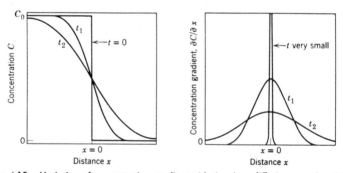

Fig. 4-15. Variation of concentration gradient with time in a diffusion experiment (4).

the boundary becomes progressively more gradual with time until, in the limit, the concentrations of solute in the two compartments will approach each other.

Theory. The net transport of solute across a plane of area A may be described by *Fick's first law* (28):

(4-40) $$\Delta G = -DA \frac{dC}{dx} \Delta t$$

SIZE, SHAPE, AND ELECTRIC CHARGE OF PROTEIN MOLECULES

Here ΔG is the number of grams of solute transported in the x direction, in the time interval Δt. dC/dx is the concentration gradient of the solute at the plane. The parameter D is the *diffusion coefficient* and serves to characterize the diffusion of the solute.

A more general equation describing the diffusion of a single solute may also be derived. This is *Fick's second law:*

$$(4\text{-}41) \qquad \frac{\partial C}{\partial t} = D \frac{\partial^2 C}{\partial x^2}$$

Equation (4-41) provides the basis for the experimental determination of D. It has been solved for a number of systems and in particular for the elongated rectangular diffusion cell which has been most generally used. In order for the diffusion to be capable of a simple description, several conditions must be met. The system must be free from convective disturbances so that transport of solute occurs exclusively by diffusion. The ends of the cell must be sufficiently far from the boundary so that the initial concentrations of solute at these points persist during the course of the experiment.

In practice, a sharp initial boundary is formed between a solution of concentration C_o and pure solvent. The progress of diffusion is usually followed by the schlieren method, which yields essentially a plot of the gradient of refractive index (dn/dx) versus x, the distance from the midpoint of the boundary. As the refractive index of the solution is directly proportional to the concentration, dn/dx is directly proportional to the gradient of concentration dC/dx. Thus a series of plots at increasing times of β (dC/dx) versus x is obtained, where β is the product of various instrumental constants.

The integration of equation (4-41) has been carried out and gives the following expression for the concentration as a function of x and t.

$$(4\text{-}42) \qquad C(x) = \frac{C_o}{2}\left\{1 - \frac{2}{\pi^{1/2}} \int_0^{x/2(Dt)^{1/2}} e^{-y^2} dy\right\}; \quad y = \frac{x}{2(Dt)^{1/2}}$$

Here C_o is the initial concentration. The integral in equation 4-42 is well known as the probability integral and is a function of $\dfrac{x}{2(Dt)^{1/2}}$. In order to analyze the experimental curves of concentration gradient as a function of distance, the derivative of equation 4-42 is needed.

$$(4\text{-}43) \qquad \frac{dC}{dx} = \frac{C_o}{2(\pi Dt)^{1/2}} e^{-x^2/4Dt}$$

As Fig. 4-15 shows, the progress of diffusion predicted by equation (4-43) is reflected by the flattening and broadening of the curve of dC/dx versus x, which has initially the form of a very sharp spike.

The area (A) under the experimental curve will be equal to

$$\beta \int_{-\infty}^{\infty} (dC/dx) \, dx$$

or βC_o. From equation (4-43) the maximum value of dC/dx will be at $x = 0$ and will be equal to $\dfrac{C_o}{2(\pi Dt)^{1/2}}$. The maximum height ($h_m$) observed experimentally is equal to $\beta C_o/2(\pi Dt)^{1/2}$. By combining this with the expression for the area, the unknown constant may be eliminated to yield an expression for D.

(4-44)
$$\frac{A}{h_m} = (4\pi Dt)^{1/2}$$

This provides a convenient means of evaluating the diffusion coefficient of a monodisperse substance. The value of D obtained in this way is moderately dependent upon concentration. The value found by extrapolation of D versus C to zero concentration (D_o) is the quantity useful for determinations of the size and shape of proteins. The diffusion coefficient is inversely proportional to the solvent viscosity (η_o) and directly proportional to the absolute temperature (T). It is usually reduced to standard conditions (H_2O, 20°C) by multiplying by the ratio of the value of (T/η_0) for standard conditions to that for the conditions of measurement.

Hydrodynamic theory has established that D_o is equal to kT/f, where f is the *frictional coefficient* and k is the Boltzmann constant (28, 29, 30). To the extent that the real molecule can be approximated by an ellipsoid of revolution, the apparent axial ratio may be computed from the frictional coefficient by the use of equation 4-20, subject to the usual reservations implicit in the hydrodynamic methods.

However, the most important application of diffusion is in computing molecular weights, in conjunction with measurements of the sedimentation coefficient (section 4-9).

Technique (31, 32, 33). The measurement of diffusion coefficients requires careful attention to a number of experimental details, including the avoidance of convection and vibration and the precise control of temperature. All methods depend upon the formation of an initially sharp boundary between solution and solvent. Some investigators have used a thin partition between solvent and solution, which can be withdrawn by the turning of a screw at the beginning of the experiment. Others have employed a sliding joint to superimpose the solvent upon the solution. In particular, an electrophoresis cell of the Tiselius type (Fig. 4-24) has often been used. In this case, the solvent and solution compartments are filled separately and then superimposed (section 4-10).

SIZE, SHAPE, AND ELECTRIC CHARGE OF PROTEIN MOLECULES

The majority of measurements have been made using a schlieren optical system to monitor the concentration gradient as a function of distance. The change with time of the curves of concentration gradient can be analyzed in the manner described in the theoretical section.

In the last few years the schlieren approach has been largely replaced by interferometric methods. An adequate description of these lies outside the scope of this book, and the reader should make use of the references cited at the end of this chapter (28, 31).

4-9 METHODS YIELDING BOTH THE MOLECULAR WEIGHT AND A MEASURE OF THE HOMOGENEITY: ULTRACENTRIFUGATION

General remarks. The settling, or *sedimentation,* of a coarse suspension produced by the action of gravity is a familiar phenomenon. If a suspension of chalk dust in a glass cylinder of water is allowed to stand, the initially opaque mixture slowly develops a clear band of pure solvent at the top, which enlarges as sedimentation progresses.

If the gravitational field is replaced by the centrifugal field obtainable with the ordinary laboratory centrifuge, the sedimenting force is increased by a factor of several thousand, permitting a greatly enhanced rate of sedimentation. Very much higher centrifugal fields are required to produce a reasonable rate of migration for protein molecules. Fields of up to 200,000 times gravity are necessary.

The centrifugal field, which is equal to the force (directed radially outwards) upon unit mass, is given by w^2r, where w is the angular velocity of rotation and r is the radial distance from the axis of rotation. Thus the only feasible way to attain the fields necessary is to increase the speed of rotation. The *ultracentrifuge* is a device for producing a very high angular velocity (34, 35). In practice, the solution is contained in a cell which is placed in a rotor driven at high speed (Figs. 4-16 and 4-17).

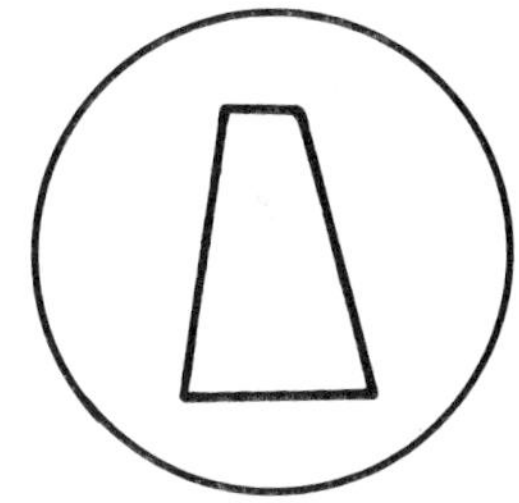

Fig. 4-16a. Cross-section of centerpiece of ultracentrifuge cell, in which the solution is contained.

There are two important variants of the ultracentrifugal method. In the first of these, the *sedimentation velocity* method, the rotor is driven at a velocity high enough to bring about essentially complete sedimentation of the solute, which is ultimately concentrated into a narrow layer at the outer periphery (bottom) of the cell.

The progress of sedimentation, as observed by a schlieren optical

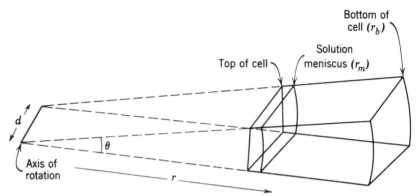

Fig. 4-16b. Schematic representation of an ultracentrifuge cell (4).

system, is accompanied by the emergence from the meniscus of a concentration gradient peak (Figs. 4-12 and 4-18). This corresponds to a boundary between solution and pure solvent, which forms a layer on the inner side of the boundary. As sedimentation progresses, the boundary moves toward the base of the cell, becoming broadened by diffusion.

If two or more solutes are present, more than one peak will appear. The area under each peak is directly proportional to the concentration of the corresponding component. In this manner, sedimentation analysis can provide a measure of the composition of a protein mixture.

Fig. 4-17. Ultracentrifuge cell, assembled and disassembled, and ultracentrifuge rotor.

The basic parameter of interest in sedimentation velocity experiments is the *rate* of movement of the peak. The rate of migration of the midpoint of the boundary corresponding to a particular species is equal to the average rate of movement of the individual molecules of that species. The speed of migration can be used to compute the molecular parameters characterizing the component, in conjunction with supplementary information from other measurements.

The other variant of the ultracentrifugal method, *sedimentation equilibrium*, represents a basically different approach. In this case the rotor is driven at a speed which is low enough so that the transport of solute under the influence of the centrifugal field can be countered by diffusion. If the speed is properly chosen, the distribution of solute in the cell will approach with time an equilibrium state for which the concentration of solute has a finite value in all parts of the cell. The distribution of solute may be regarded as reflecting a compromise between the effects of sedimentation, which tends to produce a concentration gradient, and of diffusion, which tends to erase it.

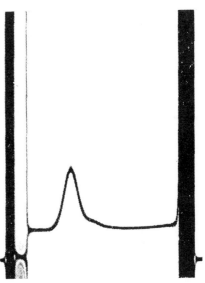

Fig. 4-18. Sedimentation diagram for bovine serum albumin (35).

The distribution of the protein at equilibrium may be observed by means of a schlieren optical system. From a knowledge of the variation of concentration with radial distance, the molecular weight of the protein may be computed.

Theory for sedimentation velocity (34-40). The fundamental parameter of sedimentation velocity is the *sedimentation coefficient* (S) which is defined as follows:

(4-45) $$S = \frac{1}{w^2 r} \frac{dr}{dt}$$

The unit is the *svedberg* (10^{-13} seconds).

Here w is the angular velocity of rotation of the rotor and is equal to 2π times the number of revolutions per second. The quantity r is the radial distance of the midpoint of the boundary (or the maximum of the peak) from the axis of rotation. dr/dt is the rate of movement of the boundary with time (t). The sedimentation coefficient is dependent only upon the molecular characteristics of the solute and the viscosity of the solvent, being independent of the rotor speed. It is customary to reduce sedimentation coefficients to standard conditions (water at 20° C), by multiplying by $\eta_o/\eta_{H_2O,20°}$ where η is the viscosity of the actual solvent at the temperature of measurement.

$$(4\text{-}46) \qquad S_{H_2O,20°} = S \frac{\eta_o}{\eta_{H_2O,20°}}$$

The sedimentation coefficient is a function of the molecular weight, density, and frictional coefficient of the solute. For movement at a constant, or terminal, velocity to occur, the centrifugal force (F_c) acting upon the protein molecule must be opposed by an equal and opposite force (F_v) arising from the viscous resistance of the solvent.

Using Archimedes' law, we may write for F_c in a solvent of density ρ_o:

$$(4\text{-}47) \qquad F_c = w^2 r M (1 - \overline{V}\rho_o)$$

The product of w^2r, which is the centrifugal force per unit mass, and the molecular weight (M) gives the force per mole of solute for a medium of zero density. This quantity, multiplied by the buoyancy factor $(1-\overline{V}\rho_o)$ gives the force per mole in the given solvent. Here $\overline{V}$, the *partial specific volume*, is equal to the volume increment per gm of solute or the reciprocal of the density of solute.

We have also:

$$(4\text{-}48) \qquad F_v = N_o f \frac{dr}{dt}$$

where N_o is Avogadro's number and f is the frictional coefficient per molecule. $N_o f$ is equal to the frictional coefficient per mole of solute.

For constant rate of movement

$$(4\text{-}49) \qquad F_v = F_c$$

$$N_o f \frac{dr}{dt} = w^2 r M (1 - \overline{V}\rho_o)$$

$$M = N_o f \frac{dr}{dt} \bigg/ w^2 r (1 - \overline{V}\rho_o)$$

or,

$$(4\text{-}50) \qquad M = S f N_o / (1 - \overline{V}\rho_o)$$

It has already been noted that $D = kT/f = RT/N_o f$. Combining this with equation (4-36), we obtain:

$$(4\text{-}51) \qquad M = \frac{SRT}{D(1 - \overline{V}\rho_o)}$$

In this manner, simultaneous measurement of the sedimentation and diffusion coefficients permits evaluation of the molecular weight. In practice, both S and D are usually concentration dependent, so that it is necessary to extrapolate both to zero concentration. In addition, both S and D must be converted to the values for standard conditions, $S_{H_2O,20°}$ and and $D_{H_2O,20°}$.

For the less asymmetric proteins, the variation of sedimentation coefficient with concentration is relatively gradual and a linear extrapolation versus concentration is generally adequate to obtain the limiting value at zero concentration. If the particles are very asymmetric (axial ratio > 20), the variation of S with concentration may show curvature. In this case, a reciprocal plot of $1/S$ versus concentration often provides a linear extrapolation (35).

Equation (4-51) has been used for the determination of the molecular weights of a large number of proteins. In addition, if the frictional coefficient is computed from equation (4-50), it may be used to compute f/f_o, the frictional ratio. Thus both the molecular weight and an index of the shape of the protein may be obtained from combined sedimentation and diffusion measurements.

It should be emphasized that the molecular weight computed from equation (4-51) is the value for the *anhydrous* protein and is not influenced by the hydration of the molecule.

Theory for sedimentation equilibrium (36, 37). If the solute consists of only one species, then the molecular weight may be computed from the values of the concentration and concentration gradient at any point of the cell (Fig. 4-19). The following equation has been derived:

$$(4\text{-}52) \qquad M = RT \frac{dC}{dr} \bigg/ C(1 - \overline{V}\rho_o)w^2 r$$

Here dC/dr and C are both measured at the same radial distance r.

If the solute consists of more than one species, then the distribution within the cell will be different for the components of different molecular weight. Thus, the apparent molecular weight, as computed from equation (4-52) will be dependent upon the position within the cell. However, a well-confined average value of the molecular weight can be obtained from the relationship:

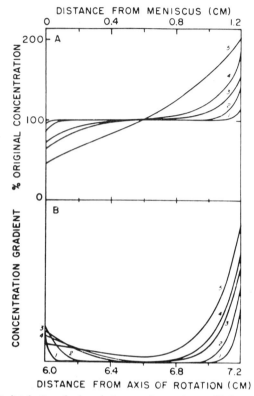

Fig. 4-19. Redistribution of solute during a sedimentation equilibrium experiment (4).

$$(4\text{-}53) \qquad M_w = \frac{2RT \int_{r_m}^{r_b} \frac{dC}{dr}\,dr}{(1 - \overline{V}\rho_o)w^2 C_o (r_b{}^2 - r_m{}^2)}$$

where r_b and r_m are the radial distances at the base and meniscus of the cell, respectively; C_o is the concentration of the original solution; and M_w is the weight-average molecular weight.

The apparent values of molecular weight computed from equations (4-52) or (4-53) must be extrapolated to zero concentration for reasons similar to those discussed for the cases of light-scattering and osmotic pressure. The molecular weight obtained in this way is that of the anhydrous protein and is not influenced by hydration.

The attainment of sedimentation equilibrium throughout the cell generally requires prolonged centrifugation and is therefore often inconvenient. However, there are two points where equilibrium is effectively attained at once. These are the meniscus and base. In the *Archibald* variant of sedimentation equilibrium equation (4-52) is applied to these points (36). The concentrations at the meniscus (C_m) and base (C_b) may be computed from:

$$(4\text{-}54) \qquad C_m = C_o - \frac{1}{r_m^2} \int_{r_m}^{r_p} r^2 \frac{dC}{dr} dr$$

$$C_b = C_o + \frac{1}{r_b^2} \int_{r_p}^{r_b} r^2 \frac{dC}{dr} dr$$

Here r_p is the (arbitrary) radial distance to a point in the cell where C is equal to C_o. It follows that the application of this method is confined to centrifugation times which are sufficiently short so that a plateau of concentration persists in the interior of the cell. As in the case of the other variant of sedimentation equilibrium, a weight average value of molecular weight is obtained.

Technique. The instrumental problems encountered in ultracentrifuge design fall into two categories. Very high centrifugal fields must be reproducibly obtained, and the redistribution of the solute within the ultracentrifugal cell must be monitored by some type of optical system.

A metallic rotor is made to rotate about an axis at a high and controlled velocity (up to 60,000 rpm). The drive may be either an electric motor or a gas turbine. A small sector-shaped cell (Fig. 4-16 and 4-17) with transparent windows, which contains the solution to be observed, is placed in the rotor. The centrifugal field is directed outward, in a radial direction which is perpendicular to the axis of rotation. The cell chamber containing the solution is positioned within the rotor so that the edges of the sector point radially outward, thereby avoiding convective disturbances.

While in the rotor, the cell is illuminated from below by a narrow beam of intense light which intercepts it once for each complete revolution. The emergent beam enters a schlieren optical unit which produces an image of the variation of refractive index gradient (and hence of concentration gradient) with radial distance. (The speed of rotation is high enough so that no perceptible "flicker" arises from the intermittent illumination of the cell.) The pattern of dn/dr, where r is the radial distance from the axis of rotation, may be photographed and analyzed.

In sedimentation velocity measurements, the sedimentation coefficient of each component may be computed from the rate of movement of the corresponding peak. The relative concentration of each species present is directly proportional to the area under its peak.

In the case of polyelectrolytes, such as proteins and nucleic acids, the sedimentation coefficient is reduced at low ionic strengths by nonspecific charge effects. These may be eliminated by working at a moderately high level of supporting electrolyte. In practice, a concentration of 0.1 M of uni-univalent electrolyte is sufficient to suppress these effects and permit application of equation 4-51.

Interpretation of the ultracentrifuge patterns of many macromolecules is complicated by the high degree of particle interaction present in solution at the concentrations usually employed in the ultracentrifuge. This is reflected by a very pronounced dependence of the sedimentation coefficient upon concentration, which complicates the extrapolation of sedimentation coefficient to zero concentration. The problem is rarely acute for proteins, but becomes very important in the case of deoxyribonucleic acid (Chapter 9).

These difficulties can be circumvented by making measurements at extreme dilutions. However, schlieren optics are insufficiently sensitive to be useful at solute concentrations much below 0.1%, thereby precluding studies at very low concentrations.

If the material absorbs visible or ultraviolet light strongly, this property may be utilized to follow its sedimentation. The optical system is relatively simple, consisting of a light source, a filter, and a set of lenses. These produce a parallel beam of monochromatic light, which passes through the ultracentrifuge cell. An image of the illuminated cell is projected upon the photographic plate. If the cell is filled with a solution which absorbs almost all the light traversing it, then the amount of light reaching the plate will be so small that the cell image is very faint. Sedimentation of the solute produces a transparent solvent region at the top of the cell so that the image of that portion of the cell is much darker (Fig. 4-20). The degree of absorption varies continuously

Fig. 4-20a. The progress of ultracentrifugation, as followed by absorption (35).

across the boundary. This is reflected by a progressive change in the blackening of the plate at the position corresponding to the boundary (Fig. 4-20a).

SIZE, SHAPE, AND ELECTRIC CHARGE OF PROTEIN MOLECULES   117

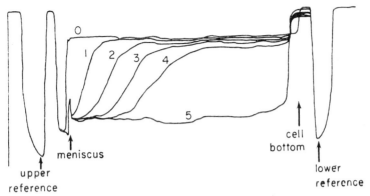

Fig. 4-20b.   Densitometer tracings of the above diagrams (35).

By densitometer measurements of the degree of darkening, curves showing the variation of absorbancy with distance, and hence of relative concentration with distance, can be obtained for various times (Fig. 4-20b). Thus the absorption technique produces curves of concentration, rather

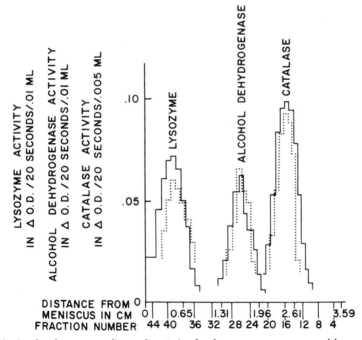

Fig. 4-20c. Results of sucrose gradient sedimentation for three enzymes, as measured by enzymatic activity. The ordinate refers to colorimetric assay units.

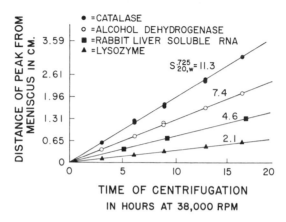

Fig. 4-20d. Demonstration of linearity of sedimentation in a sucrose gradient for four proteins.

than concentration gradient, as a function of distance. The sedimentation coefficient may be computed from the rate of movement of the midpoint of the boundary.

While less precise than schlieren methods, the absorption technique permits extension of measurements to concentrations an order of magnitude lower—down to 0.01% or less for strongly absorbing materials. It has been extensively used for deoxyribonucleic acid solutions.

Sucrose gradient centrifugation. A new form of ultracentrifugal analysis has recently been developed which has proved to be of great value for the characterization of mixtures, particularly when the component of interest is present in low concentration. While the underlying principles are of course similar, the technique of sucrose gradient centrifugation is altogether different from that of the conventional form of sedimentation velocity.

The method dispenses with optical measurements during the course of centrifugation. The analytical cell described in the preceding sections is replaced by an ordinary plastic centrifuge tube, which is shaped like a test tube.

Basically, the method depends upon the stabilization of concentration gradients by the presence of a density gradient contributed by sucrose. This permits removal of the centrifuge tube at the conclusion of sedimentation and the direct sampling of successive layers of solution. Without the sucrose, the concentration gradients would suffer extensive disruption by such handling in the absence of a centrifugal field. A rotor of the "swinging bucket" type is used, with hinged tube holders which swing out at right angles to the axis of rotation when the rotor is in motion. The centrifuge tube is thus oriented in the direction of the centrifugal field, and sedimentation proceeds radially outward, from top to bottom.

A *linear* sucrose gradient is produced by careful mixing in a plastic block containing two chambers joined at the bottom by a channel with a stopcock. An outflow tube extends from one chamber. The chambers are then separately filled with two sucrose solutions of different concentration (usually 5% and 20%). The more dense solution is placed in the mixing chamber, which is connected to the outflow tube.

The channel connecting the chambers is opened and the sucrose solution allowed to drain slowly into the centrifuge tube through the outflow tube. The concentration of the mixing chamber, which is continuously stirred, changes progressively as the more concentrated solution is diluted by the less concentrated.

In this manner, a linear gradient of sucrose concentration can be produced from the top to the bottom of the centrifuge tube, with the concentration greatest at the bottom and least at the top. A small volume of protein solution is then layered on top of the sucrose gradient. The centrifuge tube is then placed in a "swinging bucket" and spun at 30,000 to 40,000 rpm for several hours.

The rotor is then stopped, the tube removed, and a puncture carefully made in its bottom with a needle. The solution is allowed to drain dropwise into a series of collecting test tubes, with the centrifuge tube in the vertical position. Usually about ten drops are collected for each fraction.

In this manner a series of fractions corresponding to consecutive layers within the centrifuge tube are obtained. The first fraction collected comes from the bottom of the tube and the subsequent fractions from progressively higher layers. The fractions are then analyzed for protein content or enzymatic activity. The concentration may be measured by the absorption of light at 280 mμ.

In the moving rotor the centrifuge tube is perpendicular to the axis of rotation, with its bottom at the periphery. Sedimentation of the components of the mixture thus occurs from the top toward the bottom of the tube.

If the partial specific volume of each component is less than about 0.80, the distance sedimented, and hence the numerical order of the fraction in which it attains its highest concentration, is linear with respect to time. The ratio of the distances traveled for two different proteins is equal to the ratio of their sedimentation coefficients. By including a *marker* protein of known sedimentation coefficient, the value for an unknown material may be computed. From a knowledge of the volume of each fraction and the dimensions of the tube, the average distance traveled may be computed from the distribution among the fractions.

Figure 4-20c shows a representative pattern for a mixture of proteins. It should be particularly noted that, in contrast to standard sedimentation velocity, two components of differing sedimentation rate can be separated completely. The method may also be used for enzymes whose concentration is too low to permit accurate measurements of optical activity. In this case, the enzymatic activity (see Chapter 7) may be used as a direct measure of concentration.

4-10 METHODS YIELDING A MEASURE OF HOMOGENEITY: ELECTROPHORESIS

General remarks. As a consequence of the presence of ionizable groups in the side-chains of their constituent amino acids, proteins in solution are charged polyelectrolytes, whose state of ionization and net charge are dependent upon the pH and electrolyte concentration of the solution. Their charged character endows proteins with the property of migrating in an electrical field. The velocity of migration, or *mobility,* is a function primarily of the magnitude of the net charge and, to a lesser extent, of the size and shape of the molecule.

The dependence of mobility upon the molecular characteristics of the protein may be utilized for the analysis of protein mixtures. By the technique of *electrophoresis,* the relative concentrations of the various charged species may be determined. Electrophoresis is one of the most commonly employed criteria of protein homogeneity and is often used to monitor the progressive purification of a protein in the course of a preparative procedure.

If the mobility of a particular protein is observed as a function of pH, the behavior is qualitatively as follows. In the limit of highly alkaline pH (>13) the molecule will have lost almost all its dissociable protons, and its negative charge will have its maximal value. The ionizable groups will be either negatively charged (in the case of the carboxyl, tyrosyl, and sulfhydryl groups) or uncharged (in the case of the amino and histidyl groups). The mobility will have its maximum negative value, the direction of movement being toward the anode. As the pH is lowered, the net negative charge is reduced by one for each proton bound. This is paralleled by a decrease in magnitude of the mobility. Ultimately a pH is reached at which the net charge, and hence the mobility, is zero. The pH at which the mobility is zero is referred to as the *isoelectric point.* It will of course depend upon the nature and concentration of the electrolytes present in the solution.

As the pH is decreased beyond the isoelectric point, both the charge and the mobility change signs, becoming positive. The magnitudes of both increase with further decrease in pH, the limiting value usually being approached by pH 2.

Theory of protein ionization (41, 42). A complete theoretical analysis of the ionization of proteins, which may contain a hundred or more ionizable groups, is a formidable problem which has yet to be rigorously solved. The experimental curve of the number of moles of hydrogen ion bound (or dissociated) as a function of pH is referred to as the *hydrogen ion titration curve*.

At present, it is possible to account for the complete titration curve of a protein, in terms of the number and *intrinsic* dissociation constants of its ionizable sites, only in a semiquantitative manner. However, the existing theory suffices to permit analysis of titration data to yield the number of groups of each type characterized by a particular pK (section 2-4).

The most important systematic factor which modifies the intrinsic ionization constants is the electrostatic field of the protein. Since the positions of the individual charged groups are not known, some form of approximation must be made. The most generally satisfactory procedure has been to approximate the actual discrete charges by a continuous charge which is uniformly "smeared" over the surface of the protein, which is approximated by a simple geometrical shape, usually a sphere. For the purposes of computing the electrostatic influence of the net charge on the protein upon its ionization, this model thus assumes it to be equivalent to that of a uniformly charged body whose charge is equal to that of the protein.

In order to assess the magnitude of the electrostatic factor, it is necessary to know the pH at which the net charge is zero, that is, where the numbers of positively and negatively charged groups are equal. In practice, the *isoelectric point* is generally used. It may of course be influenced by the binding of ions other than H^+ or $\overline{OH}$. Since there is usually no accurate information as to ion binding, the net charge has generally been taken as equal to the number of protons bound (or dissociated) upon altering the pH from the isoelectric point to the given value.

The ionizable groups of a typical protein consist of several sets of identical sites. To each class of site, such as the carboxyl, tyrosyl, or ϵ-amino, an *intrinsic* pK may be assigned (see section 2-4). This is equal to the pK which a single isolated group of this class would have in the absence of all electrostatic (or other) influences.

The objective of theories of the hydrogen ion titration curves of proteins is to account for the experimental curves quantitatively in terms of the intrinsic constants and the net electrostatic charge of the protein. If the experimental curve cannot be fitted in a self-consistent way by this kind of approach, it is likely that some complicating factor is present, such as a structural change of the protein with pH.

The treatment of the hydrogen ion titration curves of proteins to be discussed here is based upon the following assumptions:

(1) All the groups of each class are identical.

(2) The sole source of perturbation of their ionization properties is the electrostatic influence of the protein charge.

(3) The electrostatic field present at each point on the surface is that of a uniformly charged sphere whose volume is the same as that of the protein and whose total charge is equal to the net charge of the protein.

(4) The structure of the protein does not change with pH.

(5) The number of hydrogen ions which have been bound (or dissociated) upon adjusting the pH from the isoelectric point is equal to the net charge of the protein.

In terms of the above model, the following relationship has been derived to account for the purely electrostatic displacement of the mean equilibrium constant for the binding of hydrogen ions by a set of identical groups.

(4-55) $$pK = pK_{\text{intrinsic}} - 0.868wZ_e$$

Here Z_e is the net charge (the algebraic difference between the total numbers of positive and negative charges) and w is given by:

(4-56) $$w = \frac{N_o\epsilon^2}{2DRT}\left(\frac{1}{c} - \frac{K_o}{1 + K_od}\right)^{1/2}$$

where ϵ = electronic charge; D = dielectric constant; R = gas constant; N_o = Avogadro's number; T = absolute temperature; c = radius of sphere; d = sum of radii of protein and counterion. The quantity K_o is the *Debye-Huckel constant* and is defined by:

(4-57) $$K_o = \left(\frac{4\pi N_o^2\epsilon^2}{1000DRT}\right)^{1/2}(\Sigma m_iZ_i^2)^{1/2}$$

where m_i, Z_i are the molar concentration and charge, respectively, of the ith ionic species. The summation $\Sigma m_iZ_i^2$ is equal to 2Γ, where Γ is defined as the *ionic strength*.

The relationship between pH and the fraction, α, of groups of a single class which have dissociated protons is given by:

(4-58) $$pH = pK_{\text{intrinsic}} + \log(\alpha/1 - \alpha) - 0.868wZ_e$$

The approximate nature of this treatment should be recognized. In practice distortions of the form of the hydrogen ion titration curve often occur which cannot be accounted for in terms of the generalized electrostatic factor discussed above.

If the number of groups of each type are independently known then the number of each kind which have dissociated protons at a given pH may be computed from α, as predicted by equation 4-58. By summation of the number of hydrogen ions dissociated by the various kinds of site at each pH a theoretical titration curve may be computed and compared with experiment (Fig. 4-21). In the absence of complicating factors, such as a change in the structure of the protein with pH, it is usually possible to

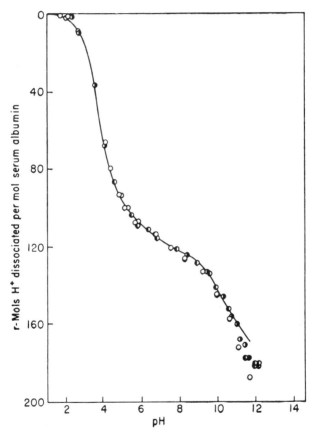

Fig. 4-21. Hydrogen ion titration curve of serum albumin. The ordinate is equal to the number of moles of hydrogen ion per mole of protein which have been dissociated upon adjusting the pH from a value in the acid plateau region (<2) to the indicated value. This quantity is equal to the difference between the total number of moles of $\overline{OH}$ which have been added and the accompanying decrease in the total number of moles of H^+ present in the unbound state (or the number of moles of H^+ which have been neutralized or titrated). The concentration of free H^+ is usually measured by means of a glass electrode.

fit the observed data fairly well with the treatment discussed above. A major departure from the predicted titration curve is, in fact, often an indication of the occurrence of a structural transition.

Theory for electrophoresis (43, 44). The possession of a *net charge* confers upon proteins the property of migrating in an electric field. The net charge is the algebraic difference between the total numbers of positive and negative charges per molecule. While the charges are contributed primarily by the ionizable groups, bound ions also make a contribution.

The velocity of movement of a protein subject to the influence of a unit electric field (1 volt per cm) is defined as the *mobility*, u. If the protein moves Δx centimeters in Δt seconds in a field of E volts per cm, then

$$(4\text{-}59) \qquad u = \frac{\Delta x}{E \Delta t}$$

The mobility of a charge particle is directly proportional to the net charge and hence is a function of the pH and the solvent composition. The constancy of the electric field within the cell results in the mobility being independent of position within the cell.

The electric field strength, E, is related to the current, i; the cross-sectional area of the cell, A; and the specific conductance, K_s, of the solution by:

$$(4\text{-}60) \qquad E = \frac{i}{AK_s}$$

Thus equation 4-59 may be rewritten as:

$$(4\text{-}61) \qquad u = \frac{\Delta x A K_s}{i \Delta t}$$

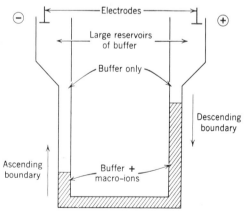

Fig. 4-22. Schematic diagram of an electrophoresis cell.

SIZE, SHAPE, AND ELECTRIC CHARGE OF PROTEIN MOLECULES

The mobility of a protein is a complex function of the net charge, the nature and concentration of electrolyte, and the size and shape of the molecule. For a spherical protein of net charge Z_e and radius c, in a solvent of viscosity η_o the theoretically derived *Henry equation* states:

(4-62) $$u = \frac{Z_e \epsilon}{6\pi\eta_o c} \frac{\phi(K_o d)}{1 + K_o d}$$

Here K_o is the *Debye-Huckel parameter*, defined by equation 4-57.

The quantity d is the distance of closest approach of the centers of the counterions, and the protein and is equal to the sum of their radii. The function $\phi(K_o d)$, the *Henry function*, increases from a limiting value of unity for low values of $K_o d$ to a value of 3/2 for very large values of $K_o d$.

Since no actual protein is likely to have a perfectly spherical shape, it is not to be expected that equation 4-62 will predict the observed mobilities quantitatively. Nevertheless, the existing data are in reasonable accord with it.

In practice, the mobility appears in general to be directly proportional to the number of hydrogen ions bound or dissociated only in the isoelectric region (Fig. 4-23). The deviations from linearity become more

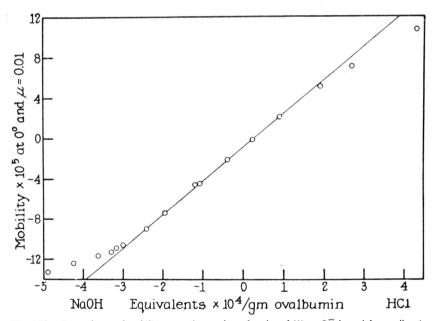

Fig. 4-23. Dependence of mobility upon the number of moles of H^+ or $\overline{OH}$ bound for ovalbumin.

important with increasing separation from the isoelectric point, possibly because of enhanced binding of ions other than H^+ or $\overline{OH}$.

Technique. As in the case of sedimentation velocity, the moving boundary method is practically universally used at present. The movement occurs in a glass cell of rectangular cross-section, in which the initial boundaries are formed by the mutual displacement of its sections (Fig. 4-22 and 4-24).

As Fig. 4-24 shows, the electrophoresis cell consists of three sections, two of which may be moved independently. First, the bottom section is filled with an excess of protein solution so that some solution extends into the two middle channels. The bottom section is then sealed off by being shifted with respect to the middle section. The left channel of the middle section is then filled with an excess of protein solution and the right channel flushed out with solvent and filled with an excess of solvent. The middle section is then shifted so as to be isolated from both top and bottom sections. Finally, the upper section is flushed and then completely filled with solvent.

At the start of the experiment the three sections are mutually aligned. This serves to produce two boundaries between solution and solvent. As one of these moves upward and the other downward, they are referred to as the *ascending* and *descending* boundaries, respectively (Fig. 4-25).

The upper section is in contact with the electrode vessels. The electrodes themselves are normally made of silver, coated with silver chloride. The electrode vessels must be sufficiently removed from the electrophoresis cell to prevent any contamination of the latter with the decomposition products formed at the electrodes.

The movement of the boundaries has usually been followed by schlieren optics, as described in section 4-5. Two diagrams are obtained, corresponding to the ascending and descending boundaries (Fig. 4-25). The two small stationary boundaries at the initial boundary position, the δ-

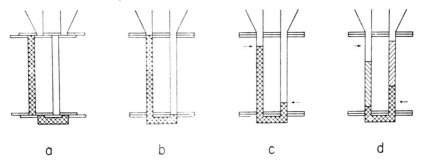

Fig. 4-24. Technique of boundary formation in an electrophoresis cell.

SIZE, SHAPE, AND ELECTRIC CHARGE OF PROTEIN MOLECULES

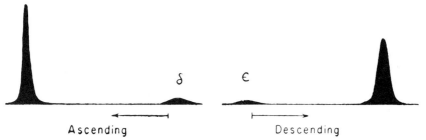

Fig. 4-25. A typical electrophoresis diagram of a pure material.

and ϵ-boundaries, are artifacts arising from the existence of a small gradient of electrolyte concentration across the boundary. The rate of migration of each moving boundary is equal to the velocity of transport of the corresponding protein component and may be used to compute the mobility from equation 4-59. Since the electrophoresis cell is of constant cross-sectional area, there is no dilution accompanying movement of the boundary, in contrast to the case of the ultracentrifuge.

To minimize the convective effects arising from the electrical heating of the solution, it is usual to maintain the electrophoresis cell at a temperature between 0° and 1°.

Finally, it is of course very important that no important gradient of pH exist within the cell. Thus the protein solution must be strongly buffered. The usual buffer concentration is 0.1 to 0.2 M. Uni-univalent buffers are used to avoid artifacts arising from the presence of polyvalent electrolytes. It is important to minimize any difference in salt concentration between solution and solvent. This is accomplished by prolonged dialysis versus solvent.

TABLE 4-1. MOLECULAR WEIGHTS ($\times 10^{-4}$) OF SEVERAL PROTEINS FROM VARIOUS TECHNIQUES

Protein	M (sedimentation velocity)	M (sedimentation equilibrium)	M (light-scattering)
Thyroglobulin	63	65	65
Myosin		6.1	
Fibrinogen	34		34
Serum albumin	6.5	6.8	7.0
Ovalbumin	4.4	4.4	
Pepsin	3.6	3.9	3.7
Chymotrypsin	2.2		2.5
Trypsin	2.4		2.4
Lysozyme	1.4	1.5	
Ribonuclease	1.3	1.3	
Myoglobin	1.7	1.8	

TABLE 4-2. VALUES OF IMPORTANT CONSTANTS IN C.G.S. UNITS

Constant	Symbol	Units	Numerical Value
Gas constant	R	erg degree^{-1} mole^{-1}	8.31×10^7
Boltzmann constant	k	erg degree^{-1}	1.38×10^{-16}
Avogadro's number	N_o	—	6.02×10^{23}
Electrostatic unit of charge		absolute electrostatic units	4.8×10^{-10}

TABLE 4-3. UNITS OF IMPORTANT QUANTITIES

Quantity	Symbol	C.G.S. Units	Other Units
Osmotic pressure		dynes cm^{-2}	cm Hg; atm
Reduced intensity	R_θ	cm^{-1}	
Diffusion coefficient	D	cm^2 sec^{-1}	
Sedimentation coefficient	S	sec	svedbergs (1 svedberg = 10^{-13} sec)
Viscosity	η	gms cm^{-1} sec^{-1}	poise (1 poise = 1 C.G.S. unit)
Intrinsic viscosity	$[\eta]$	cm^3 gms^{-1}	
Frictional coefficient	f	gms sec^{-1}	
Mobility	u	cm^2 volts^{-1} sec^{-1}	

TABLE 4-4. VALUES OF SEVERAL IMPORTANT QUANTITIES

Quantity	Units	Numerical Value
Viscosity of water at 20°C	poise	1.0×10^{-2}
Dielectric constant of water at 20°C	—	80.0
Difference between °K and °C	degrees	°K = °C + 273.2

GENERAL REFERENCES

1. J. T. Edsall in *The Proteins*, H. Neurath and K. Bailey (eds.), vol. 1, chap. 7, Academic Press, New York (1953).
2. *Analytical Methods of Protein Chemistry*, P. Alexander and R. Block, vol. 3, Pergamon Press, London (1960).
3. *Colloid Science*, A. Alexander and P. Johnson, Oxford University Press, New York (1947).
4. *The Physical Chemistry of Macromolecules*, C. Tanford, Wiley, New York (1961).

SPECIFIC REFERENCES

Electron microscopy

5. Reference 2, p. 3.
6. *Introduction to Electron Microscopy*, C. Hall, McGraw-Hill, New York (1953).
7. *Practical Electron Microscopy*, V. Cosslett, Butterworth, London (1951).

Osmotic pressure

8. Reference 2, p. 23.
9. G. Scatchard, *J. Am. Chem. Soc.*, **68,** 2315 (1946).
10. R. Wagner in *Physical Methods of Organic Chemistry*, A. Weissberger (ed.), vol. 1, part 1, p. 487, Interscience, New York (1949).

Light scattering

11. *Light Scattering in Physical Chemistry*, K. Stacey, Butterworth, London (1956).
12. E. Geiduschek and A. Holtzer, *Advances in Biological and Medical Physics*, vol. 6, p. 431, Academic Press, New York (1958).
13. P. Debye, *J. Phys. Chem.*, **51,** 18 (1947).
14. P. Doty and J. T. Edsall, *Advances in Protein Chemistry*, vol. 6, p. 35, Academic Press, New York (1951).
15. P. Doty and R. Steiner, *J. Chem. Phys.*, **18,** 1211 (1950).
16. B. Zimm, *J. Chem. Phys.*, **16,** 1093 (1948).

Schlieren optics

17. Instruction manual, Spinco ultracentrifuge, section V-A (1953).

Viscosity

18. A. Einstein, *Ann. Physik.*, **19,** 289 (1900).
19. A. Einstein, *Ann. Physik.*, **34,** 591 (1911).
20. R. Simha, *J. Phys. Chem.*, **44,** 25 (1940).
21. References 3, Chap. 13.
22. J. Kirkwood and P. Auer, *J. Chem. Phys.*, **19,** 281 (1951).
23. Reference 2, p. 173.
24. J. Yang, *J. Phys. Chem.*, **62,** 894 (1958).

Fluorescence polarization

25. G. Weber, *Biochem. J.*, **51,** 145, 155 (1952).
26. G. Weber, *Advances in Protein Chemistry*, vol. 8, Academic Press, New York (1953).
27. R. Steiner and H. Edelhoch, *Chem. Rev.*, **62,** 457 (1962).

Diffusion

28. L. Gosting, *Advances in Protein Chemistry*, vol. 11, p. 429 Academic Press, New York (1956).
29. H. Neurath, *Chem. Rev.*, **30,** 357 (1942).
30. J. W. Williams and L. Cady, *Chem. Rev.*, **14,** 171 (1934).
31. L. Longsworth, *J. Am. Chem. Soc.*, **74,** 4155 (1952).
32. A. Geddes in *Physical Methods of Organic Chemistry*, A. Weissberger, (ed.), vol. 1, part 1, p. 551, Interscience, New York (1949).
33. L. Longsworth, *J. Am. Chem. Soc.*, **69,** 2510 (1947).

Ultracentrifugation

34. *The Ultracentrifuge,* T. Svedberg and K. Pederson, Oxford University Press, New York (1940).
35. *Ultracentrifugation in Biochemistry,* H. Schachman, Academic Press, New York (1958).
36. W. Archibald, *J. Phys. and Colloid Chem.*, **51**, 1204 (1947).
37. R. Goldberg, *J. Phys. Chem.*, **57**, 194 (1953).
38. L. Gosting, *J. Am. Chem. Soc.*, **74**, 1548 (1952).
39. R. Baldwin, *Biochem. J.*, **65**, 490 (1957).
40. J. Williams, K. van Holde, and R. Baldwin, *Chem. Rev.*, **58**, 715 (1958).

Hydrogen ion titration

41. J. Steinhardt and E. Zaiser, *Advances in Protein Chemistry,* vol. 10, p. 152, Academic Press, New York (1955).
42. R. Cannon, A. Kibrick, and A. Palmer, *Ann. N. Y. Acad. Sci.*, **41**, 243 (1942).

Electrophoresis

43. *Electrophoresis,* M. Bier, Academic Press, New York (1959).
44. R. Alberty in *The Proteins,* H. Neurath and K. Bailey (eds.), vol. 1, chap. 6, Academic Press, New York (1953).

PROBLEMS (See Tables 4-2 to 4-4)

1. What is the osmotic pressure of a 1% solution of a protein of molecular weight 50,000? Assume ideal behavior.

2. What is the reduced intensity of a 0.5% solution of a protein of molecular weight 90,000 when dissolved in water (refractive index = 1.33)? The refractive increment is 0.190 (ml/gm). The wave-length of light used is 4360 Å. Assume ideal behavior and that the dimensions of the protein are negligibly small in comparison with the wave-length of light.

3. What is the intrinsic viscosity expected for an unhydrated prolate ellipsoidal particle of axial ratio 10 and density 1.3? The solvent is water.

4. What is the frictional ratio expected in the above case?

5. Repeat problems 3 and 4, assuming a hydration of 0.20.

6. What is the rotational relaxation time for a spherical particle of molecular weight 30,000 and density 1.3 when dissolved in water at 20°C? Assume zero hydration.

7. Repeat problem 6, assuming a hydration of 0.20.

8. What is the diffusion coefficient predicted for the system of problem 6?

9. Discuss and criticize the kind of approximation involved in assuming an ellipsoidal model for the shape of an actual protein.

10. Compute the expected degree of ionization for a set of groups of intrinsic pK equal to 9.0 at pH 9.0, when the groups are attached to a protein of molecular weight 30,000 and density 1.3. The net charge on the protein is −20. The groups acquire a positive charge upon binding hydrogen ion. The solvent is 0.1 M KCl at 20°C.

5

The Spatial Organization of Proteins

5-1 GENERAL REMARKS

A knowledge of the purely chemical aspects of protein structure, such as the amino acid composition and sequence, is by no means sufficient to explain many of their properties (1-5). The most conspicuous conclusion of the extensive physical studies which have been carried out on natural proteins is that their molecular characteristics are very different from those expected for unorganized polymers of the same size and composition. The evidence is overwhelming that the polypeptide chains of each protein molecule must be folded to form a highly specific three-dimensional pattern (1, 2, 3). Indeed, the geometry of natural proteins appears in general to be as sharply defined as their primary structure. Disruption of this molecular organization by heat, or by such agents as detergents or urea, can result in drastic changes in the properties of the protein.

Thus the determination of the spatial organization of proteins must be regarded as a separate and very important phase of the analysis of their structure. The problem subdivides naturally into that of determining the configuration of the polypeptide backbone (the *secondary structure*) and that of describing the residual structure stabilized by secondary bonds involving the side-chains (the *tertiary* structure).

Generalization about the tertiary structure is difficult at this time. However, the secondary structures of a number of proteins of widely divergent structure and properties can be discussed from a unified point of view. To anticipate, the polypeptide backbones of most of the proteins which have been examined to date appear to exist wholly, or in part, in a helical conformation which corresponds closely to the α-*helix* originally proposed by Pauling, Corey, and co-workers.

This structure is stabilized by intramolecular hydrogen bonding (see section 5-2) involving the C=O and N—H groups of the peptide linkage. The α-helix was initially proposed on stereochemical grounds and has the advantage of permitting the formation of the maximum number of hydrogen bonds without requiring any excessive distortion of bond distances or angles.

The experimental demonstration of the occurrence of the α-helix in proteins and synthetic polypeptides* has involved the use of several very different techniques. The most powerful approach is undoubtedly that of x-ray diffraction. The analysis of fiber diffraction patterns has indicated that a number of synthetic polypeptides can exist largely in this form in the solid state.

The application of X-ray diffraction to proteins is attended with difficulties so formidable that a complete structural analysis has been approached in only two instances. This and the restriction of this technique to crystalline proteins have led to the search for supplementary methods for assessing the helical content of proteins in solution. Those which have been developed to date are of an indirect and essentially empirical nature.

Examination of models of the α-helix permits the conclusion that such a structure must be essentially completely rigid, as any important degree of torsional bending would require an intolerable distortion of bond distances. Thus it can be confidently predicted that any polypeptide existing as a perfect α-helix would have the shape of a rigid rod (Fig. 5-1). From the known pitch and radius of the α-helix, it is possible to compute the expected dimensions of a polypeptide of known molecular weight which exists entirely in this form.

The other extreme of possible polypeptide configurations is that of a completely unorganized random coil. This type of configuration occurs frequently for synthetic polymers in good solvents. As was discussed in the preceding chapter, the random, or Gaussian, coil is a relatively open structure which is permeable to solvent.

Investigation of the solution properties of synthetic polypeptides has revealed that their configuration is strongly dependent upon the nature of the solvent. In solvents favorable to the formation of intramolecular hydrogen bonds, several polypeptides have been shown to have dimensions corresponding closely to those expected for perfect α-helices. In solvents which competitively disrupt internal hydrogen bonding, the same polypeptides exist as random coils.

The availability of systems of known conformation has assisted the development of methods for assessing the helical content of polypeptides

* For a discussion of the preparation of synthetic polypeptides, see Appendix C.

in solution. The optical rotatory properties of polypeptides in the helical and coiled forms have been found to be dramatically different. This has provided the basis for the use of optical rotation as a structural probe for helical regions in proteins.

Studies of the properties of proteins in solution by the physical methods described in the preceding chapter have revealed that the majority do not have the dimensions expected for perfect α-helices. On the other hand, their hydrodynamic properties indicate that their structures are much too compact and impermeable to solvent to be consistent with a randomly coiled configuration (Fig. 5-1).

A considerable body of evidence suggests that proteins of this type, which are called *globular proteins,* generally contain both helical and nonhelical regions (Fig. 5-2). The former are too short to endow the protein

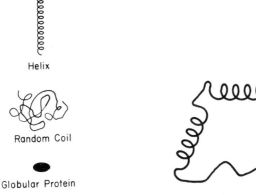

Fig. 5-1. Schematic representation of several protein configurations. The helix is drawn on a much larger scale than the other two.

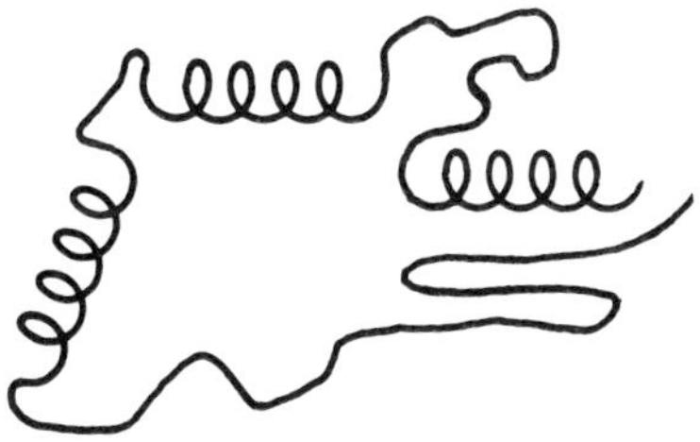

Fig. 5-2. Schematic configuration of a globular protein with a fractional helical content.

with the rod-like character of the α-helix. To account for the compact nature of globular proteins it has been postulated that, in addition to the helical regions, they are stabilized by interactions involving the sidechains. Structural elements of this kind, which are still imperfectly understood, are often referred to as the *tertiary structure.*

5-2 FORCES INVOLVED IN THE STABILIZATION OF PROTEIN STRUCTURE (6-9)

Coulombic interaction. A basic principle of electrostatics states that the force (F) exerted between two point charges of magnitude ϵ and ϵ', imbedded in a medium of dielectric constant D and separated by a distance r is given by:

$$(5\text{-}1) \qquad F = \frac{\epsilon\epsilon'}{Dr^2}$$

If ϵ and ϵ' have the same sign, the force is repulsive; if the signs are opposite, it is attractive.

This results in a net interaction energy (U) of the form

$$(5\text{-}2) \qquad U = \int_r^\infty F\, dr = \frac{\epsilon\epsilon'}{Dr}$$

The interaction energy represents the difference in potential energy of the pair when infinitely far apart and when separated by a distance r. For attraction its sign is negative and for repulsion positive.

Since proteins are charged polyelectrolytes, it is to be expected that coulombic forces would have a definite influence upon their stability. The mutual repulsion of like charges is of course an important destabilizing factor. Conversely, the existence of "salt linkages" resulting from the juxtaposition of oppositely charged sites may assist in the stabilization of the structures of some proteins.

Two features of coulombic interactions deserve comment.

(1) Coulombic forces decay relatively gradually with distance compared to the other types of interaction discussed in this section and thus are less dependent upon a small separation of the interacting sites.

(2) There is no dependence upon orientation. Hence coulombic forces cannot have an important directional effect upon interacting groups.

Dipole interactions. The generalization can be made that, for almost all molecules composed of two different kinds of atom, the centers of gravity of the positive and negative charges will not coincide (6, 7). (An exception must be made in the cases of such symmetrical molecules as methane and carbon tetrachloride.) This effect arises from the partially ionic character of most covalent bonds, which results in an unequal distribution of electric charge among the bonded atoms. Thus a bond of the type A—B may be regarded as fluctuating between the forms:

$$\text{A : B}$$
$$\text{A}^+ \text{ B}^-$$
$$\text{A}^- \text{ B}^+$$

If atom B is more electronegative than A, the $A^+ B^-$ species will be more important than $A^- B^+$, and atoms A and B will possess on the average a net positive and negative charge, respectively, giving rise to a permanent dipole moment.

The magnitude of the dipole moment (μ) of such a pair of atoms is defined as the product of the absolute magnitude (ϵ) of the (equal) positive and negative charges and their average separation (d)

$$(5\text{-}3) \qquad \mu = \epsilon d$$

Physically, the dipole moment must be regarded as a vector quantity, whose properties depend upon its direction. It may be represented by a vector whose direction is that of the line joining the two atoms and whose magnitude is given by equation 5-3.

The net dipole moment of an entire molecule or group is equal to the *vector* sum of the dipole moments of its constituent bonds. If the dipole corresponding to each bond is canceled by another of equal magnitude and opposite direction, as in the case of carbon tetrachloride, the net dipole moment of the molecule as a whole will be zero.

If two dipoles are placed side by side in a head-to-tail arrangement of the type

$$+ \longrightarrow - \qquad + \longrightarrow -$$

then each charged site will be closer to the oppositely charged site of the other dipole than to the similarly charged site. As a consequence, the negative coulombic terms in the expression for the energy of interaction will predominate, resulting in a net attraction. The interaction energy for two dipoles may be obtained by summation of the coulombic terms for the four charge elements involved. In terms of the separation (r) of the centers and the angles (θ_A, θ_B) formed by the dipoles with the line joining the centers, the result is (for dipoles lying in the same plane):

$$(5\text{-}4) \qquad U = -\frac{\mu_A \mu_B}{Dr^3}(2 \cos \theta_A \cos \theta_B - \sin \theta_A \sin \theta_B)$$

Maximum attraction $\left(U = -\dfrac{2\mu_A\mu_B}{Dr^3}\right)$ occurs when the dipoles are aligned in the head-to-tail position:

$$\begin{cases} \theta_A = 0 \\ \theta_B = 0 \end{cases} \text{ or } \begin{cases} \theta_A = \pi \\ \theta_B = \pi \end{cases}$$

Maximum repulsion $\left(U = \dfrac{2\mu_A\mu_B}{Dr^3}\right)$ occurs when the dipoles are in the head-to-head or tail-to-tail position:

$$\begin{cases} \theta_A = 0 \\ \theta_B = \pi \end{cases} \text{ or } \begin{cases} \theta_A = \pi \\ \theta_B = 0 \end{cases}$$

A dipole also interacts with the electric field of a point charge. The interaction energy is given by:

$$(5\text{-}5) \qquad U = -\frac{\epsilon\mu \cos \phi}{Dr^2}$$

where ϕ is the angle formed by the dipole and the line joining its center with the charge.

Dispersion forces. Intermolecular forces of this kind, which are often referred to as *London* forces, are of universal occurrence. In the case of molecules for which electrostatic or chemical interactions are absent, such as the inert gases helium and argon, they represent the only type of intermolecular attraction present and are responsible for the formation of the liquid state at low temperatures.

The origin of the dispersion forces lies in the fact that the nuclei and electrons of every molecule, including the monatomic gases, must be regarded as undergoing a continuous oscillation with respect to each other. As a consequence of this oscillation, the centers of gravity of the positive nuclear charge and the negative electronic charge will not in general coincide at any particular instant. This mutual displacement of the electronic and nuclear charge produces the equivalent of a transient dipole moment, whose magnitude fluctuates according to the charge distribution at each moment. Thus every molecule may be regarded as an oscillating dipole, irrespective of whether it possesses a permanent dipole moment. If a permanent dipole is absent, the varying dipoles will of course yield a resultant of zero when averaged over a long period of time. Nevertheless, despite their transient character, they are capable of giving rise to a nonvanishing interaction energy between two different molecules.

This arises from the capacity of an oscillating dipole to induce in a nearby molecule a dipole of opposite direction which oscillates *in phase* with it, thereby producing a resultant interaction which is basically analogous to that of two permanent dipoles. The interaction energy for a pair of isotropic molecules is of the form

$$(5\text{-}6) \qquad U = -\frac{A}{r^6}$$

where r is the intermolecular separation and A is the product of various molecular constants.

The dependence of the interaction energy upon the reciprocal sixth power of the intermolecular separation renders the attraction extremely sensitive to the mutual positions of the molecules. Dispersion forces between two different molecules are therefore very dependent upon their shapes and mutual orientations. Any complementarity of shape, which permits the mutual contact of large areas in a "lock and key" arrangement, would be expected to enhance interactions of this kind. As a consequence, dispersion forces may have an especial significance in controlling the specificity of many biological interactions.

The hydrogen bond. Since the hydrogen bond is primarily responsible for the stabilization of many of the protein conformations to be discussed in this chapter, it is desirable to preface a detailed description of these structural forms by an account of its properties. Under appropriate conditions a hydrogen atom may be attracted by two different atoms instead of one, thereby forming a bond between them (8, 9). No instances are known of the linkage of more than two atoms.

Since the hydrogen atom contains only a single stable orbital it cannot of course form more than one covalent bond. The forces stabilizing the hydrogen bond are usually of a predominantly electrostatic character. A hydrogen bond between the two atoms X and Y may be represented as X—H . . . Y, if the hydrogen atom is attached to atom X by a covalent bond. Formation of a hydrogen bond is favored if the resonance patterns of the molecules to which these atoms are attached are such as to place a net positive charge upon the hydrogen atom and a negative charge upon Y. A positive hydrogen ion is a bare proton stripped of its electron shell. Such a cation of negligible radius can simultaneously attract two anions and form a linkage between them of the type $X-H^+Y^-$. In general, only the most electronegative atoms form hydrogen bonds. Fluorine, oxygen, nitrogen, and chlorine atoms have this capacity, the extent diminishing in this order.

Several structural aspects of the hydrogen bond are quite general. It can be predicted on theoretical grounds that the X—H . . . Y bond will attain maximum stability when the arrangement of the three atoms is close to linearity. In practice, the angle between the internuclear lines X—H and X—Y is generally quite small.

In the vast majority of cases the hydrogen atom is closer to one of the two linked atoms than to the other. For example, in the case of two hydrogen-bonded oxygen atoms in ice, the proton is 1.00 Å from one oxygen and 1.76 Å from the other.

The presence of a hydrogen bond is usually reflected by a shortening of the X . . . Y interatomic separation to a value smaller than the sum of the two van der Waals radii, which are determined from measurements of the distance of closest approach in compounds where such bonding is absent. This has provided a criterion for the detection of such bonds in crystals.

The strength of the hydrogen bond is much less than that of normal covalent bonds. As a consequence, hydrogen bonds may be ruptured reversibly at elevated temperatures, or under conditions of pronounced electrostatic, or other, stress. Because of the low magnitudes of its bond energy and the energies of activation for its formation and dissociation, the hydrogen bond is particularly well adapted to figure in reactions occurring at normal temperatures.

It is difficult to overstate the importance of hydrogen bonding in the structure and interactions of biological molecules. Its occurrence is so universal that it may without exaggeration be said to have a significance more general than that of any other single structural feature of biopolymers.

5-3 POLYPEPTIDE CONFORMATIONS

The α-helix. Of the large number of configurations which have been proposed for natural and synthetic polypeptides, only a few are of more than historical interest today. The most important of the configurations for which there is direct experimental evidence is the α-helix proposed originally by Pauling and co-workers (10, 11, 12, 13).

The search for a plausible model for the structure of polypeptides centered upon the problem of finding the conformation of minimal potential energy. This amounted to choosing a model permitting the maximum possible formation of hydrogen bonds, while requiring the least distortion of valence bonds and angles. It should be emphasized that the theoretical proposal of the α-helical model preceded and stimulated the numerous experimental studies which verified its presence in real systems. As in the case of the Watson-Crick structure for DNA, a hypothesis based on experience accumulated from simpler systems provided an indispensable guide for experiment.

The detailed criteria invoked by Pauling and co-workers for a satisfactory model for a polypeptide conformation involving only intramolecular hydrogen bonds were as follows (10, 11, 12, 13).

(1) The peptide group is *planar*, as is invariably the case for simple peptides. This is a consequence of the partial double bond character of the C—N linkage. Any important deviation from planarity would be energetically unfavorable.

(2) The maximum possible number of hydrogen bonds between $\overset{\overset{\displaystyle O}{\|}}{C}$ and N—H groups is formed.

(3) The bond lengths of the —CH—C(=O)—NH— repeating unit (with R on CH) are the same as in the case of small molecules. Any distortion of bond angles would incur a severe energetic penalty.

(4) The hydrogen-bonding H atom does not deviate by more than 30° from the straight line joining the nitrogen and oxygen atoms involved.

(5) The orientations about the C—C and N—C single bonds are close to the potential energy minima for rotation about these bonds.

The only structure which was found to satisfy the above requirements was the α-helix. All other models would either not permit maximum hydrogen-bond formation or would require some distortion of bond angles or distances from their optimal values.

The α-helix is shown in Figs. 5-3 and 5-4. The structure is stabilized

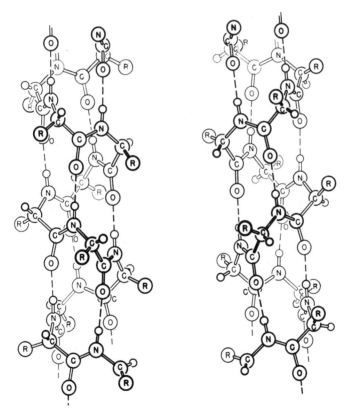

Fig. 5-3. Left- and right-handed α-helices.

140 THE CHEMICAL FOUNDATIONS OF MOLECULAR BIOLOGY

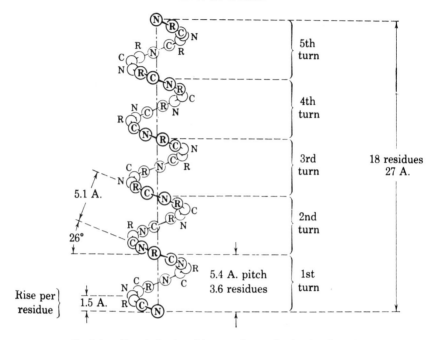

Fig. 5-4a. The α-helix viewed from another angle, showing dimensions.

by C=O . . . H—N hydrogen bonds which are roughly parallel to the axis of the helix. Only the polypeptide backbone is involved in the structure. The configuration of the side-chains is not specified.

Each peptide nitrogen is hydrogen bonded to the carbonyl oxygen of the third nearest residue. That is, if the residues are numbered 1, 2, 3, . . . i, $i + 1$, . . . , and so forth, the nitrogen of the ith residue is hydrogen bonded to the carbonyl of the $i + 3$ th residue. Thus each hydrogen bond spans three residues. This fact is of crucial importance to theories of the helix-coil transition of polypeptides.

The nature of the hydrogen bonding is such as to prevent all free rotation about the bonds of the polypeptide backbone and thus to endow the helix with essentially complete structural rigidity. As a consequence, it can be predicted that a polypeptide which exists in solution as a perfect α-helix will have the shape of a thin rigid rod. Moreover, at least three successive hydrogen bonds must be broken before *any* freedom of rotation is introduced. This is a consequence of the bridging of three residues by each hydrogen bond and may be verified by examining molecular models.

The feature of the α-helix which represented the greatest departure from expectations aroused by known helical systems was its *non-integral* character. There are 3.6 amino acid residues per each turn of the helix. An alternative name sometimes used for the α-helix is the *3.6 residue helix.*

The *pitch,* or distance between successive turns, is 5.4 Å. The separation, in the direction of the fiber axis, of adjacent like atoms is 1.5 Å. The α-carbon atoms, to which the side-chains are attached, are at a radius of 2.3 Å.

It should be noted that the hole running down the center of the α-helix is too narrow to allow penetration by solvent. This is probably a stabilizing factor in the case of hydrogen-bonding solvents.

There is now extensive evidence, obtained by X-ray diffraction and other techniques, for the occurrence of the α-helix in polypeptide systems in the solid state. The problem of detecting its presence in polypeptides in solution requires a quite different experimental approach, which is of an indirect nature.

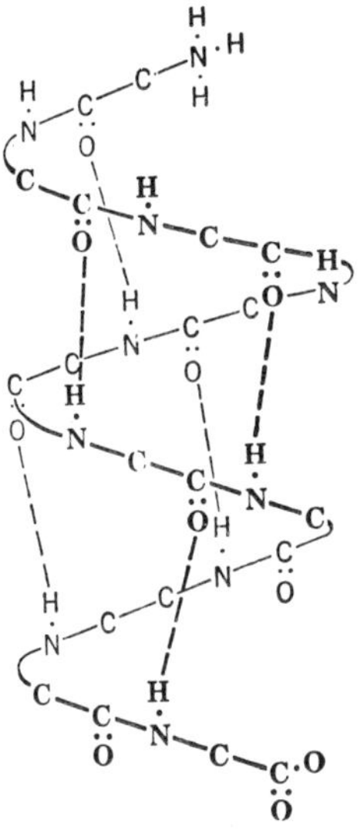

Fig. 5-4b. Hydrogen bonding of the α-helix (5).

The β-conformation. A variety of evidence indicates that the α-helix is the most stable of the polypeptide conformations which are stabilized by *intramolecular* hydrogen bonds. However, structures stabilized by *intermolecular* bonds between different chains are also possible and are believed to occur in some cases.

The completely extended polypeptide chain (Fig. 5-5) is referred to as the β-configuration. Such a maximally stretched form cannot be stabilized by hydrogen bonds of the intramolecular type. However, the potential hydrogen bonding groups (C=O and N—H) all lie in a plane and extend in a direction roughly perpendicular to the direction of the chain. As a consequence, adjacent chains can be linked by lateral hydrogen bonds which serve to stabilize the β-form. In this manner the polypeptide chains can be united to form infinite planar sheets (Fig. 5-6). This is the *pleated sheet* structure of Pauling and Corey (Fig. 5-6).

Two alternative orientations are possible for the pleated sheet con-

142 THE CHEMICAL FOUNDATIONS OF MOLECULAR BIOLOGY

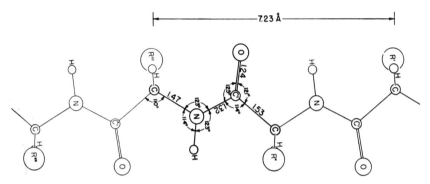

Fig. 5-5. A completely extended polypeptide chain (β-configuration).

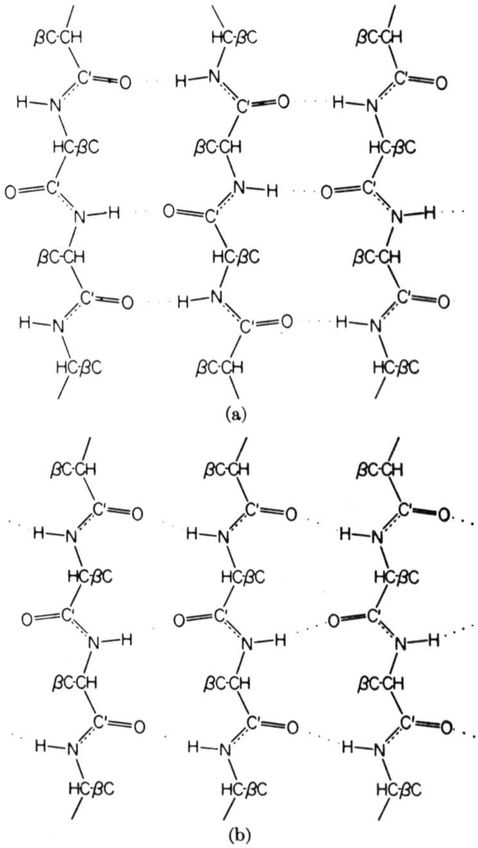

Fig. 5-6. Parallel (b) and anti-parallel (a) pleated sheet structure. Note positions of inter-chain hydrogen bonds.

figuration. In the *parallel* form all the chains have the same direction, while in the *anti-parallel* variant the chains run alternately up and down (Fig. 5-6).

5-4 THE DETERMINATION OF THE CONFORMATION OF POLYPEPTIDES IN THE SOLID STATE

X-ray diffraction. A detailed discussion of the principles of X-ray diffraction is beyond the scope of this book. However, in view of the central importance of this technique in current studies of the structure of proteins and nucleic acids, it is important to give some account of its capabilities (14-19).

X-rays are a form of electromagnetic radiation produced by the bombardment of a metallic target by a stream of electrons. While the wave-lengths of X-rays range from 0.01 Å to 1000 Å, most crystallographic work has been done with wave-lengths of 1.5 Å.

The method is only useful for studying systems which have a definite regularity in their spatial geometry. Systems which have a completely ordered three-dimensional organization are called crystals. A crystal is generated by the indefinite repetition of a small three-dimensional unit, called the *unit cell*. This consists of a small integral number of molecules. Each atom of the molecule is part of a space lattice made up of the collection of such atoms occurring in all molecules of the crystal. If the molecule contains s atoms, then the crystal may be regarded as consisting of s overlapping lattices, of identical geometry, each of which consists of only one kind of atom.

The objectives of X-ray crystallography are twofold. The first, and much the easiest, is the determination of the geometry of the repeating scheme and the dimensions of the unit cell. The second phase, which is more important and much more difficult, consists of the determination of the mutual positions of the atoms making up the unit cell, or, in other words, of the structure of the molecules of the crystal.

If a protein (or other) crystal is mounted in the path of a thin beam of monochromatic X-rays and rotated about some axis perpendicular to the direction of the beam, then a photographic plate placed behind the crystal will register (in addition to the transmitted beam itself) the beams resulting from diffraction or scattering by the atoms of the crystal (Fig. 5-7). A representative result is shown in Fig. 5-8. The pattern does not consist of a diffuse disk or halo, *but rather of a series of discrete spots of varying spacing and intensity.* This indicates that the diffraction of X-rays by a crystal occurs only in certain sharply defined directions.

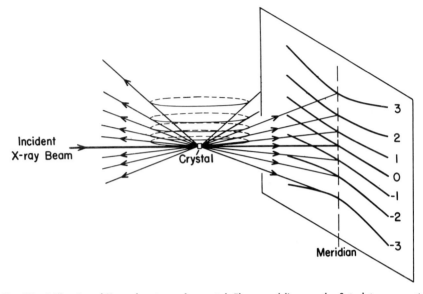

Fig. 5-7. Diffraction of X-rays by atoms of a crystal. The curved lines on the flat plate represent the loci of spots, or *layer lines.*

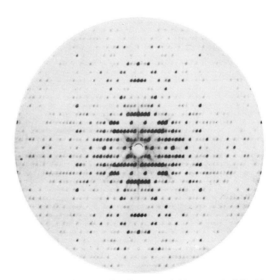

Fig. 5-8. An X-ray diffraction photograph of a myoglobin crystal (19). Note regularity and symmetry of pattern.

Even a cursory glance at Fig. 5-8 reveals several regularities. Thus the upper and lower halves of the photograph are identical, as are the right and left halves. The spots are not distributed at random, but fall on a series of horizontal lines. The intensities of the spots on a given line vary in a regular manner.

The X-ray crystallographer has basically two kinds of experimental information to work from. These are the positions of the spots and their intensities. The problem is to relate this information to the arrangement of the atoms in the unit cell.

Any crystal, by virtue of its generation by repetition of a simple unit, will consist of a large number of planar layers of similar atoms (Fig. 5-9). The diffraction characteristics of the crystal arise from the mutual

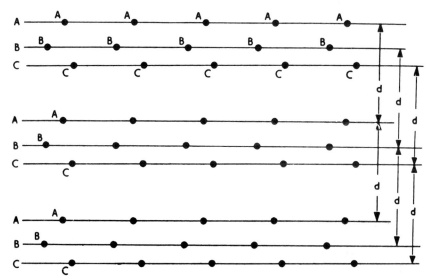

Fig. 5-9. Layers of similar atoms in a crystal (18). The interplanar separation of layers of like atoms is constant.

interference, or reinforcement of beams reflected from different layers. For reinforcement to occur, the beams reflected from a set of parallel planes must be in phase. This is achieved if the difference in path-lengths to the point of observation of rays reflected from successive layers is an integral multiple of the wave-length λ (Fig. 5-10). This will be the case only for certain angles (θ) of incidence (Fig. 5-10). The conditions for reinforcement may be simply stated in the *Bragg equation:*

(5-7) $$n\lambda = 2d \sin \theta$$

where $d =$ the spacing of the planes.

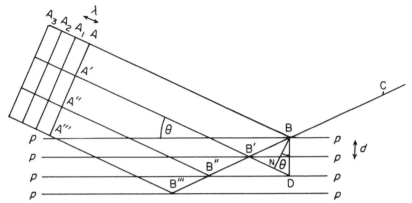

Fig. 5-10. Reinforcement of X-rays reflected by layers of atoms (fulfillment of the Bragg condition).

The derivation of this equation may be written down easily from Fig. 5-10 and is left as an exercise for the reader.

Thus strong diffracted beams will occur only in certain directions for which the Bragg condition is satisfied. In other directions, the beams will be out of phase and destructive interference will greatly reduce their intensity. Every spot on the diffraction photograph represents the fulfillment of the Bragg condition for a particular set of planes.

The *positions* of the spots are determined by the shape and dimensions of the unit cell. Their *intensities* are determined by the mutual positions of the atoms of the unit cell and hence by the structure of the molecule.

If the structure of the molecule is known, it is possible to predict its diffraction pattern by lengthy, but basically straightforward, calculations. Working backwards to obtain the structure from the pattern is much more difficult. This is because one necessary piece of information is a knowledge of the phases of the diffracted rays. These cannot be determined directly from the photograph. In favorable cases they can be estimated by examining the effects of selective chemical modification of the protein upon the diffraction pattern. At best the process is exceedingly demanding and laborious and has been carried to completion in only two cases, which will be discussed in Chapter 6.

An alternative procedure is to make a reasonable guess as to the structure, compute a diffraction pattern, and compare the observed and predicted patterns. Deviations are progressively reduced by making refinements in the original model. This approach has been applied with conspicuous success to the case of DNA (Chapter 9).

Analysis of the X-ray diffraction patterns of fibers of a number of synthetic polypeptides has revealed them to exist, at least to a major extent, in the α-helical conformation. Two examples are poly-γ-methyl-L-glutamate and poly-L-alanine.

5-5 THE ESTIMATION OF HELICAL CONTENT FOR PROTEINS AND POLYPEPTIDES IN SOLUTION

Conformation and molecular dimensions. It is clear from the discussion of section 5-2 that very different physical properties would be expected for a linear polypeptide in the helical and coiled configurations. Doty and co-workers have found that the hydrodynamic properties of poly-γ-benzyl glutamate are profoundly dependent upon solvent. In solvents relatively favorable to the formation of intramolecular hydrogen bonds, such as dimethylformamide, the exponential dependence of intrinsic viscosity upon molecular weight is typical of rod-shaped molecules. Moreover, the molecular dimensions computed from intrinsic viscosity correspond closely to those expected for essentially perfect α-helices.

In contrast, in solvents which competitively disrupt hydrogen bonding, as dichloroacetic acid, both the magnitude of the intrinsic viscosity and its dependence upon molecular weight are characteristic of random coils. On the basis of these results and parallel optical rotation studies it was concluded that poly-γ-benzyl glutamate can, depending upon the solvent, exist in either the α-helical or the randomly coiled configuration.

Subsequent work has confirmed these results and extended them to other synthetic polypeptides. A number of instances have been observed of a transition between the two states which is mediated by variation of the solvent composition or the temperature.

The problem of the conformation of proteins in solution is complicated by the usual presence of more than one polypeptide chain and by the occurrence of important interactions involving the side-chains. It is possible to group proteins roughly into two categories. The class of *fibrous* proteins includes many of those whose function is structural or contractile, such as myosin, tropomyosin, collagen, and fibrinogen. The X-ray diffraction patterns of fibers or films are indicative of a high helical content. The hydrodynamic properties are those of very elongated and asymmetric particles. However, the molecular dimensions are rarely those of a single α-helix, suggesting that they may consist of cables of two or more polypeptide strands or that the individual strands are partially folded.

The class of *globular* proteins is characterized by very compact shapes in solution, which have no dimension comparable in magnitude to that expected for an α-helix of similar molecular weight. Hydrodynamic and light-scattering evidence indicate that the spatial extensions of proteins of this class are generally much more compact than those of typical random coils of the same molecular weight. Moreover, the available fluorescence polarization data are consistent with a high degree of internal rigidity. The most realistic model for the typical globular proteins appears to be that of condensed "supercoils" which are structurally rigid and impermeable to solvent. The helical portions of such molecules must consist of short segments separated by nonhelical zones which endow the polypeptide chains with sufficient flexibility to be folded into a compact configuration (Fig. 5-2).

The determination of the spatial organization of proteins and polypeptides by the X-ray diffraction methods discussed earlier is extremely laborious and is limited by the nature of the technique to the crystalline state. The conformation of proteins is not necessarily the same in solution as in the solid state. Moreover, a variety of evidence indicates that the conformation in solution is not always constant and may vary with such parameters as pH, temperature, and solvent composition.

Thus it is useful to have secondary criteria for the helical content of proteins in solution. The limitations of the available techniques should be recognized. They do not attempt to do more than assess the over-all fraction of the polypeptide chain which is in the helical conformation. In particular, they cannot locate the helical regions within the protein or differentiate between a single long helix and several shorter helices. Moreover, the information obtained is approximate at best.

Nevertheless, the methods described in this section remain the only available structural probes which can assess the helical content of proteins in solution. Their relative simplicity and applicability to non-crystalline proteins has rendered them the only source of information for most proteins.

Optical rotation. All natural proteins are optically active. The observed optical rotation is of a composite character. Since proteins are formed exclusively from amino acids in the L-configuration, a major component of the rotation arises from the summed contributions of the individual amino acids. This *intrinsic* optical activity persists under conditions where the structural organization of the protein is partially, or completely, lost. Since the L-amino acids are laevo-rotatory, the sign of this component of the total activity is negative.

However, in the case of proteins with an important degree of α-helical content, a second major contribution to the optical rotation arises from the intrinsic asymmetry of the α-helix itself (20-24). The α-helix can exist in either a right-handed or a left-handed form (Fig. 5-3). The two forms are mirror images and cannot be interconverted by rotating the helices in space. In the case of synthetic polypeptides it has been predicted on theoretical grounds and verified experimentally that polypeptides composed of L-amino acids form right-handed helices preferentially and *vice versa*. While it is not as yet completely certain that this is the case for proteins, it is entirely reasonable as a working hypothesis.

As a consequence of its spatial asymmetry, the α-helix will have a definite optical activity which is quite distinct from that arising from the individual amino acids. In the case of right-handed helices, the two components are comparable in magnitude and opposite in sign.

The problem of relating optical activity to helical content quantitatively has been approached both theoretically and empirically. The present discussion will be confined to the empirical approach.

The *specific rotation*, $[\alpha]$, has already been defined as

$$(5\text{-}8) \qquad [\alpha] = \frac{\alpha}{dc}$$

where α is the observed rotation; d is the path-length in decims; and c is the concentration in gms per ml.

In dealing with polypeptides it is often convenient to define a new parameter, the molar rotation ($[m']$), in terms of the molar concentration of residues:

$$(5\text{-}9) \qquad [m'] = \frac{M_o[\alpha]}{100}$$

where M_o is the *average* molecular weight of the amino acid residues making up the polypeptide.

In comparing the rotation of polypeptides in different solvents, it is important to take account of the nonspecific effect of refractive index (n). A systematic dependence of specific rotation upon refractive index exists even if the structure is invariant to the solvent. Electromagnetic theory predicts that the rotation should be proportional to $n^2 + 2$. Thus it has become the practice to normalize optical rotation to the value it would have for a solvent of unit refractive index by multiplying by $3/(n^2 + 2)$.

$$(5\text{-}10) \qquad [m] = \frac{3[m']}{n^2 + 2} = \text{corrected molar rotation}$$

It is not as yet possible to relate $[m]$ or $[\alpha]$ quantitatively to the α-helical content of a protein or synthetic polypeptide. Certain semiquantitative relationships have however emerged from studies upon the synthetic systems. An increase in helical content is (for polymers of L-amino acids) generally accompanied by a *decrease* in laevo-rotation. At a wave-length of 5890 Å, which is that of the frequently used sodium "D" line, a polypeptide in the randomly coiled state usually has a value of $[\alpha]$ close to $-100°$. A completely helical polypeptide has a value near $0°$. However there is definite evidence for the existence of nonspecific solvent effects upon the specific rotation.

A more quantitative basis for the estimation of helical content is provided by the wave-length dependence, or *dispersion*, of optical rotation. A number of equations have been proposed to account for this dependence. The most widely used of these has the following form:

(5-11) $$[m] = \frac{a_o \lambda_o^2}{\lambda^2 - \lambda_o^2} + \frac{b_o \lambda_o^4}{(\lambda^2 - \lambda_o^2)^2}$$

Here λ is the wave-length, and a_o, b_o, and λ_o are characteristic constants. For many substances b_o is zero and equation 5-11 reduces to the *Drude equation*.

(5-12) $$[m] = \frac{a_o \lambda_o^2}{\lambda^2 - \lambda_o^2}$$

If the single-term Drude equation is obeyed, the wave-length dependence, or dispersion, is said to be *simple*. If a second term is required to fit the data, the dispersion is said to be *anomalous*.

Equations 5-11 and 5-12 may be placed in the alternative forms:

(5-13) $$(\lambda^2 - \lambda_o^2)[m] = a\lambda_o^2 + \frac{b_o \lambda_o^4}{(\lambda^2 - \lambda_o^2)}$$

(5-14) $$\lambda^2[m] = \lambda_o^2[m] + a_o \lambda_o^2$$

If the dispersion is simple, $\lambda^2[m]$ varies linearly with $[m]$, and the slope is equal to λ_o^2. If the dispersion is anomalous, a value of λ_o which will produce a linear plot of $(\lambda^2 - \lambda_o^2)[m]$ versus $\frac{1}{(\lambda^2 - \lambda_o^2)}$ is found by trial and error. The parameter b_o is found from the slope, which is equal to $b_o \lambda_o^4$.

Thus three different parameters characterize the dispersion of optical rotation: a_o, b_o, and λ_o. The quantity a_o is strongly influenced by the solvent and hence is not useful for the estimation of helical content. However, b_o appears to be determined primarily, at least in the case of polypeptides, by the fraction of the chain which exists in the helical conformation.

From pilot studies upon synthetic polypeptides which are known to exist as 100% helical or randomly coiled forms, the limiting values of

b_o appear to be close to $-630°$ for the helical species and to $0°$ for the nonhelical (Fig. 5-11). Thus, if it is assumed that the variation is linear, the fraction of the polypeptide which is in the helical conformation is equal to $-b_o/630$.

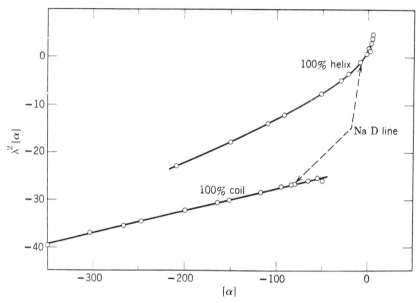

Fig. 5-11. Optical rotatory dispersion for poly-γ-benzyl glutamate in helix and coil forms. (From J. Yang and P. Doty, J. Am. Chem. Soc., 79, 761 [1957].)

The extension of this relationship to proteins has been tested in the case of the protein myoglobin, for which X-ray diffraction has provided a virtually complete determination of the three-dimensional structural pattern. Estimates of the fractional helical content obtained from rotatory dispersion are in agreement with those predicted from the more complete structural determination.

Table 5-1 cites several examples of the determination of helical content from b_o for several proteins in aqueous solution.

Deuterium exchange. An altogether different approach to the problem of helical content determination involves the use of "heavy water" in which all the hydrogen is the isotope of mass number 2, or deuterium (D). To differentiate it from ordinary water, this is usually referred to as deuterium oxide (D_2O).

A protein molecule in water will have virtually all its hydrogen as the ordinary isotope (H), since the natural occurrence of deuterium is less than 1%. If the protein is placed in D_2O solution, its H atoms will be replaced by D atoms at rates which depend upon the atoms to which they are attached and upon their environment. In general, H atoms attached to nitrogen or oxygen exchange very rapidly, while those attached to carbon do not exchange at all except at elevated temperatures.

Thus, in the absence of other factors, it would be expected that the H atoms attached to nitrogen or oxygen, including those of the CONH groups of the peptide backbone, would exchange very rapidly, while the remainder of the H atoms, which occur as C—H, would not exchange at appreciable rates at ordinary temperatures. On this basis the theoretically expected number of rapidly exchangeable hydrogen atoms can be computed (25, 26, 27).

The experimental procedure is to dissolve the protein in D_2O and to keep it in contact with this solvent for a sufficiently long time to permit quantitative replacement of its exchangeable hydrogen atoms with deuterium. The protein solution is then frozen and all water removed by exhaustive evacuation. The dry, D_2O-free protein is redissolved in H_2O. The exchange is followed by observing the rate of appearance of deuterium atoms in the H_2O.

In the case of most proteins it has been found that only a fraction of the theoretical number of exchangeable H atoms equilibrate rapidly. The balance exchange at rates which are very much slower. It has been postulated that the class of slowly exchangeable H atoms arises from the involvement of the peptide CONH groups in the helical regions of the molecule. The reason for the impaired exchangeability is uncertain. It is possible that the hindrance is primarily steric, reflecting the inaccessibility of the helical zones to solvent.

Calculations of the fractional helical content have been made on this basis and appear to be in fair agreement with those based on optical rotation.

5-6 THE HELIX-COIL TRANSITION FOR POLYPEPTIDES

Experimental observations. As has been discussed in earlier sections, many synthetic polypeptides can, depending upon the solvent, exist in either the completely helical or the randomly coiled form. The two states have dramatically different hydrodynamic and optical rotatory properties.

The intrinsic stability of the α-helix is strongly dependent upon conditions. It is to be expected that reagents which competitively rupture hydrogen bonds will tend to destabilize the structure. Since the formation of hydrogen bonds is exothermic, an increase in temperature will favor conversion of the helix to a random coil. If the side-chains of the polypeptide contain similar ionizable groups, then the helix will be less stable at pH's where the groups are ionized, since the groups will be more closely spaced in the helical than in the randomly coiled form. This results in a greater electrostatic repulsion between similarly charged groups in the former case.

By varying a single experimental parameter it is often possible to observe the transition between the two states in a single solvent. An example of a helix-coil transition which is mediated by a change in ionization is furnished by polyglutamic acid in water-dioxane mixtures. In this case, the side-chains contain carboxyl groups which become negatively charged upon ionization.

At pH's below 4, where the carboxyls exist in the un-ionized (COOH) form, polyglutamic acid has optical rotatory and hydrodynamic properties typical of the α-helical conformation. At pH's above 6, where ionization of the carboxyls is complete, it has the characteristics of a random coil. If its properties are monitored as a function of pH, a major change in optical rotation and viscosity is observed to occur in the vicinity of pH 6 (Fig. 5-12). The limiting values of $[\alpha]_D$ at acid pH ($+6°$) and at

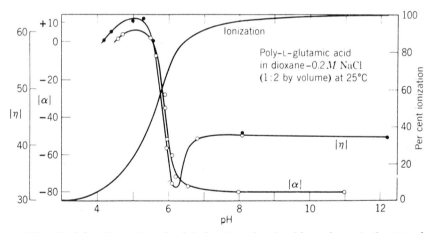

Fig. 5-12. The helix-coil transition of poly-L-glutamic acid, induced by a change in the state of ionization. Note parallel changes in different experimental parameters. (From P. Doty, A. Wada, J. Yang, and E. Blout, J. Polymer Sci., 23, 851 [1957].)

alkaline pH ($-80°$) are typical of α-helical and of randomly coiled polypeptides, respectively.

Thus ionization of the side-chain carboxyls imposes in this case an intolerable electrostatic stress upon the helical form. This can be partially relieved by a transition to the more open and irregular coiled form.

The viscosity of polyglutamic acid actually passes through a minimum (Fig. 5-12). The initial drop arises from the loss of rigidity accompanying disruption of the helix when the ionization has attained a critical value. The subsequent partial recovery results from the expansion of the coil under electrostatic stress as the degree of ionization approaches completion.

The helix-coil transition may also be brought about by varying the temperature. Such a transition occurs in the case of poly-γ-benzyl glutamate (Fig. 5-13) in ethylene dichloride-dichloroacetic acid. Optical rotation is a convenient index of the extent of conversion, the change in rotation being roughly proportional to the fraction of residues which have undergone the transition.

A characteristic feature of the helix-coil transition for polypeptides of high molecular weight is the dramatically sharp character it often assumes. In the case of poly-γ-benzyl glutamate (Fig. 5-13) of high molec-

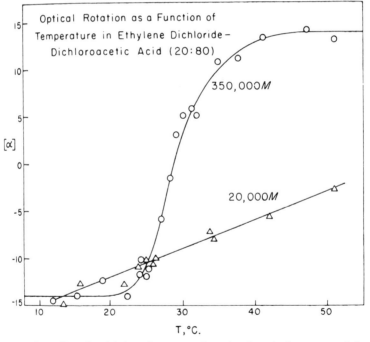

Fig. 5-13. Thermally induced helix-coil transition for poly-γ-benzyl glutamate, as followed by optical rotation. The solvent is ethylene dichloride—dichloroacetic acid. Observe the contrast between the *sharp* thermal profile for the high molecular weight sample and the *broad* profile for the low molecular weight sample.

ular weight, the process attains completion over a range of about 5°. In view of the low (−5 KCal) heat change accompanying the formation of hydrogen bonds, the abruptness of this transition cannot be accounted for by any simple equilibrium treatment.

Mechanism of the transition. The sharpness of the thermally induced helix-coil transition is in fact of central importance in clarifying the mechanism of the process. Elaborate statistical mechanical theories have been developed which can account quantitatively for the temperature dependence of the helical content (30, 31). However, the basic principles involved can be readily understood from a purely qualitative exposition.

The high degree of structural rigidity characteristic of the α-helix confines the groups of the polypeptide backbone to a single configuration. Thus the transition from a random coil to the α-helix reduces the number of configurations which the chain can assume from a large number to only one. The transition is from a state in which the relative positions of two residues cannot be precisely specified to one in which they are entirely predictable. Alternatively, this process, which is analogous to the crystallization of a liquid, may be said to be from a *disordered* to an *ordered* state.

In the language of thermodynamics (Appendix B), such a process is said to involve a decrease of *entropy*. In the absence of other compensating factors, processes of this kind do not occur naturally, for they correspond to the transition from a state of greater to one of lesser probability. A simple analog might be a glass jar containing black and white marbles arranged in alternate layers which are all black or all white. Such an ordered, predictable state is easily disrupted by shaking the jar and is not likely to be reformed by subsequent shakings.

In the case of the α-helix the compensating factor which permits formation of the ordered structure is the decrease in total energy accompanying the formation of $C=O \cdots H-N$ hydrogen bonds. When the stabilizing hydrogen bonds are ruptured thermally, the rigidity conferred by the helical conformation is lost and the molecule reassumes the randomly coiled configuration.

Each $C=O \cdots H-N$ hydrogen bond spans a total of three residues and holds them in a rigid configuration irrespective of the state of the other two hydrogen bonds. The situation may be represented as follows:

$$\begin{array}{c} \text{H}\cdots\cdots\cdots\cdots\cdots\cdots\text{O} \\ | \phantom{\text{XXXXXXXXXXXXXXXX}} \| \\ -\text{N}-(\text{CO}-\underset{\text{R}}{\text{CH}}-\text{NH})_3-\text{C}- \end{array}$$

In order for any free rotation, and hence any loss of ordered character (or gain of entropy), to occur, three *consecutive* hydrogen bonds must be broken. The first two bonds must be ruptured without any compensating increase of entropy.

Once a sequence of three broken bonds has been created, the zone of random residues can be enlarged with ease, because for each residue added the loss of a hydrogen bond is compensated for by a gain of entropy arising from the addition of one unbridged residue with freedom of rotation.

In other words, the nature of the hydrogen bonding which stabilizes the α-helix is such as to render energetically unfavorable the presence of numerous junctions between helical and random regions and thereby to minimize the number of such junctions. Thus short helical zones tend to coalesce into extended sequences. At a temperature in the middle of the thermal transition the system may be visualized as consisting of a small number of alternating helical and random zones of considerable length. It may be shown by the methods of statistical thermodynamics (30, 31) that the thermal dependence of the distribution of residues between the two states is greatly sharpened for such a system.

Indeed, the helix-coil transition may be regarded as somewhat similar to a one-dimensional phase transition. (Strictly speaking, however, a *true* phase transition cannot occur for a one-dimensional system.) The extended helical and random regions may be thought of as analogous to the crystalline and liquid phases which are at equilibrium at the melting point of a pure solid. The imperfect character of the analogy is indicated by the finite breadth of the helix-coil transition, as contrasted with the discontinuous nature of a true phase transition.

The helix-coil transition for a polypeptide is shown schematically in Fig. 5-14.

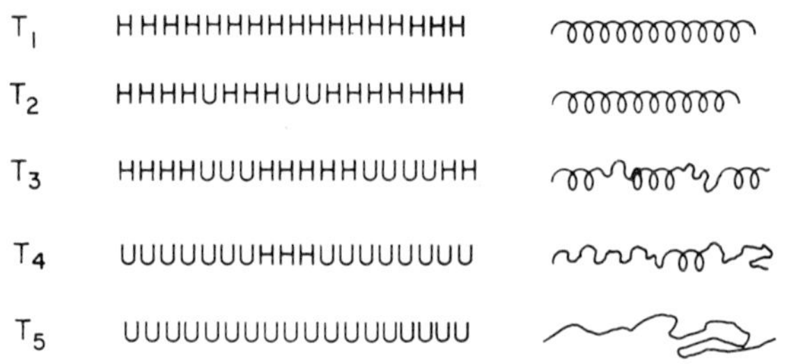

T_1	HHHHHHHHHHHHHHHH
T_2	HHHHUHHHUUHHHHHHH
T_3	HHHHUUUHHHHHUUUUHH
T_4	UUUUUUUHHHUUUUUUUU
T_5	UUUUUUUUUUUUUUUUUU

Fig. 5-14. Schematic representation of a helix-coil transition. The temperature increases from T_1 to T_5, with T_3 at about the midpoint of the transition. "H" represents a hydrogen-bonded residue; "U" represents an unbonded residue. Note that three "U's" in sequence are needed to disrupt part of the helix.

Effect of chain length. The preceding discussion requires modification for relatively short helices, a hundred residues long or less. In this case an alternative mechanism becomes important. At the terminus of each α-helix there are two residues which are not bridged by hydrogen bonds. These provide a "built-in" nucleus of random residues which can be enlarged relatively easily.

For very short helices, unraveling from the ends is the dominant mechanism for the thermal transition. The longer the chain, the more important is the alternate mechanism involving breaks in the interior. For chains longer than about 1000 residues the latter mechanism is predominant.

The stability of the α-helix decreases with decreasing chain-length. This is reflected by a decrease in the transition temperature. The stability of the very short helical regions found in globular proteins is probably conferred by other structural elements.

The helix-coil transition in proteins. In the case of proteins the situation is complicated by the presence of important interactions involving the side-chains, as well as by disulfide cross-links. In the case of globular proteins the helical regions are relatively short and may differ in stability.

In general the disruption of the native structure of globular proteins, or *denaturation*, follows from a primary event which does not involve the polypeptide backbone directly. While the helical content is usually reduced or eliminated by denaturation, this appears to be in general a consequence of the loss in molecular organization rather than the cause of it. In other words the short helical sequences present in globular proteins are stabilized by their molecular environment. Once this is lost the unfolding of the helical regions proceeds spontaneously.

The mechanism of the stabilization is still a matter of controversy, but it seems likely that a dominant factor is the shielding of the helix from water, which would otherwise tend to break competitively its hydrogen bonds.

5-7 THE TERTIARY STRUCTURE OF PROTEINS

It is clear from the available information for natural proteins that the structural forms discussed in earlier sections by no means account for all aspects of their molecular organization in most cases. The helical content of most proteins is only fractional and in many instances can account for only a minor portion of the molecule. Nevertheless, such proteins appear to have configurations which are no less sharply defined than those of the highly helical systems.

It has become customary to group arbitrarily all aspects of the spatial organization of proteins other than the helical forms into the classification of "tertiary structure." This is presumably stabilized by interactions involving the *side-chains* rather than the peptide backbone.

At this point, generalization becomes impossible, and each protein must be regarded as a special case. Indeed, in no case can a particular structure be regarded as established beyond question. The models which will be described in subsequent sections should be regarded as no more than plausible guides to further investigation.

Hydrophobic bonds. The most familiar example of the kind of adhesive force responsible for bonds of this type is furnished by solutions of the detergents. These consist of long hydrocarbon chains concluding in an ionic group. When dissolved in a polar solvent, such as water, they combine to form *micelles* in which the hydrocarbon chains occupy the interior and the ionic groups are on the periphery. The geometry of the micelles is such as to minimize the area of contact of the hydrocarbon with the water.

The reasons for this behavior are not difficult to understand. Water has a high degree of internal hydrogen bonding. This results in the formation of a partially ordered lattice. Since hydrocarbons are ineffective formers of hydrogen bonds, the existence of hydrocarbon chains in direct contact with water would serve to reduce the total number of hydrogen bonds, thereby producing a situation which is energetically unfavorable. Such hydrocarbon zones would resemble "holes" in the water lattice.

In contrast, the ionic groups can be easily accommodated by the water structure. The energy of interaction of charged groups with the polar H_2O molecules is more than sufficient to compensate for any blocking of hydrogen-bond formation.

The formation of micelles results in a partial shielding of the hydrocarbon portions of the detergent molecules from the solvent. The micelle structure is such as to replace hydrocarbon-H_2O contacts as much as possible by hydrocarbon-hydrocarbon contacts.

Since the side-chains of many amino acids are of a hydrocarbon nature, it is natural to suppose that preferential hydrocarbon-hydrocarbon contacts might make a contribution to the stability of protein structure. Such cohesive forces have been designated as *hydrophobic bonds*. The importance of such forces would be expected to vary from protein to protein, as does the proportion of nonpolar side-chains.

Hydrogen bonds between side-chains. There has been much speculation that hydrogen bonds may exist between polar groups of side-chains and that such bonds may assist in the stabilization of the molecular organization of proteins. In particular, hydrogen bonds between tyrosine and carboxylate ions have often been postulated.

$$\overset{-}{\underset{\underset{O}{\overset{\|}{C}}}{\overset{|}{-C}}}\!\!-\!\overset{-}{O}\cdots H\!-\!O\!\!-\!\!\!\bigcirc\!\!\!-$$

Such linkages were originally invoked to explain certain anomalies in the hydrogen ion titration behavior of a number of proteins. Histidine-carboxylate bonds have also been proposed occasionally.

At present there is little direct evidence either for or against such bonds. Their existence must be regarded as quite speculative.

Other linkages. If two oppositely charged groups are placed in close juxtaposition, the electrostatic interaction of the two (section 5-2) will tend to maintain them at a short separation and hence to stabilize the protein configuration. Dipole-dipole interactions, as between the hydroxyl groups of two serine residues, and charge-dipole interactions (section 5-2) may also occur in particular instances.

At present it is not possible to assess the relative importance of these types of interaction. Figure 5-15 represents schematically the various possibilities.

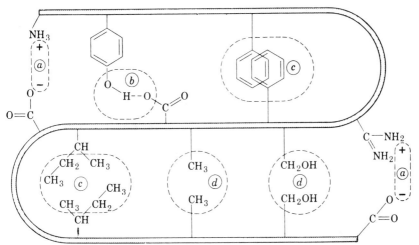

Fig. 5-15. Some possible types of side-chain interaction for globular proteins: (a) electrostatic (ion-pair), (b) hydrogen bond (tyrosyl-carboxyl), (c) hydrophobic, and (d) dipole-dipole (from reference 9, Chapter 2).

TABLE 5-1. COMPUTED HELICAL CONTENTS FOR SEVERAL
REPRESENTATIVE PROTEINS

Protein	b_o	% helix
Serum albumin	$-290°$	46
Ribonuclease	$-100°$	16
Insulin	$-240°$	38
Tropomyosin	$-650°$	100

GENERAL REFERENCES

1. *Physical Chemistry of Macromolecules*, C. Tanford, Wiley, New York (1961).
2. *Biophysical Science*, J. Oncley (ed.), Wiley, New York (1960).
3. *Biophysical Chemistry*, J. Edsall and J. Wyman, Academic Press, New York (1958).
4. *The Nature of the Chemical Bond*, L. Pauling, Cornell University Press, Ithaca (1960).
5. *Synthetic Polypeptides*, C. Bamford, A. Elliott, and W. Hanly, Academic Press, New York (1956).

SPECIFIC REFERENCES

Intermolecular forces

6. *Theoretical Chemistry*, S. Glasstone, chap. IX, Van Nostrand, Princeton (1944).
7. Reference 1, p. 129.

The hydrogen bond

8. Reference 4, chap. 12.
9. *The Hydrogen Bond*, G. Pimentel and A. McClellan, Freeman, San Francisco (1959).

The α-helix

10. Reference 4, p. 498.
11. Reference 5, chap. IV.
12. L. Pauling, R. Corey, and H. Branson, *Proc. Nat. Acad. Sci. U. S.*, **37**, 205 (1951).
13. Reference 1, p. 51.

X-ray diffraction

14. *Chemical Crystallography*, C. Bunn, Oxford University Press, New York (1945).
15. *X-ray Crystallography*, Wiley, New York (1942).

16. *The Optical Principles of the Diffraction of X-rays*, R. James, Bell, London (1948).
17. Reference 1, chap. 2.
18. A. Stokes in *Progress in Biophysics and Biophysical Chemistry*, **5,** 140 (1955).
19. F. Crick and J. Kendrew in *Advances in Protein Chemistry*, vol. 12, p. 133, Academic Press, New York (1957).

Optical rotation

20. Reference 1, p. 119.
21. *Optical Rotatory Dispersion*, C. Djerassi, McGraw-Hill, New York (1960).
22. C. Schellman and J. Schellman, *Compt. rend. trav. lab. Carlsberg*, Ser. chim., **30,** 463 (1958).
23. J. Yang and P. Doty, *J. Am. Chem. Soc.*, **79,** 761 (1957).
24. W. Moffett, D. Fitts, and J. Kirkwood, *Proc Nat. Acad. Sci. U. S.*, **43,** 723 (1957).

Deuterium exchange

25. Reference 1, p. 665.
26. K. Linderstrom-Lang, *Chem. Soc.* (London), Spec. Publ., no. 2, **1** (1955).
27. A. Hvidt and K. Linderstrom-Lang, *Biochem. et Biophys. Acta*, **18,** 306 (1955).

Helix-coil transition

28. Reference 1, p. 509.
29. P. Doty, A. Wada, J. Yang, and E. Blout, *J. Polymer Sci.*, **23,** 851 (1957).
30. T. Hill, *J. Chem. Phys.*, **30,** 383 (1959).
31. J. Gibbs and E. DiMarzio, *J. Chem. Phys.*, **30,** 271 (1959).

PROBLEMS

1. Compute the length of an α-helical polypeptide 1000 amino acid units long.
2. What is the axial ratio of the polypeptide of problem 1? Compute its intrinsic viscosity from Simha's equation (Chapter 4).
3. Make a plot of $\lambda^2[m]$ versus $[m]$ from 6000 Å to 3500 Å for a protein for which $a_o = -600°$
   $\lambda_o = 2120$
   $b_o = 0$
4. Repeat problem 3 for
   $a_o = 0$
   $\lambda_o = 2120$
   $b_o = -630°$

6

Structure and Function of Certain Important Proteins

6-1 GENERAL REMARKS

The number of different proteins which have been isolated from biological systems is very large. For perhaps a hundred of these there is available reasonably reliable information as to the more basic physical parameters. In only a dozen or so cases is there really definitive information as to structure. The proteins to be discussed in this chapter have been selected from this small, but expanding, group.

It must be acknowledged that, at the present time, the complete structure of no protein is completely known. In the cases of the two proteins, hemoglobin and myoglobin, for which knowledge of the spatial organization approaches completion, areas of uncertainty remain in the primary structure. Proteins, such as insulin and ribonuclease, for which there is complete information as to amino acid sequence, have not been examined with comparable success by X-ray diffraction.

Thus in no case have both the primary structure and all details of the three-dimensional structural pattern been completely worked out. There is every reason to believe, however, that this situation is temporary and that the future will witness first one, and then many, examples of the complete structural description of individual proteins.

Much of the discussion of the following sections will center about the relationship of structure to biological function. Here too, the information is fragmentary, and gaps exist in most of the models to be described. However, progress in this area has been dramatic in recent years, and there can be little doubt that ambiguities will be progressively eliminated in the next few years (1-5).

6-2 STRUCTURAL PROTEINS: COLLAGEN

Occurrence and function. Collagen is, together with elastin, the primary fibrous constituent of connective tissue. As such, it is among the most abundant proteins known, forming a major part of skin, cartilage, ligament, tendon, and bone. Its function appears to be primarily structural. In the form of filaments it can possess extraordinary tensile strength, approaching steel wire in this respect.

Collagen appears to have developed rather early in evolutionary time and is found in a wide variety of living species ranging from the most primitive to the most complex. While quantitative differences exist, the general properties of collagens from different species are remarkably similar.

Collagen occurs in biological systems as bundles of individual linear *fibrils*, whose diameter varies with the source (Fig. 6-1). The bundles, which may be examined with the electron microscope, may be randomly

Fig. 6-1a. Electron microscope photograph of native collagen fibrils from skin, showing the 700 Å band spacing (8).

arranged, as in skin, or aligned in a parallel manner, as in tendon (6-10).

Electron microscopic examination of collagen fibrils from a number of sources have revealed the presence of a definite banded pattern (Fig. 6-1a and 6-1b). The spacing of the bands is quite regular and constant,

Fig. 6-1b. Regenerated collagen fibrils of native form.

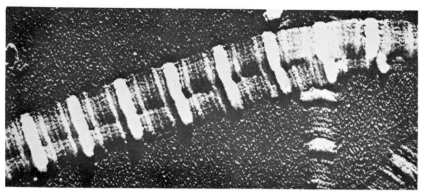

Fig. 6-1c. Regenerated fibrils of FLS form.

being close to 700 Å. For a long time it was believed that this spacing was related to the dimensions of the fundamental molecular unit of collagen. However, it now appears to be a consequence of the nature of the arrangement of the molecular units within the fibril, as will be discussed below (10).

Properties of collagen fibrils. A distinctive characteristic of macroscopic collagen fibers, which consist of many fibrils arranged in parallel, is the abrupt and dramatic contraction which occurs on heating. The onset of this phenomenon, which resembles a phase transition, is remarkably

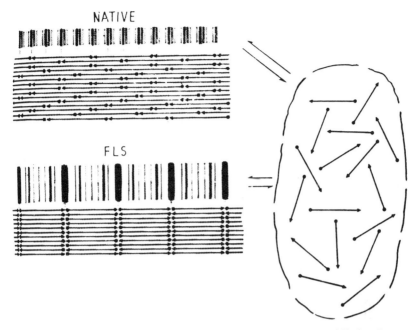

Fig. 6-1d. Schematic version of structures of two types of regenerated fibrils (10).

sudden and usually goes to completion over a rather narrow temperature range, whose midpoint is close to 60° for most mammalian collagens. The amount of shortening can be very large—down to one-third to one-quarter of the original length. The shrunken fiber has mechanical properties which are quite different from those of the original; it displays rubber-like elasticity and has a greatly reduced tensile strength.

Collagen fibrils are highly insoluble in water but will dissolve in dilute acetic acid at room temperature. Neutralization of the solution results in the reappearance of the fibers. Electron microscopic examination of fibrils regenerated in this way shows them to have the characteristic banded appearance of native collagen with the same constant 700 Å spacing (Fig. 6-1b).

Fibrils of different appearance may be obtained by varying the conditions of reformation. The addition of certain glycoproteins to acid collagen solutions produces fibrils in which the 700 Å spacing is replaced by a long spacing of 2800 Å (Fig. 6-1c). This variant is often referred to as the *fibrous long spacing* (FLS) form. The difference between the native and FLS forms is believed to arise from the different manners in which the individual molecular units are aligned within the fibril. However, a discussion of this point should be prefaced by an account of the *molecular* characteristics of collagen.

Molecular structure. It has already been mentioned that certain forms of collagen, including those of rat-tail tendon and carp bladder, may be dissolved in acid buffers. The resultant solutions contain individual collagen molecular units as well as aggregates of varying size. Removal of the aggregates by high speed centrifugation permits physical studies upon the residual, molecularly dispersed material. This is known as *tropocollagen*, or sometimes *procollagen*.

Much of the information available as to the molecular characteristics of tropocollagen has been obtained for *ichthyocol*, the collagen of carp bladder. What is known of the size and shape parameters of other tropocollagens indicates that they are similar to those of ichthyocol.

Light scattering measurements upon ichthyocol solutions, using the Zimm grid extrapolation method described in Chapter 4, have revealed its molecular weight to be close to 350,000. Light scattering and viscosity determinations agree in assigning a highly asymmetric and rodlike shape to the molecule. Doty has estimated from a combination of data obtained by these two methods that the best model for the molecular unit of ichthyocol is a rigid rod, or cylinder, about 3000 Å long and 14 Å in

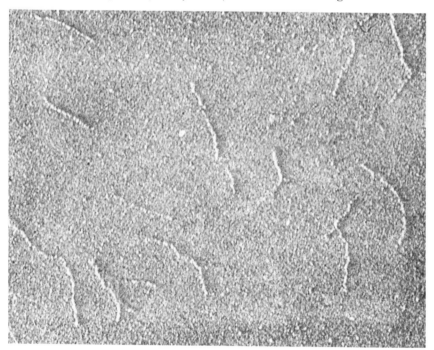

Fig. 6-2a. Electron microscope photograph of tropocollagen (10).

diameter (9). These dimensions have been confirmed by direct electron microscopic examination (Fig. 6-2).

It is clear from this result that the 700 Å spacing observed in native collagen fibrils cannot correspond to individual molecular units. However, there is an encouraging similarity of the long dimension of tropocollagen to the long spacing of the FLS form of regenerated collagen fibrils. This will be discussed further in the next section.

The extreme asymmetry and rigidity of tropocollagen are of course suggestive of a helical conformation. This has indeed proved to be the case. However, analysis of the X-ray diffraction patterns produced by collagen fibers has shown that the helical form is not the α-helix described in Chapter 5.

This appears to be a consequence of the unusual amino acid composition of collagen (Table 6-1). While the figures vary somewhat for collagens from different sources, the mole fractions of glycine, which makes up about one-third of the collagen molecule, and of proline and hydroxyproline, which together account for about one-quarter, are invariably high.

The high proline and hydroxyproline content is probably responsible in large measure for the unique structural characteristics of collagen. It will be recalled (Chapter 2) that these are the only two amino acids which have an imino, rather than an amino, group attached to the α-carbon. The formation of a peptide linkage involves the loss of the single hydrogen of the imino group. Thus the peptide group is in this case left without a potential hydrogen-bond donor. As a consequence, the type of hydrogen bonding which stabilizes the α-helix is not possible. Hence, proline and hydroxyproline cannot be accommodated within an α-helical

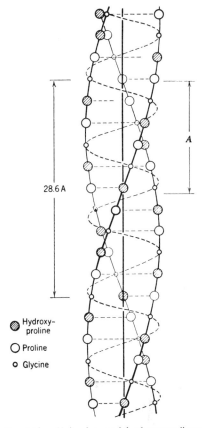

Fig. 6-2b. Molecular model of tropocollagen (10).

THE CHEMICAL FOUNDATIONS OF MOLECULAR BIOLOGY

TABLE 6-1. AMINO ACID COMPOSITION OF SEVERAL IMPORTANT PROTEINS

Amino Acid	Moles per 100,000 Gms			Residues per Mole
	Collagen	Myosin (rabbit)	Fibrinogen (human)	Serum Albumin (bovine)
Alanine	107	73	42	48
Glycine	363	25	75	17
Valine	29	22	35	35
Leucine	—	119	54	65
Isoleucine	43		37	14
Proline	131	17	50	29
Phenylalanine	15	26	28	24
Tyrosine	5	19	30	19
Tryptophan	0	4	16	2
Serine	32	41	67	28
Threonine	19	43	52	29
Cystine/2	0	12	19	28
Cysteine	0	—	3	1
Methionine	5	23	17	4
Arginine	49	42	45	23
Histidine	5	15	17	18
Lysine	31	81	63	60
Aspartic acid	47	67	99	65
Glutamic acid	77	150	99	78

structure. In view of the high frequency of occurrence of these residues, it is not surprising to find that the conformation of collagen is not based on the α-helix.

Collagen owes its characteristics to its molecular organization as well as to its chemical composition. The structure of collagen has been solved by several independent groups of X-ray crystallographers (11). The currently accepted model of the collagen molecule pictures it as consisting of a cable of *three* polypeptide strands (Fig. 6-2b). Each chain is twisted into a left-handed helix (Fig. 6-3), and the three helices are wrapped around each other to form a right-handed superhelix (Fig. 6-3).

The three chains are held together by *inter-chain* hydrogen bonds. These probably join the carbonyl oxygens and amino nitrogens (of residues other than proline and hydroxyproline) of peptide groups located in different chains (Fig. 6-3). The character of the stabilizing hydrogen bonding is thus altogether different from that of the α-helix.

Arrangement of collagen units within the fibrils. Recognition that the true length of the tropocollagen molecule is close to 3000 Å has led to abandonment of the idea that the 700 Å spacing of native collagen fibrils has any direct relation to the dimensions of the molecular unit.

STRUCTURE AND FUNCTION OF IMPORTANT PROTEINS

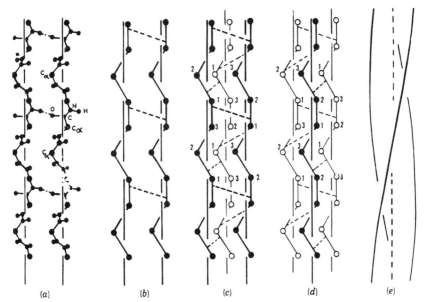

Fig. 6-3. Assembly of the molecular model of tropocollagen (5). Parts (a) and (b) show two polypeptide backbones lying side by side; (c) and (d) show two alternative ways of assembling three chains to form a triply stranded molecule; (e) shows the deformation of the triply stranded structure to form the collagen model.

The 700 Å band spacing is comprehensible only if the tropocollagen units are not completely regular and display appreciable variations in thickness along their length (Fig. 6-1d). These might arise from a clustering of bulky residues in certain zones to form nodules. To explain the difference in band spacing between the native and FLS fibrils, it is sufficient to postulate that the two ends of the tropocollagen unit are distinguishable from the rest of the molecule and from each other (Fig. 6-1d).

One model which has been proposed for native collagen represents it as composed of *overlapping* tropocollagen units (Fig. 6-1d). The inevitable blurring occurring in electron microscope photographs causes the rows of junction zones between tropocollagen units to appear continuous (Fig. 6-1a), thereby creating the 700 Å banded pattern.

In the case of fibrils of the FLS type, there is no reason to doubt that the 2800 Å spacing corresponds to the molecular length of tropocollagen. Thus the *parallel nonoverlapping* alignment of the collagen molecular units within the regenerated fibril would account for banding of this type (Fig. 6-1d). The fine structure (Fig. 6-1d) between the major bands may arise from lesser irregularities in the middle of the tropocollagen units, which form bands when lined up in rows.

Thermal denaturation. When heated in solution, the tropocollagens undergo a very dramatic structural change at a temperature which is close to 40° in the case of ichthyocol. The molecular weight, as monitored by light scattering (Fig. 6-4a), falls by a factor of two to three. Both the radius of gyration and the intrinsic viscosity show a major drop. The change is rapid and abrupt, going to completion over a temperature interval of several degrees (9, 12).

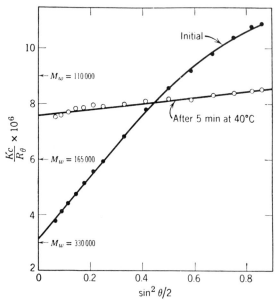

Fig. 6-4a. Light-scattering data on the thermal denaturation of ichthyocol. The curves were obtained by extrapolation of Zimm grids to zero concentrations (9).

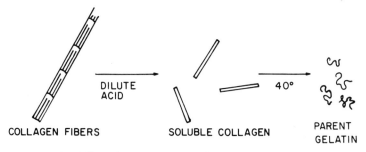

Fig. 6-4b. Schematic version of collagen denaturation.

The molecular events associated with the thermal transition of tropocollagen are readily interpretable in terms of the triply stranded model discussed in an earlier section. The process is a helix-coil transition and is accompanied by a partial splitting into the individual polypeptide strands. Once the rigidity conferred by the helical structure is lost, the separated chains assume a randomly coiled configuration, with a concomitant drop in radius of gyration and intrinsic viscosity (Fig. 6-4b).

It is of course very likely that the thermal shrinkage shown by collagen fibers reflects the same basic molecular events as the process occurring in solution. The contraction in length and the appearance of rubberlike elasticity are both consistent with this view.

The product of the thermal transition in solution is called *parent gelatin*, of which ordinary gelatin is a degraded derivative. There is evidence that the collagen-parent gelatin transition is at least partially reversible upon prolonged standing at low temperatures (12).

6-3 CONTRACTILE PROTEINS: MYOSIN AND ACTIN (13-18)

Occurrence and function. Myosin is one of the principal protein components of muscle tissue and, together with the fibrous protein actin, appears to form an essential part of the contractile system, which endows muscle fibers with the capacity to perform mechanical work.

Muscle tissue is very diverse in appearance and structure. The muscle tissues of mammals may be roughly grouped into three categories, which share the property of contractibility (16).

Smooth muscle cells are normally spindle-shaped and have a homogeneous appearance. They do not possess any distinct enclosing sheath. They are found in the walls of the digestive and respiratory tracts, in the skin (together with connective tissue fibers), and in the walls of blood vessels.

Skeletal muscle is by far the most commonly occurring form in mammals, making up a greater proportion of the body weight than any other tissue. Skeletal muscles are invariably striated, consisting of bands of different refractive index. A typical muscle is encased in a sheath, the *perimysium*, and may be infiltrated extensively by fat deposits and connective tissue. The contractile tissues occur as *fibers* 0.01 to 0.1 mm thick, each of which is enclosed in a membrane, the *sarcolemma*. The fibers are further split into *fibrils* about 10,000 Å thick. In the electron microscope these are seen to consist of bundles of thinner threads, called *filaments* (Fig. 6-5).

Cardiac muscle, the muscle of the heart, has striations which are less

172   THE CHEMICAL FOUNDATIONS OF MOLECULAR BIOLOGY

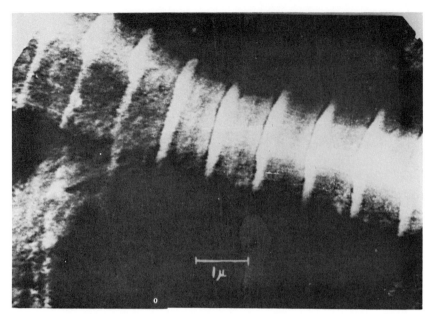

Fig. 6-5a.   Electron microscope photograph of myofibril, showing sarcomeres and individual filaments (16).

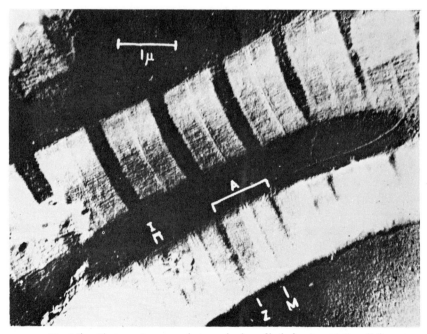

Fig. 6-5b.   Electron microscope photograph of myofibril, showing striations (16).

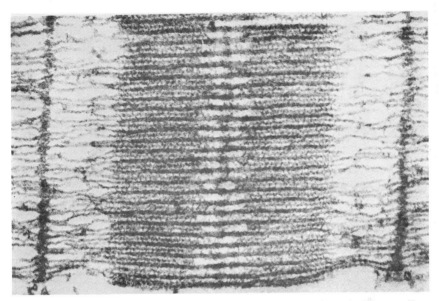

Fig. 6-5c. High resolution electron microscope photograph of striated muscle filaments. (From H. Huxley, J. Biophys. Biochem. Cytol., 3, 631 [1957].)

distinct than those of skeletal muscle. The fibers are surrounded by a very thin sheath which is difficult to isolate. Cardiac muscle is intermediate in properties to smooth and skeletal muscle.

Many of the classical investigations upon muscle fibers have centered about their interaction with the nucleotide *adenosine triphosphate*, generally abbreviated as ATP. (For a discussion of the structure and properties of this and related compounds, see Chapter 8.) ATP is of central biological importance as a carrier of energy in available form (Chapter 13) and as such figures in a host of biological processes.

Much work has been done with preparations of rabbit psoas muscle which have been dehydrated by treatment with glycerol. Such fiber preparations have been shown to undergo a dramatic contraction in the presence of ATP. Observations of this kind led early to speculation that the basic event of muscle contraction is the interaction of ATP with the protein components of muscle fibrils. The precise nature of the role of ATP remains however the subject of intense controversy, despite an enormous volume of experimental work.

Molecular properties of myosin. Myosin may be prepared from skeletal muscle fibers by extraction with 0.6 M KCl, buffered at a pH between 7 and 8. The resultant preparation is heavily contaminated with actin, from which it may be separated by subsequent fractionation. Actin-free myosin is electrophoretically homogeneous at pH 9. It is quite soluble in 0.5 M KCl at neutral pH but becomes relatively insoluble at lower salt concentrations.

The solution properties of myosin are those of a highly asymmetric, rodlike protein. The molecular weight is still not known with precision, but the true value appears to be close to 600,000, as determined from sedimentation equilibrium and light scattering. The molecular length, obtained from light scattering measurements, is about 1500 Å. The combined application of light scattering and the hydrodynamic techniques has permitted the assignment of an approximate model for the over-all size and shape of myosin. The best approximation, in terms of a geometrically simple shape, is a thin, rigid cylinder, or prolate ellipsoid, about 1500 Å long and 25 Å thick (17, 18).

Myosin belongs to the class of fibrous proteins of relatively high helical content. The helical fraction, as estimated from optical rotatory dispersion, accounts for a major portion of the molecule.

The amino acid composition of myosin is cited in Table 6-1. It has no particularly exceptional features.

Myosin has enzymatic properties and catalyzes the hydrolysis of adenosine triphosphate (ATP) to adenosine diphosphate (ADP). The terminal phosphate of ATP is split off during the process.

Molecular properties of actin. The other major protein component of muscle fibers is actin, which may be extracted from dried preparations with water and subsequently purified.

The basic molecular unit of actin is a globular protein of low asymmetry, whose molecular weight is close to 65,000. The monomeric form of actin, which is called *globular* actin or *G-actin*, is stable only at very low ionic strengths at neutral pH. In the presence of moderate concentrations of electrolyte, such as 0.1 M KCl or 0.1 M NaCl, a spontaneous polymerization of G-actin occurs to form extended linear filaments several thousand angstroms or more in length and of the order of one G-actin unit in diameter. The polymeric form of actin is known as *fibrous actin*, or *F-actin* (16, 17).

The interconversion of G-actin and F-actin is basically reversible. F-actin may be depolymerized to monomer units by high concentrations of urea (6 M) or potassium iodide (0.5 M). The depolymerization is reversible provided that ATP is present. The polymerization of G-actin thus does not appear to involve the formation of primary chemical bonds.

The action of electrolyte in the conversion of G-actin to F-actin appears to be nonspecific in nature. The stability of G-actin at low ionic strengths appears to stem primarily from the strong electrostatic repulsion present between the similarly charged molecular units under these conditions. The addition of electrolyte, in accord with the predictions of the Debye-Huckel theory, serves to reduce the inter-particle repulsion, thereby permitting association.

Interaction of myosin and actin. If solutions of purified myosin and F-actin are mixed, a molecular combination of the two proteins occurs to form a complex species whose composition reflects the ratio of the two proteins in the mixture. This polydisperse molecular complex, which is called *actomyosin*, appears to consist of bundles of overlapping myosin and F-actin molecules in side-by-side combination. Like its constituents, actomyosin is highly asymmetric, attaining lengths of several thousand angstroms or more.

The most interesting properties of actomyosin center about its capacity to form threads upon extrusion of a concentrated solution through a narrow orifice into a very dilute salt solution, in which it is insoluble. Such macroscopic threads display a dramatic *contraction* in the presence of ATP.

The action of ATP upon actomyosin is not confined to macroscopic threads. The addition of ATP to actomyosin solutions in 0.6 M KCl results in a partial dissociation of the complex.

The observation of the contractibility of actomyosin threads naturally stimulated speculation that this process might be a model for molecular contraction *in vivo* and that the actomyosin complex might be the actual contractile unit. However, as will be discussed in subsequent sections, this view is oversimplified and the molecular events accompanying muscle contraction in living systems do not permit so simple a description.

Molecular organization of the myofibril. The discussion of this section will refer particularly to skeletal muscle, for which the most detailed information is available. It will be recalled that skeletal muscle fibers consist of microscopic fibrils (or *myofibrils*) about 10,000 Å in diameter, which are separated from each other by a sarcoplasmic gap of about 5000 Å. The myofibrils are the ultimate morphological units of muscle and are the smallest units visible by ordinary microscopy, using visible light. The much higher resolution obtainable with electron microscopy has revealed that the myofibrils are composed of bundles of filaments. The distinction between fibers, fibrils, and filaments should be kept clearly in mind.

A conspicuous feature of the myofibrils is their characteristic *striated* appearance (Fig. 6-5). When properly stained or shadowed, two alternating broad bands—the A and I bands—are clearly visible. The A and I bands are both bisected by narrower bands, called H (or M) in the former case and Z in the latter.

The A and I bands correspond in reality to periodicities in bundles of filaments. Since the latter are of molecular dimensions, it is clear that the structural periodicity reflected in the visible bands must arise somehow from the nature of the arrangement of the protein components of the myofibril within the filaments.

The determination of the substructure of the myofibril on a molecular level has been a most difficult problem, which still cannot be regarded as completely solved. One basic approach has been to observe the effects of selective extraction of the myofibril by solvents which preferentially dissolve myosin. In recent years an alternative approach which utilizes selective immunological staining has also been applied. Both lines of attack have reached basically similar conclusions. Myosin appears to be concentrated in the A band (Fig. 6-6), while the I band consists largely

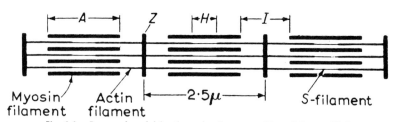

Fig. 6-6. Proposed model for the molecular composition of the myofibril.

of actin. The Z bands are generally believed to consist of continuous membranes which bisect the bundle of filaments and upon which individual filaments terminate. The Z lines thus divide the myofibril into discrete sections, or *sarcomeres* several microns in length (Fig. 6-5b).

A detailed model for the myofibril has been proposed by Hanson and Huxley (13). Its main features may be summarized as follows:

(1) Striated muscle is built up from two distinct sets of filaments which extend longitudinally through the sarcomere.

(2) The actin filaments extend from the Z line through the I band and into the A band (Fig. 6-6), where they terminate at the edge of the H zone.

(3) The myosin filaments occur in the A bands and extend throughout their length.

(4) The H (or M) band corresponds to the gap between the ends of

the actin filaments. To account for some of the mechanical properties of muscle it has been postulated that the gap is bridged by very thin filaments of unknown composition, called the S-filaments (Fig. 6-6).

While a considerable amount of evidence in favor of the above model has been accumulated, a number of features remain obscure. Thus the dimensions of the myosin filaments are greatly in excess of those of individual myosin units and could only exist as longitudinal polymers of these. The organization of the myosin units within the filaments remains conjectural.

Relation of myofibril structure to muscle contraction. From electron microscopic studies of the contraction process as reflected by changes in in the sarcomere, Hanson and Huxley (13) have proposed the following general picture of the events at the molecular level (Fig. 6-7).

When muscle contracts, the actin filaments are drawn into the A bands. The over-all length of the actin filaments remains constant until the H zone is filled up. Subsequent shortening must involve some degree of folding of the actin filaments. The length of the A bands, which corresponds to the length of the myosin filaments, remains virtually constant until an advanced stage of contraction.

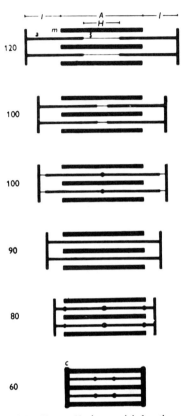

Fig. 6-7. Hanson-Huxley model for the contractile process in one sarcomere. The left-hand column gives the degree of contraction (or stretching) from the resting length.

The most salient feature of this model (and the major point of divergence from earlier models) is the *sliding* of actin filaments past myosin filaments. Such movement is possible only if *permanent* actin-myosin linkages are absent. On this basis contraction is visualized as a *stepwise* process, the actin filament sliding in each step along a distance equal to the separation of adjacent globular subunits on the actin polymer.

Such lateral mobility requires that any myosin-actin linkages be weak and rapidly reversible, as a considerable proportion, or all, of them would have to be broken and reformed at each step of the process.

Problems of muscle contraction. The above discussion has been inevitably centered about the mechanics of the contractile process—about *what* happens rather than *how* it happens. The model described above, which has won wide, but not universal, acceptance accounts for only part of the process.

Whatever the detailed mechanism of the sliding process and the chemical nature of the sites involved, it is clear that a continuous supply of external energy is essential. Indeed, the myofibril has been described as a kind of *transducer* which converts chemical energy into external mechanical work.

Two principal areas of the contractile mechanism remain highly controversial. The first of these is the problem of exactly how the chemical energy supplied by cellular metabolism is made available to overcome the frictional and other forces which oppose the mutual slippage of the filaments. Agreement is general that ATP, which is a substrate for the enzymatic sites of myosin, is involved in the process, but whether by virtue of its hydrolysis to ADP, its binding to a constituent of the myofibril, its role in actin polymerization, or by some indirect mechanism remains conjectural. A discussion of the numerous ingenious theories lies outside the scope of this book.

The second area of uncertainty is the mechanism whereby muscular contraction is triggered by nerve impulses. Here there is even less detailed information available, and any description on a molecular level lies far in the future.

6-4 ANTIBODIES (19-23)

Occurrence and function. The body has adopted a plural approach to the problem of protecting itself from the invasion of foreign infectious agents. The over-all defense against bacterial or viral attack can draw upon many elements including cellular agents, as the *phagocytes;* the substances collectively known as *complement,* which promote lysis of bacteria; and a number of other factors. However, the *specific* defense mobilized against particular harmful organisms is dependent upon the antibody response (20).

Antibodies do not arise solely to counter microorganisms. They are produced by animals in response to the injection of some foreign material, or *antigen*. This may be a protein, polysaccharide, or some other biopolymer; whole cells or tissue fragments; or particles of a nonbiological nature. While small molecules are known to have antigenic properties, the most effective antigens are usually of large size.

The versatility of the antibody response is astonishing. The only general requirement is that the material be foreign to the circulation of the animal. Loosely speaking, the more foreign it is, the stronger its antigenic power, as measured by the speed and quantity of the antibody response. Thus an animal does not produce antibodies to its own tissue or plasma proteins. The response to the injection of proteins from another source generally becomes stronger with increasing taxonomic separation of the two.

Antibodies are globular proteins present in the blood plasma and belonging to the class of plasma proteins known as the γ-globulins. In over-all chemical and physical properties they resemble normal γ-globulin and each other. The sole striking difference is the capacity of antibodies to *combine selectively* with the antigenic agent responsible for their production. It is this specific combining ability which is responsible for the capacity of antibodies to neutralize bacteria and viruses and facilitate their destruction by other, preformed elements of the body's defensive mechanism.

The combining ability of antibodies is conferred by the presence of specific sites, localized in nature, upon their surface. There is no evidence for any prosthetic group, and the sites are believed to consist of specific arrangements of amino acids attained by specific folding of the polypeptide chain.

The primary site of antibody synthesis is the *reticulo-endothelial* system. This is a particular class of cells found in the spleen, liver, and lymphatic glands, as well as located circulating freely in the blood. Antibody formation may be regarded as a modification of the normal synthesis of globulins. At some stage of biosynthesis the antigen interferes with globulin formation so as to produce molecules which are adapted in a complementary manner to the determinant group of the antigen molecule.

The mechanism of antibody formation is very far from being completely understood. The existing theories fall into two principal categories.

According to the classical "instructive" theory, the specific combining site of the antibody acquires its pattern by being synthesized in contact with the antigenic determinant. The antigen is believed to enter the cell and to exert its action as the polypeptide chain is in the process of being folded into globular form. At this folding stage the globulin is in intimate contact with the antigen, which serves as a mold for the formation of the complementary pattern.

The alternative "selection" theory holds that antibody synthesis does not differ essentially from that of other proteins and does not depend upon the arrival in the cell of information from outside. Instead, the body is regarded as containing a finite number of cells capable of synthesizing antibodies to each of the thousands of possible foreign antigens. Each specialized cell, or group of cells, is supposed to be capable of making its specific antibody even if the complementary antigen is wholly absent. The antigen is believed to function by stimulating selectively the proliferation of the appropriate specialized cells, thereby enhancing the production of its antibody. In some versions of this theory the difficult question of how such a wide variety of specialized cells arises is met by postulating that a very high rate of mutation exists for the antibody-synthesizing cells in the early stages of embryonic life. In this manner all possible types of specialized cells are created in a random fashion. Each mutant cell becomes through division the ancestor of a small group, or *clone*, of cells which specialize in the formation of one, or a few, types of antibody. Later in embryonic life the mutation rate of immunological cells slows to a normal level. While some recent evidence appears to favor the selection theory, the issue must be regarded as still in doubt.

Among the best known of the phenomena associated with the interaction of antigen and antibody *in vitro* is the *precipitin* reaction, which is easily observed with solutions of antibody and protein antigen. If, to a series of solutions of rabbit antibody of constant concentration, there is added increasing amounts of protein antigen, a visible precipitate is formed in the region of antibody excess. This increases in magnitude as more antigen is added, passing through a maximum at an antigen:antibody ratio which depends upon the nature of the particular system. At higher proportions of antigen the amount of precipitate decreases again and finally approaches zero.

Efforts to gain a detailed understanding of the factors responsible for antibody specificity are handicapped in the case of protein antigens by the unknown character of the antigenic site, which consists of a patch of amino acids. On the other hand, small organic molecules whose structures are well understood do not ordinarily function as antigens.

The classical work of Landsteiner was decisive in overcoming this difficulty. This investigator found that polar organic molecules, such as the aminobenzoic acids or the aminophenylarsonic acids, *when chemically coupled to a protein*, became antigenic determinants in their own right. Injection of an animal with a protein labeled with such a *hapten* group results in the production of antibodies directed solely against the group (as well as other, different antibodies directed against the protein, if it is foreign). If the protein carrier is a *homologous* plasma protein (obtained from animals of the same species), then only antibodies to the hapten will be produced.

Studies of antibodies to artificial haptens have been extremely useful in clarifying the factors involved in antibody specificity.

Molecular structure of antibodies. No clear-cut differences in over-all amino acid composition or physical parameters have been detected between antibodies and nonimmune γ-globulin of the same species. Both appear to be somewhat inhomogeneous by the criterion of electrophoretic mobility (19, 20, 22).

The size and shape parameters of antibodies of varying specificity appear to be similar and to resemble those of normal γ-globulin. The molecular weight is close to 160,000. The frictional ratio (1.4) and intrinsic viscosity (6.0) are larger than is usual for globular proteins and are suggestive of a moderate degree of molecular asymmetry. Formal application of the equations relating viscosity and frictional ratio to molecular shape (section 4-5) yields an axial ratio of about 7, as computed for a prolate ellipsoidal model (assuming a hydration of 0.3) (19, 20, 22).

The molecular organization of the γ-globulins, both immune and normal, displays some rather unusual features. The α-helical content, as measured by optical rotatory dispersion, is vanishingly small. Thus the molecular configuration must be largely tertiary in nature and stabilized by side-chain interactions.

At extremes of pH, γ-globulin undergoes a partially reversible change in structure which is reflected by an increase in viscosity and frictional ratio and probably corresponds to an expansion of the molecule.

Perhaps the most important generalization which emerges from comparative studies upon the chemical and physical properties of antibodies and normal γ-globulins is that the differences between antibodies of varying specificity must reside in relatively *subtle* structural factors, which do not involve *gross* changes in the over-all composition or molecular organization of the protein. These presumably arise from a *localized* alteration in the nature of the folding of the polypeptide chain, possibly accompanied by an alteration in amino acid sequence. The question of how the subtle and limited nature of the structural variations can be reconciled with the extraordinary versatility of the antibody response is a challenging one.

The antibody combining sites. The specific combination of antibody with the antigenic receptors does not occur at loci distributed uniformly over the antibody surface. It is rather confined to certain localized areas of restricted extent, which are called combining *sites*. These are undoubtedly complex in nature and are composed of a number of amino acid residues maintained in correct mutual position by the tertiary organization of the protein.

The number of combining sites per antibody molecule remained a point of intensive controversy for a surprisingly long time. However, in recent years electrophoretic and ultracentrifugal studies of the composition of soluble antigen-antibody complexes under conditions of saturation by antigen have provided evidence that the overwhelming majority of antibody molecules must be *bivalent,* containing two sites per molecule (19). Whether antibodies of different valence are produced in minor quantities is still uncertain, and there is some evidence for the existence of "incomplete" or univalent antibodies.

In many ways, the combining sites of antibodies present an analogy to the catalytic sites of enzymes, extending to the rather poorly understood nature of both. In particular, the high degree of specificity of the antibody sites parallels the substrate specificity of enzymes. In both cases, the combination of site and receptor is reversible and is mediated by secondary forces. The specificity shown by antibodies is even more remarkable since it is attained by variations in the nature of folding (and possibly by variations in amino acid sequence) of polypeptide chains of similar over-all composition.

The poorly characterized nature of the antigenic sites of natural proteins has led to the use of artificial haptens of simple and known structure in efforts to gain a closer look at the antibody combining site. The usual procedure has been to immunize animals against a plasma protein

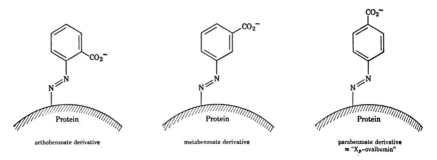

Fig. 6-8. Hapten groups coupled to a protein carrier by diazotization. (See Fig. 6-10.)

of the same species coupled with an organic hapten, such as an aminosulfonic acid. If the free hapten contains an amino group, coupling may be accomplished by diazotization (Fig. 6-8).

In this manner antibodies directed against a single hapten may be obtained. If solutions of antibody and the protein conjugate of its hapten are mixed in the proper ratio a specific precipitate is obtained, whose magnitude serves as a semiquantitative indicator for the reaction (20, 22).

If a set of protein conjugates are prepared with a series of chemically related haptens, their cross-reaction with antibodies to the original hapten may be tested for by the precipitin reaction. Alternatively, the competitive inhibition of the reaction between the original conjugate and its antibody by unconjugated hapten may be used as an index of cross-reaction. In this manner the rigidity of the specificity requirements of the antibody may be examined. In practice, the haptens tested have usually been aromatic molecules with charged substituents.

Studies of this kind have revealed that antibody specificity, while high, is not in general absolute and that cross-reactions do occur (Fig. 6-9). The following generalizations have emerged:

(1) The influence of charged groups in the hapten tends to be decisive. In general, any change in the nature or position of the charged site results in complete, or almost complete, abolition of the reaction. This is particularly true if the polar group is acidic in nature. Thus antibodies to a sulfanilic acid-conjugated protein show little reaction with haptens in which the sulfonic acid group is replaced by a carboxyl group (Fig. 6-9).

Antisera made with	Tested against antigens made with					
	 NH_2-C$_6$H$_5$	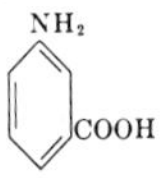 NH_2-C$_6$H$_4$-COOH	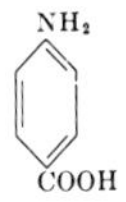 NH_2-C$_6$H$_4$-COOH	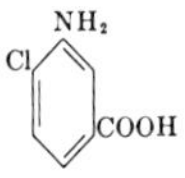 NH_2,Cl-C$_6$H$_3$-COOH	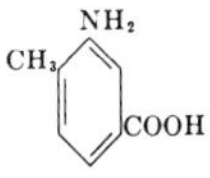 NH_2,CH$_3$-C$_6$H$_3$-COOH	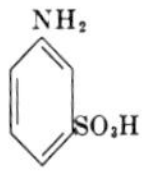 NH_2-C$_6$H$_4$-SO$_3$H
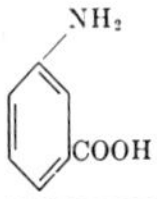 NH_2-C$_6$H$_4$-COOH	0	+++	0	++++	+++	+
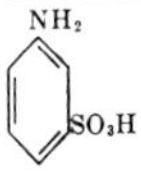 NH_2-C$_6$H$_4$-SO$_3$H	0	0	0		0	++++

0, no reaction; +, positive reaction; ++++, very strong reaction.

Fig. 6-9. Cross-reaction between antibodies to related haptens (22).

(2) Introduction of uncharged groups, as methyl or halogen, has less influence. A great many cross-reactions have been observed among such compounds.

(3) The spatial configuration of the hapten has an important effect. Thus the position of a polar group in a benzene ring influences the extent of reaction strongly (Fig. 6-9). Antibodies to one of the ortho, meta, or para phenylarsonic acids have no difficulty in distinguishing between the different isomers.

Speculation as to the origin of the specificity of the combining site has stemmed from the prevalent picture of the nature of the forces which cause antigens and antibodies to unite. It is supposed that the antibody sites consist of characteristic arrangements of amino acids in certain areas on the surface of the molecule. Electrostatic forces, van der Waals forces, and hydrogen bonding may all be important in individual cases. Since all of these decay rapidly with distance, the close approach of the elements of the antigenic and antibody sites is important for a strong interaction. This can be achieved only if a certain complementarity of shape exists between the two.

One of the simplest models of how this might be attained (Fig. 6-10)

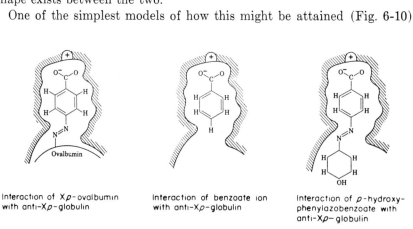

Fig. 6-10. Schematic version of one model for the combining sites of antibodies. (See Fig. 6-8.)

visualizes the antibody combining site in the form of a cavity into which the hapten or other antigenic determinant fits. The fit is supposed to be a close one and as such would be very sensitive to changes in the configuration of a hapten.

This is the famous "lock and key" model of Pauling, which has guided thinking in this area and has provided a framework for the interpretation of observations upon hapten specificity.

The antigen-antibody reaction. When a protein antigen and its antibody are mixed in solution at neutral pH, room temperature, and ionic strength of the order of 0.1, a molecular combination occurs (Fig. 6-11). Depending upon the ratio of antigen to antibody, either soluble complexes or a visible precipitate may be formed.

The molecular combination of antigen and antibody may be profitably discussed in a generalized and formal manner, without reference to the specific characteristics of the individual sites. All, or almost all, antibody molecules appear to have a valency of two; that is, they contain two combining sites per molecule. In contrast, protein antigens behave as multivalent reactants, whose effective valency increases with the size of the molecule— from 5 for ovalbumin (molecular weight = 44,000) to 40 for thyroglobulin (molecular weight = 650,000). If the antigenic determinant is a hapten attached to a protein, then the effective valency will be equal to the (usually large) number of hapten groups per molecule.

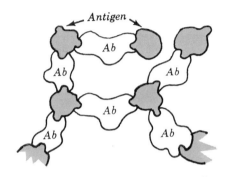

Fig. 6-11a. Schematic model of the interaction of antigen and antibody.

Thus a maximum of two antigen molecules may be bound by the antibody, while the limiting number of antibodies which can combine with one antigen is much larger. Let us consider the reaction between a bivalent antibody and a protein antigen of valency x. It is convenient to consider the effects of increasing the ratio of the molar concentration of antibody (A) to that of antigen (G).

(1) If the antigen to antibody ratio is very high (> 10), so that antigen is present in large excess, the antibody will be effectively *saturated* with antigen according to the equation.

(6-1)
$$A + G \rightleftarrows AG$$
$$AG + G \leftrightarrows AG_2$$

No precipitation occurs under these conditions. Ultracentrifugal examination of the solution reveals only two peaks, due to free antigen and the AG_2 complex species (21) (Fig. 6-11b).

(2) If the ratio of antibody to antigen is increased, then the antibody ceases to be saturated, so that a single antigen may combine simultaneously with two or more antibodies, forming a bridge between them.

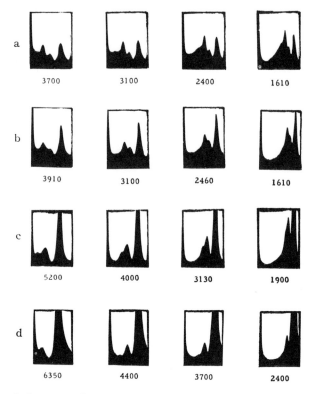

Fig. 6-11b. Sedimentation diagrams of a mixture of bovine serum albumin and its rabbit antibody as a function of antigen: antibody ration (21a). Sedimentation proceeds to the left. The peaks observed, beginning at the right, are free antigen, AG_2 complex, A_2G_2 (or A_2G_3), and higher complexes. The numbers indicate sedimentation times: (a) A:G = 0.56, (b) A:G = 1.0, (c) A:G = 1.6, (d) A:G = 5.9.

(6-2)
$$A + G \rightleftarrows AG$$
$$AG + A \rightleftarrows A_2G$$
$$A_2G + G \rightleftarrows A_2G_2$$
$$A_2G_2 + G \rightleftarrows A_2G_3, \text{ etc.}$$

In this manner a series of complexes containing more than one molecule each of antigen and antibody is built up. A sedimentation diagram of a solution with an $A:G$ ratio of the order of unity shows, in addition to the peaks representing free antigen and the AG_2 species, a considerable fraction of poorly resolved, more rapidly sedimenting material, which corresponds to a mixture of complex species of high molecular weight (21). As the ratio of antibody to antigen increases, the amount of the AG_2 species present declines, with a corresponding increase in the concentration of the higher species (Fig. 6-11).

(3) As the molar ratio of antibody to antigen is increased beyond unity, the average size of the complexes increases to the point where visible precipitation occurs. In the *equivalence* zone, virtually all the antigen and antibody is incorporated into large fragments of a three-dimensional network.

(4) A further increase in the antibody-antigen ratio results in an increase in the fraction of antibody in the precipitate. In the case of rabbit antibodies precipitation persists even at very high $A:G$ ratios. Apparently, the intrinsic solubility of the rabbit γ-globulin is too low to permit the formation of soluble complexes of high antibody content.

The reactions discussed above are entirely reversible. The antigen-antibody precipitate formed in the equivalence zone can be redissolved in the presence of excess antigen to yield a solution whose composition is identical to that produced by direct mixing of the components.

The interaction of antibody and protein antigen in solution is formally analogous to a reversible condensation reaction between a bivalent and a multivalent reactant and can be analyzed on this basis (22). A statistical analysis of the process in terms of this kind of model, assuming equal reactivity of all sites, has been developed by Goldberg (21b).

Active antibody fragments. In a dramatic recent development, Porter has shown that rabbit anti-hapten antibodies may be partially digested by the proteolytic enzyme papain to produce three fragments, referred to as I, II, and III, which have approximate molecular weights of 50,000, 50,000 and 80,000, respectively. Fragments I and II retain the capacity to combine with hapten, while fragment III is inactive (23).

Most interestingly, I and II have closely similar amino acid compositions and physical properties although they must come from different parts of the original molecule. Both contain one combining site.

The most plausible picture of the mutual positions of the three fragments represents them as arranged in linear sequence, the two similar segments being joined to a third of a quite different structure (Fig. 6-12).

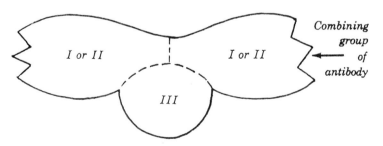

Fig. 6-12. Schematic model of possible manner of attachment of antibody fragments.

Subsequent work has shown that γ-globulin consists of two similar long polypeptide chains (called A chains by Porter) and two shorter chains (called B chains). The four chains are joined by disulfide bridges in the sequence B-A-A-B. Papain fragments I and II consist of a B chain and part of an A chain, while III consists of the greater part of two A chains (23).

6-5 FIBRINOGEN

Occurrence and function. Fibrinogen is one of the major components of mammalian blood plasma and accounts for about 4% of the total plasma protein. Its biological importance arises from its central role in the process of blood clotting.

The coagulation of whole blood is one of the simplest biological processes to observe *in vitro*. If fresh blood is withdrawn from a blood vessel and placed in a test tube, it remains fluid only for a short interval which is terminated by an abrupt gelation to form a semirigid clot. This is the basis for the self-sealing properties of blood. Traumatic injury to the blood vessel is followed by local coagulation of the blood to form a clot which plugs the wound and stops the loss of blood.

The molecular event corresponding to clot formation is the incorporation of fibrinogen, in altered form, into a three-dimensional fibrous network called *fibrin* (Fig. 6-13). The complete mechanism is quite complicated. Its numerous steps may be conveniently grouped into three principal stages: (1) the formation of the fibrinogen-activating enzyme *thrombin* from its inactive precursor *prothrombin;* (2) the action of thrombin upon fibrinogen to form *activated* fibrinogen or *profibrin;* and (3) the polymerization of molecular units of activated fibrinogen to form the tangle of elongated strands of which fibrin is composed.

The first stage of the clotting process is much less completely understood than the latter two, about which the discussion of this section will be centered. The final step of the initial stage is the activation of prothrombin to form thrombin. This represents the culmination of an intricate series of reactions which begins when blood leaves the circulation and comes in contact with a foreign surface, such as the wound surface or a test tube wall.

Blood is a fluid two-fifths of whose volume is composed of formed elements (red cells, white cells, and so on) and the remainder of the plasma, a concentrated solution of a great many different proteins. Among the formed elements are the *platelets*, which are disk-shaped bodies about 2μ in diameter. These contain enzymatic or other factors essential to the activation of prothrombin.

STRUCTURE AND FUNCTION OF IMPORTANT PROTEINS 189

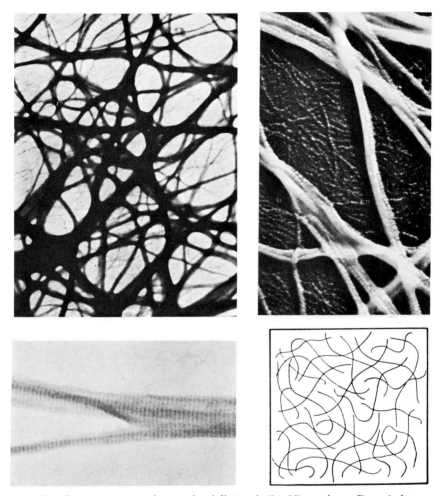

Fig. 6-13. Electron microscope photographs of fibrin gels (24, 25) are shown. Figure in bottom row at left shows a single strand stained with phosphotungstic acid, to bring out the striations. Figure in bottom row at right shows a schematic picture of a fibrin clot.

The platelets are fragile and easily ruptured upon contact with a foreign surface. Breakup of the platelets releases a series of substances. Some of these proceed to interact with factors already present in plasma to produce agents, presumably enzymes, which activate prothrombin. At least six poorly characterized factors are involved in the over-all process.

Stages (2) and (3) are beginning to be fairly well understood at the molecular level. This is largely a consequence of work with purified preparations of thrombin and fibrinogen. The addition of small amounts of thrombin to solutions of purified fibrinogen results in a progressive increase in viscosity, culminating in gelation. Examination of the gel by electron microscopy shows it to consist of a network of strands similar in appearance to those formed by whole blood or plasma (24, 25).

The diameter of the strands is variable, the average value depending upon the pH and ionic strength at which polymerization is carried out. That a high degree of order is present in the strands is indicated by their regular, banded appearance when stained by phosphotungstic acid (Fig. 6-13). Definite cross-striations, whose spacings are 120 Å and 240 Å, appear on the wider strands (Fig. 6-13).

The incorporation of fibrinogen into the fibrin network proceeds essentially to completion, irrespective of the thrombin concentration, although the *rate* of conversion does of course depend on the thrombin level. Thrombin itself is not incorporated into the clot and may be recovered in unchanged form from the fluid squeezed out of a compressed clot. It thus is clear that the action of thrombin must be of an *enzymatic* nature and serve to convert fibrinogen to an altered form which is capable of polymerization.

A gel formed by *purified* fibrinogen may easily be dissolved in high concentrations of urea (6 M) to produce a solution in which the profibrin units exist in molecularly dispersed form, as shown by ultracentrifugal analysis. Since the action of urea is confined to the rupture of *secondary* bonds, the forces holding the profibrin units together in the intact strands cannot involve true chemical, or *primary*, bonds (24, 26).

The above should be qualified by mentioning that whole blood, or plasma, contains a poorly characterized clotting factor which, in conjunction with Ca^{++}, serves to introduce new bonds of a more permanent nature into the fibrin network so as to render it insoluble in urea. However, the basic nature of the gelation process is, apart from this complication, the same for both systems.

Molecular characteristics of fibrinogen. The molecular weight of fibrinogen is difficult to determine accurately because of a pronounced tendency of this protein to aggregate in solution. The best current estimates place it close to 330,000 (24).

The high values of the intrinsic viscosity (30) and frictional ratio (2.3) observed for this protein indicate a considerable asymmetry of shape. A combination of light scattering and hydrodynamic evidence has shown that the best simple geometrical model approximating the actual shape is a prolate ellipsoid, or cylinder, about 500 Å long and about 25 Å in diameter.

In this case it has been possible to obtain a closer look at the over-all size and shape by electron microscopy. Direct observation shows fibrinogen to consist of three semispherical nodules connected by a slender filament (Fig. 6-14). The molecular length determined in this way is close to 475 Å, in remarkably good agreement with the light scattering value.

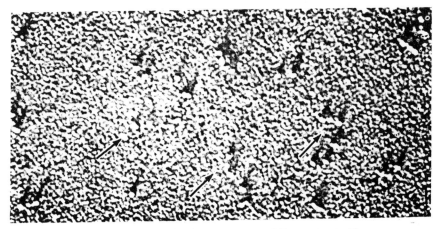

Fig. 6-14a. Electron microscope photograph of fibrinogen units (1).

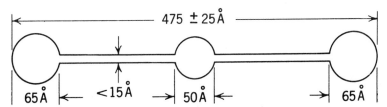

Fig. 6-14b. Schematic model of fibrinogen (1).

The amino acid composition of fibrinogen (Table 6-1) has no particularly unusual features. There are two NH_2-terminal tyrosines and two NH_2-terminal glutamic acids per molecule (24) of bovine fibrinogen.

The conversion of fibrinogen to fibrin. A comparison of the molecular parameters of fibrinogen and fibrin, when dissolved in 6 M urea, shows no important difference in molecular weight or intrinsic viscosity. It follows that the action of thrombin upon fibrinogen has not resulted in any *major* alteration in its size or shape. However, a change in electrophoretic mobility indicates that the net charges of fibrinogen and profibrin are different and hence that ionizable groups have been introduced or removed.

Chromatographic examination of the products of the action of thrombin upon fibrinogen has shown that this altered charge pattern results from the thrombin-catalyzed splitting off of two small polypeptides A and B. These have a total molecular weight close to 8000 (24). The drop in molecular weight resulting from their removal is too slight (2%) to be easily detectible.

The removal of the fibrinopeptides is accompanied by a change in the NH_2-terminal groups. Bovine fibrinogen contains two NH_2-terminal glutamic residues and two NH_2-terminal tyrosines. While the NH_2-terminal tyrosines persist in profibrin, the NH_2-terminal glutamics are replaced by NH_2-terminal glycines.

Thus the action of thrombin upon fibrinogen appears to be proteolytic in nature and to involve the hydrolysis of peptide bonds. Thrombin has, in fact, been shown to be a typical peptidase (Chapter 7) and to catalyze the hydrolysis of synthetic peptides.

The fibrinopeptides are particularly rich in glutamic and aspartic acid. The loss of the acidic groups contained in the side-chains of these residues easily accounts for the change in electrophoretic mobility accompanying the activation of fibrinogen.

Activated fibrinogen, or profibrin, has the property of polymerizing spontaneously. In water, at pH's between 6 and 10, the polymerization proceeds to the point of gelation. In the presence of certain inhibitors of gelation, including low concentrations (0.4 M) of hexamethylene glycol, the solution remains fluid. Under these conditions the presence of rodlike polymers of great length (>5000 Å) can be demonstrated by ultracentrifugation and light scattering. The polymerization is reversible and the distribution of profibrin between the monomer and polymer species approaches an equilibrium which depends upon pH, temperature, and the concentration of gelation inhibitor (26, 27).

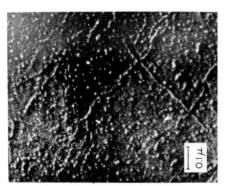

Fig. 6-15a. Electron microscope photograph of profibrin polymers formed at an early stage of polymerization (24).

The linear character of the intermediate polymer, as shown by electron microscopic examination (Fig. 6-15), suggests that the initial mode of combination of the profibrin monomer units is end to end (Fig. 6-15). However, the presence of strands of much greater width in the fibrin gel (Fig. 6-13) indicates that a considerable degree of lateral aggregation of the initial polymer must also occur.

In the absence of hexamethylene glycol the profibrin polymers increase progressively in length until gelation occurs. The later stages of polymerization are accompanied by considerable side-by-side aggregation

Fig. 6-15b. One model for the end-to-end polymerization. Note additional nodule believed to arise during activation (25).

of the linear polymers to form bundles of varying width. The ordered structure of these, as revealed by the regularity of their striations (Fig. 6-13), indicates that this process is not random and that certain structural features of the monomer units must be maintained in register. The degree of lateral aggregation depends upon the pH and decreases with increasing pH.

The final stage of the clotting process is the appearance of gel properties. This probably reflects the establishment of occasional cross-links between the elongated strands, so as to form a three-dimensional network. The nature of the cross-links is uncertain. They may be of the same type as are involved in the polymerization itself (Fig. 6-13).

Bonds formed during polymerization. Speculation as to the nature of the bonds formed between profibrin units during the polymerization reaction has been guided by the following observations (24, 28).

(1) The polymerization is reversible and hence cannot depend upon the formation of primary bonds.

(2) Polymerization is inhibited by pH's above 10 or below 5.

(3) An analysis of hydrogen ion titration data has indicated that formation of the linkages between profibrin units involves two classes of ionizable groups. One of these (X) has an intrinsic pK close to 6.1 and the other (Y) a pK close to 9.6. The binding of protons by the former and their dissociation by the latter are competitive with respect to bond formation.

(6-3) $$\text{(unbonded)} \quad \overset{H^+}{} \text{(bonded)} \quad \overset{OH^-}{} \text{(unbonded)} \\ XH^+ + HY \rightleftharpoons X\cdots HY \rightleftharpoons X + Y^-$$

A tentative model has been proposed which postulates that the linkages involve the formation of hydrogen bonds between uncharged imidazole acceptor groups and tyrosine donors. The pK's of these are in the proper range, and ionization of either group would eliminate its capacity to serve as a hydrogen-bond acceptor (in the case of imidazole) or donor (in the case of tyrosine). This explains the inhibition of polymerization at pH's above 10 or below 5.

6-6 PLASMA PROTEINS: SERUM ALBUMIN

Occurrence and function. Serum albumin is an important component of blood plasma. It is the protein species present in the highest concentration and represents about half of the total plasma protein. The serum albumins of the higher mammalian species are quite similar in composition and properties, but can be differentiated by immunological criteria.

Serum albumin is responsible for a disproportionate fraction (about 75%) of the osmotic pressure of human plasma, primarily as a consequence of its low molecular weight as compared with the other plasma proteins. No very specific physiological activity has been found as yet for serum albumin. It appears that the primary function of this protein may be to maintain the osmotic pressure of blood, thereby preserving the critical balance of electrolyte concentration.

Chemical structure. Ordinary serum albumin has been shown to contain two species of identical amino acid composition, except for the presence of a single cysteine residue in one case. This fraction, which accounts for about two-thirds of bovine serum albumin, is referred to as mercaptalbumin. Since no other structural differences have emerged as yet, the two species will be discussed in a unified manner. The amino acid composition is listed in Table 6-1.

Both human and bovine serum albumin appear to consist of single polypeptide chains, which are cross-linked by 17 disulfide bridges. The NH_2-terminal residue for both proteins is aspartic acid.

Dimerization. The presence of a single —SH group in mercaptalbumin endows it with the capacity of combining strongly with Hg^{++} to form a stable complex. If the mole ratio of Hg^{++} to mercaptalbumin is 1:1, or greater, each albumin binds a single Hg^{++} and remains in the monomer state. If, however, mercaptalbumin is in excess, then two mercaptalbumin molecules can share a single Hg^{++}, thereby forming a dimer.

(6-4) $$\text{Alb—SH} + Hg^{++} \leftrightarrows \text{Alb—S—}\overset{+}{Hg} + H^+$$

(6-5) $$\text{Alb—S—}\overset{+}{Hg} + \text{Alb—SH} \leftrightarrows (\text{Alb—S})_2\text{—Hg} + H^+$$

The addition of excess Hg^{++} results in the stoichiometric conversion of the dimer to the monomeric mercury complex.

(6-6) $$(\text{Alb—S})_2\text{—Hg} + Hg^{++} \leftrightarrows 2\ \text{Alb—S—}\overset{+}{Hg}$$

Because the mercury dimer has solubility properties somewhat different from those of uncombined albumin, this reaction has provided a basis for isolating purified mercaptalbumin.

Size and shape. The molecular weight of serum albumin is close to 67,000 as determined by sedimentation and diffusion measurements (Table 6-2). The frictional ratio and intrinsic viscosity are suggestive of a moderately asymmetric shape, the computed axial ratio on the prolate ellipsoidal model being 4.1 to 6:1 (Table 6-2). However, the asymmetry is sufficiently low so that the computed axial ratio is very sensitive to the choice of the value of hydration, as well as to any deficiencies of this kind of model.

The rotational relaxation time is twice the value predicted for an anhydrous sphere of the same molecular weight (Table 6-2). It is clear

TABLE 6-2. PHYSICAL PARAMETERS OF SEVERAL IMPORTANT PROTEINS

Protein	Molecular Weight ($\times 10^{-4}$)	Frictional Ratio	Intrinsic Viscosity	Rotational Relaxation Time ($\times 10^7$)	Isoelectric Point
Ichthyocol	34	6.8			
Fibrinogen	34	2.3	25		
Myosin	60	3.5	19		5.4
Serum albumin	6.8	1.3	4	1.4	4.7
Hemoglobin	6.8	1.1	4		6.6
Myoglobin	1.8	1.1	3		

that the molecule must possess a high degree of internal rigidity. The axial ratio computed on the prolate ellipsoidal model from the rotational relaxation time obtained from fluorescence polarization is close to 4:1 (assuming 15% hydration).

Secondary and tertiary structure. What is known of the secondary and tertiary structure of serum albumin show it to be a typical globular protein. The helical content, estimated from optical rotatory dispersion, is close to 46% (Table 5-1).

In view of the compact and rigid character of serum albumin, it is clear that the tertiary structure must be of major importance in stabilizing the molecular organization of this protein, since the fraction of the molecule in the α-helical configuration is too small to endow it with structural rigidity.

Acid structural transition. The molecular parameters of serum albumin are relatively invariant to pH between the limits pH 4-pH 11. Outside of this range definite structural changes occur, which are fairly well understood in the case of the acid transition (34).

While no change in molecular weight occurs, the intrinsic viscosity increases with decreasing pH between pH 4 and pH 2 (Fig. 6-16). The magnitude of the increase is reduced by the addition of electrolyte.

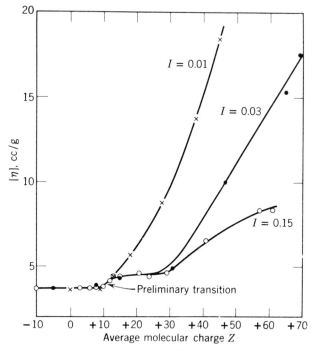

Fig. 6-16. Increase in viscosity of serum albumin at acid pH for several concentrations of electrolyte (1). The abscissa is the net positive charge and is equal to the number of protons bound per molecule in excess of the number bound at the isoelectric point (Chapter 4).

The change in viscosity is paralleled by a rise in frictional ratio and an increase in laevo-rotation (Fig. 6-17). The rotational relaxation time, as determined by fluorescence polarization, shows a major decrease. The alterations in all these quantities are completely reversed by a return to neutral pH.

The above data indicate that an important loss of helical content and of structural rigidity occur at acid pH's. They are consistent with, and suggest that, the highly organized configuration prevailing at neutral pH is progessively replaced by one which is relatively loose and much closer to the randomly coiled state.

It is of particular interest that so extensive a loss of molecular organi-

STRUCTURE AND FUNCTION OF IMPORTANT PROTEINS

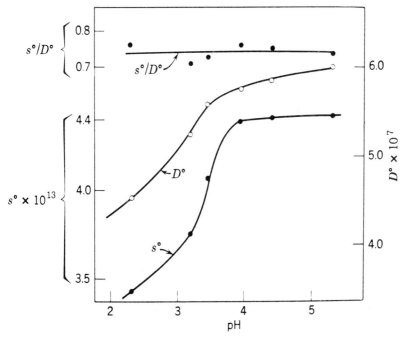

Fig. 6-17. Increase in frictional ratio of serum albumin at acid pH (1). The ratio of sedimentation coefficient to diffusion coefficient remains constant, indicating that the molecular weight is invariant to pH. The fall in the magnitude of these parameters may thus be attributed entirely to an increase in the frictional coefficient accompanying the structural transition.

zation can be reversed so rapidly and completely. It is likely that the 17 disulfide bridges assist in guiding the reformation of the original structure. This resiliency of the molecular organization of serum albumin indicates that the polypeptide chain possesses sufficient mobility to explore various competitive configurations and select that of maximum stability. This is of importance with respect to the mechanism of the formation of the secondary and tertiary structure of proteins (Chapter 10).

6-7 HEMOGLOBIN (29, 30, 31, 32)

Occurrence, function, and nomenclature. Hemoglobin is the respiratory protein of vertebrates and accounts for virtually all the soluble protein of the red blood cells (*erythrocytes*) of these species. It can be easily prepared in nearly pure form by rupture of the cell walls of the erythrocytes and centrifugal removal of the insoluble residue.

Hemoglobin is perhaps the most intensively studied of the small group of proteins which have the property of reversible combination with molecular oxygen. This endows it with the capacity to serve as the agent for the transport of oxygen by way of the blood stream to the sites of its utilization.

Hemoglobin can exist in several forms, distinguished by the state of the oxygen-binding site. The general term "hemoglobin" will be used in discussing properties for which this is immaterial.

Hemoglobin is a conjugated protein and consists of four ion-containing *heme* groups, plus a protein, *globin*. The molecular weight is close to 65,000.

The heme groups belong to the class of *porphyrins*, which are large ring compounds (Fig. 6-18). The porphyrins have the property of form-

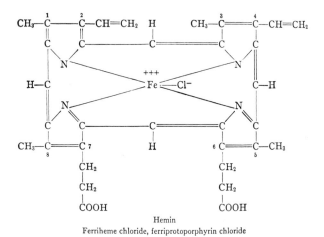

Hemin
Ferriheme chloride, ferriprotoporphyrin chloride

Fig. 6-18. Structure of iron-protoporphyrin complex (heme).

ing complexes with iron and other metals. When iron is introduced into a porphyrin it enters the center of the ring, forming bonds with four inner nitrogens (Fig. 6-18).

Both ferrous (Fe^{++}) and ferric (Fe^{+++}) ions are capable of forming a stable complex with porphyrins. The complexes formed with *protoporphyrin* (Fig. 6-18) are the *hemes*. These can exist in either oxidized (Fe^{+++}) or reduced (Fe^{++}) states. The form of hemoglobin with oxidized heme groups is referred to as *ferrihemoglobin* and the reduced form as *ferrohemoglobin*. Of these, only ferrohemoglobin has the capacity to combine with molecular oxygen to form oxyhemoglobin. It also unites with carbon monoxide to form *carbonmonoxyhemoglobin*.

The sites of attachment of the hemes to the globin are still uncertain, but there is evidence that the iron of heme is linked to the imidazole nitrogen of a histidine.

The presence of the hemes is responsible for the characteristic red color of hemoglobin. The absorption spectra of the different forms of hemoglobin vary somewhat, but in each case there is a strong absorption band between 5000 Å and 6000 Å. For ferrohemoglobin the maximum is near 5650 Å.

Hemoglobin is somewhat unusual in that a number of genetically controlled variants have been detected (for human hemoglobin), some of which are associated with a hereditary disease. The form of hemoglobin which predominates for normal adults is called *hemoglobin A*. This is homogeneous by the criterion of electrophoresis for a wide range of conditions. The amino acid composition is cited in Table 6-3.

TABLE 6-3. AMINO ACID COMPOSITION OF MYOGLOBIN AND TWO FORMS OF HEMOGLOBIN (29)

Amino Acid	Residues per Mole		
	Hemoglobin A (human)	Hemoglobin F (human)	Myoglobin
Alanine	76	74	16
Glycine	43	41	9
Valine	61	55	5
Leucine	76	79	23
Isoleucine	0	10	—
Proline	26	25	3
Phenylalanine	31	33	4
Tyrosine	11	13	2
Tryptophan	5	—	2
Serine	32	45	4
Threonine	33	42	6
Cysteine	6	—	—
Methionine	6	9	2
Arginine	13	13	2
Histidine	36	33	12
Lysine	44	49	19
Aspartic acid	55	54	8
Glutamic acid	33	35	20

In the erythrocytes of normal human unborn infants the hemoglobin content is largely of a different form. This is *fetal hemoglobin* or *hemoglobin F*. It is replaced by hemoglobin A by the end of the first year of postnatal life. The two forms differ in amino acid content (Table 6-3) and in electrophoretic mobility (29).

A third variant of human hemoglobin occurs in the blood of patients suffering from the hereditary disease *sickle cell anemia*. This receives its name from the characteristic shape of the red blood cells when oxygenated. This effect appears to arise from the properties of the hemoglobin. *Sickle cell hemoglobin* (*hemoglobin S*) differs from the A form in mobility and amino acid content (32).

A number of other human hemoglobins have been identified, primarily by virtue of their different electrophoretic properties, which arise from variations in amino acid content. Many, but not all, are symptomatic of disease (29).

Size and shape. The mammalian hemoglobins appear to have molecular weights close to 65,000. Their frictional ratios are quite low (Table 6-2) and could easily arise from hydration. Thus hemoglobin appears to belong to the class of symmetrical proteins whose shapes do not deviate greatly from the spherical.

At pH 7 the sedimentation properties of hemoglobin are those of a single homogeneous substance. There is no sign of any molecular dissociation into subunits at the lowest concentration to which measurements can be extended. Under more acidic conditions a reversible dissociation into half-molecules of molecular weight 33,000 occurs. This has been detected by combined sedimentation and diffusion measurements.

Alkaline denaturation. At high pH, denaturation of oxyhemoglobin occurs to form an irreversibly altered form of ferrihemoglobin. The process is accompanied by a change in absorption spectrum and a loss in solubility. The rates of relative change in both these parameters are consistent with a mechanism which involves several consecutive steps rather than with a single-stage process.

Secondary and tertiary structure. Hemoglobin is one of the two proteins whose spatial organization has been determined with a high degree of precision by X-ray diffraction studies upon crystals. We have already observed that the hydrodynamic properties of the molecule are indicative of a rather compact and symmetrical shape. While the surface is of course somewhat irregular, the best geometrical approximation to its shape, as deduced from X-ray studies, is that of a spheroid, 64 Å by 55 Å by 50 Å (30, 31).

The four heme groups are located on the surface of the molecule and are widely separated. The iron atoms, which lie in the centers of the protoporphyrin rings, may be regarded as lying at the four corners of an irregular tetrahedron (Fig. 6-19). The distance of closest approach between iron atoms is 25 Å.

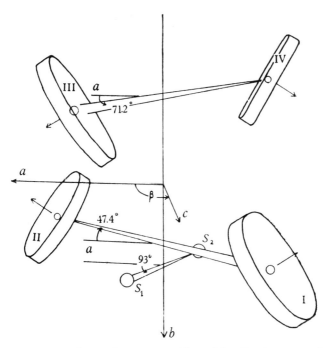

Fig. 6-19. Configuration of the four heme groups of hemoglobin. The short arrows indicate the reactive side (for O_2) of each heme. S_1 and S_2 are two sulphydryl groups of hemoglobin. The c-axis is pointed toward the reader.

Hemoglobin contains four polypeptide chains of similar length, which are grouped into two identical pairs. In terms of Figs. 6-20 to 6-22 each half-molecule consists of a "white" and a "black" chain. The contact between the chains is tenuous, suggesting a rather loose connection. Each heme is associated with a different chain.

The most striking features of the complete model are its symmetry and the complementarity of the configurations of the subunits (Figs. 6-19 to 6-22). The configurations of the two chains of each half-molecule are similar although not identical. The straight regions of the chains in Fig. 6-20 are α-helices. The over-all α-helical content is thus quite high, accounting for about 70% of the polypeptides.

It is clear that the acid dissociation of hemoglobin must involve a splitting into two half-molecules. Somewhat less obvious is the origin of the forces holding together the two chains of each half-molecule. These cannot be disulfide cross-links, since hemoglobin has no cystine.

An interesting structural transition has been found to accompany the combination of hemoglobin with oxygen. The two forms have long been

202   THE CHEMICAL FOUNDATIONS OF MOLECULAR BIOLOGY

(a)

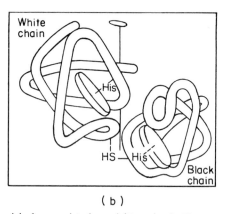

(b)

Fig. 6-20a. Schematic model of a complete hemoglobin molecule. The gray disks represent heme groups (30, 31). (Lower) Configuration of the polypeptide chains of a half-molecule of hemoglobin (31).

known to differ in solubility and crystal structure. The "white" subunits of Figs. 6-20 to 6-22 are unchanged in structure and relative arrangement. However, a striking rearrangement of the two "black" subunits occurs, resulting in an increase of over 7 Å in the separation of related structural features. This configurational change is undoubtedly related to the oxygen-combining capacity of hemoglobin and raises the question of whether a parallel change may not occur in the case of many enzymatic processes.

A surprising feature of the molecular organization of hemoglobin is the wide separation of the heme groups (Fig. 6-19), since it is known

STRUCTURE AND FUNCTION OF IMPORTANT PROTEINS

Fig. 6-20b. Complete hemoglobin molecule viewed from another angle (31).

Fig. 6-21. Model of hemoglobin molecule with one "white" chain removed (31). The angle of viewing is different from Fig. 6-20.

that the binding of one oxygen favors (increases the association constant for) the binding of subsequent oxygens. It is clear that this effect cannot reflect any short-range interaction of heme groups and must arise from some other cause.

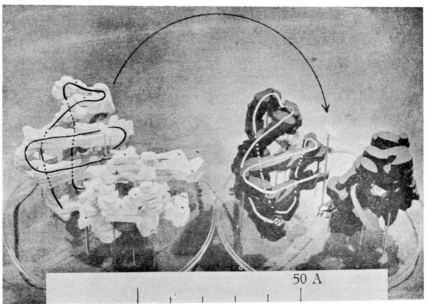

Fig. 6-22. Assembly of hemoglobin model (31).

6-8 MYOGLOBIN

Occurrence and function. Myoglobin is a heme-containing intracellular pigment which is present in both vertebrates and invertebrates. Like hemoglobin, it can combine reversibly with molecular oxygen to form *oxymyoglobin,* which serves as an oxygen reservoir within the cells.

Chemical structure. The molecular weight of myoglobin is close to 18,000. It contains only one heme and appears to consist of a single polypeptide chain, with an NH_2-terminal valine.

Like hemoglobin, myoglobin occurs in a number of variants, which differ slightly in the amino acid content of the globin portion. In the case of whale myoglobin, for which most of the detailed information is available, five distinct forms have been detected, by chromatographic analysis. Two of these, myoglobins IV and V, account for most of the material. They do not appear to differ in amino acid composition or sequence, and the basis for their chromatographic divergence remains unknown.

Myoglobin contains no cystine and hence is not internally cross-linked. The molecular configuration is thus stabilized entirely by secondary linkages.

Secondary and tertiary structure. Myoglobin shares with hemoglobin the distinction of being the first example of a protein whose spatial organization has been almost completely determined by means of X-ray diffraction. The configuration of the single polypeptide chain is remarkably similar to those of the hemoglobin chains, despite major differences in amino acid content and sequence (Fig. 6-23).

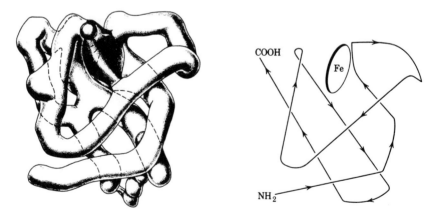

Fig. 6-23. Model of myoglobin molecule (1).

As in the case of hemoglobin, most of the residues are in the α-helical conformation. This includes the linear regions separated by bends. The molecule contains eight nonhelical zones. One of these occurs at the COOH-terminal end of the chain and the balance at the turns, or corners, between lengths of helix. These contain all of the proline residues, which do not fit into an α-helix, except at the NH_2-terminal end (33).

Again in analogy to hemoglobin, the heme group appears to be linked by its iron atom to a histidine. The whole inner part of the heme is surrounded by nonpolar side-chains.

The preceding observations refer, of course, to the crystalline state, to which structural analysis by X-ray diffraction is confined. That the basic features of the structure are preserved in solution is indicated by a comparison of the helical content estimated from optical rotatory dispersion measurements in aqueous solution with those predicted from the model derived from X-ray diffraction.

It is not possible to make rotatory dispersion measurements upon myoglobin itself because of the very strong absorption of the hemes. However, measurements upon the globin portion, after removal of the heme, give a helical content close to 73%. This is in reasonable agreement with the figure of 77% expected for the structure deduced from the X-ray studies.

GENERAL REFERENCES

1. *Physical Chemistry of Macromolecules*, C. Tanford, Wiley, New York (1961).
2. *The Proteins*, H. Neurath and K. Bailey, Academic Press, New York (1953).
3. *Introduction to Protein Chemistry*, S. Fox and J. Foster, Wiley, New York (1957).
4. *Symposium on Protein Structure*, A. Neuberger, Methuen and Co., London (1958).
5. *Biophysical Chemistry*, J. Edsall and J. Wyman, Academic Press, New York (1958).

SPECIFIC REFERENCES

Collagen

6. R. Bear in *Advances in Protein Chemistry*, **7**, p. 69 (1952).
7. W. Harrington and P. von Hippel in *Advances in Protein Chemistry*, vol. 16, p. 1, Academic Press, New York (1961).
8. J. Gross, *Sci. Amer.*, **204**, 121 (1961).
9. H. Boedtker and P. Doty, *J. Am. Chem. Soc.*, **78**, 4267 (1956).
10. F. Schmitt, *Rev. Mod. Phys.*, **31**, 349 (1959).

11. A. Rich and F. Crick in *Recent Advances in Gelatin and Glue Research*, G. Stainsby, Pergamon Press, London (1958).
12. P. von Hippel and W. Harrington in *Protein Structure and Function*, Brookhaven Symposia in Biology, no. 13, p. 213 (1960).

Muscle proteins

13. E. Hanson and H. Huxley, *Symp. Soc. Exptl. Biol.*, **9,** 228 (1955).
14. D. Wilkie in *Progress in Biophysics and Biophysical Chemistry*, vol. 4, Pergamon Press, London (1954).
15. A. Huxley in *Progress in Biophysics and Biophysical Chemistry*, vol. 7, Pergamon Press, London (1957).
16. K. Bailey in reference 2, vol. IIb, chap. 24.
17. E. Geiduschek and A. Holtzer in *Advances in Biological and Medical Physics*, vol. 6, p. 431, Academic Press, New York (1958).
18. A. Holtzer and S. Lowey, *J. Am. Chem. Soc.*, **78,** 5954 (1956).

Antibodies

19. H. Isliker in *Advances in Protein Chemistry*, vol. 12, p. 388, Academic Press, New York (1957).
20. *Fundamentals of Immunology*, W. Boyd, Interscience, New York (1956).
21a. S. Singer and D. Campbell, *J. Am. Chem. Soc.*, **77,** 3504 (1955); **74,** 1794 (1952).
21b. R. Goldberg, *J. Am. Chem. Soc.*, **74,** 5715 (1952).
22. W. Boyd in reference 2, vol. IIb, chap. 22.
23. R. Porter in *Protein Structure and Function*, Brookhaven Symposia in Biology, no. 13, p. 203 (1960); *Biochem. J.*, **88,** 220 (1963).

Fibrinogen and blood clotting

24. H. Scheraga and M. Laskowski, Jr., in *Advances in Protein Chemistry*, vol. 12, p. 1, Academic Press, New York (1957).
25. K. Laki, *Sci. Amer.*, **206,** 60 (1962).
26. S. Shulman, S. Katz, and J. Ferry, *J. Gen. Physiol.*, **36,** 750 (1953).
27. S. Shulman and J. Ferry, *J. Phys. Coll. Chem.*, **55,** 135 (1951).
28. J. Sturtevant, M. Laskowski, Jr., T. Donnelly, and H. Scheraga, *J. Am. Chem. Soc.*, **77,** 6168 (1955).

Hemoglobin

29. H. Itano in *Advances in Protein Chemistry*, vol. 12, p. 216, Academic Press, New York (1957).
30. M. Perutz et al., *Nature*, **185,** 416 (1960).
31. M. Perutz in *Protein Structure and Function*, Brookhaven Symposia in Biology, no. 13, p. 165 (1960).
32. V. Ingram, *Nature*, **792** (1956).

Myoglobin

33. J. Kendrew *et al.*, *Nature* **190,** 661 (1961).

Serum albumin

34. J. Yang and J. Foster, *J. Am. Chem. Soc.*, **76,** 1588 (1954).

7

Catalytic Proteins: The Enzymes

7-1 INTRODUCTION

Nature of enzymes. An enzyme may be defined as a protein molecule which has the property of catalyzing, or increasing the velocity of, a particular chemical reaction. Like other catalysts, an enzyme is involved in its reaction without being irreversibly altered by it and at the conclusion of the reaction returns to its original state. Thus any interaction of the enzyme and the reactants must be of a transient nature and cannot involve the formation of *permanent* chemical bonds.

Enzymes share the diversity of composition, structure, and function which is characteristic of proteins in general. Some enzymes are simple proteins containing only amino acids. Others are conjugated proteins. The limited class of enzymes for which definitive structural information exists includes fibrous, highly helical examples, as myosin, and globular proteins of vanishingly small helical content, as pepsin. Enzymes are known which can be boiled briefly without deactivation; others are so unstable as to be almost impossible to preserve under nonphysiological conditions.

The diversity of enzymes extends to their catalytic function. Some enzymes catalyze the transformation of only a single substrate; others catalyze a family of related reactions. As section 7-2 will make clear, a catalog of known enzyme substrates would contain hundreds of entries of very divergent chemical natures.

Nevertheless, sufficient grounds for generalization exist to make it worthwhile to consider enzymes in a unified manner. Enzymes of widely different properties obey similar kinetic principles and may be studied by similar means. Here, as elsewhere in molecular biology, the ultimate objective is to relate structure and function. Again in parallel to other areas of this field, this objective is only beginning to be realized.

Kinds of reaction catalyzed by enzymes. Over 600 different enzymes have been isolated from biological systems.* There is no reason to believe that the list is approaching saturation. Indeed, the rate of discovery of new enzymes appears to be accelerating, and there is little doubt that the list will ultimately contain many thousands of entries.

A detailed account of all enzymatic reactions would require many volumes and is out of the question here. Fortunately, the *types* of reaction which have been identified thus far fall into a relatively limited number of categories (1).

One generalization may be made at the beginning. The vast majority of enzymatic processes involve the *transfer* of a particular chemical group between two substrate molecules or between solvent and substrate.

(7-1) $$AB + C \rightarrow AC + B$$

Since water is the only biological solvent, enzymes which catalyze reactions of the latter kind are grouped together in the class of hydrolytic enzymes.

All hydrolytic enzymes catalyze the rupture of chemical linkages by combination with water.

(7-2) $$AB + H_2O \rightarrow AOH + BH$$

The kinds of linkage which are known to be attacked by hydrolytic enzymes may be briefly summarized. Almost all examples which have been identified fall into six categories:

(1) amide linkages (—CONH—)
(2) amine groups (—C—NH$_2$)
(3) ester linkages

    (a) carboxylic $\left(\begin{array}{c} O \\ \| \\ -C-O-R \end{array} \right)$

    (b) phosphoric $\left(\begin{array}{c} O \\ \| \\ -O-P-O-R \\ | \\ O- \end{array} \right)$

---

* The Commission on Enzymes of the International Union of Biochemistry has recently proposed a system of nomenclature for enzymes. Their recommendations are cited in the *Report of the Commission on Enzymes of the International Union of Biochemistry,* Pergamon Press, Oxford (1961). However, since the older, trivial names for enzymes continue to be used in most publications, they will be retained in this book. It is likely that the newer designations will gradually replace the traditional terminology within the next few years.

(c) sulfuric $\left(\begin{array}{c} O \\ \| \\ -S-O-R \\ | \\ O- \end{array} \right)$

(4) glycosidic (—R—O—R′, where R and R′ are sugar residues)

(5) pyrophosphate linkages $\left(\begin{array}{cc} O- & O- \\ | & | \\ -P-O-P- \\ \| & \| \\ O & O \end{array} \right)$

(6) carboxyl groups (R—$\overline{COO}$)

Class (1) includes the *peptidases*, which hydrolyze *peptide* linkages. Several examples of these have already been discussed in Chapter 3, in connection with problems of primary structure determination. The specificity of a number of peptidases is not confined to peptide bonds. Thus trypsin and chymotrypsin catalyze the hydrolysis of *amide* derivatives of amino acids.

(7-3) $\quad$ R—CO—NH$_2$ + H$_2$O → R—$\overline{COO}$ + NH$_3$ + H$^+$

Enzymes which hydrolyze simple amides are also known. Thus *urease* converts urea to ammonia and CO$_2$.

(7-4) $\quad$ H$_2$NCONH$_2$ + H$_2$O → CO$_2$ + 2NH$_3$

Category (2) contains enzymes which catalyze the conversion of primary amino groups to hydroxyl groups.

(7-5) $\quad$ —C—NH$_2$ + H$_2$O → —C—OH + NH$_3$

Enzymes of category (3) hydrolyze esters to the corresponding acidic and hydroxyl compounds. In the case of *carboxylases*, the reaction is:

(7-6) $\quad$ —CO—OR + H$_2$O → —$\overline{COO}$ + H$^+$ + HOR

An important example of a carboxylase is *acetylcholinesterase*, which has an important role in nerve function (section 7-5). Other *esterases* catalyze the hydrolysis of phosphate esters. *Phosphoesterases* are of central importance in nucleic acid chemistry. Two important examples are *ribonuclease* (Chapter 10) and *deoxyribonuclease* (Chapter 9).

Enzymes of category 4, the *glycosidases*, hydrolyze polysaccharides, such as starch and glycogen (Chapter 12). *α-Amylase* converts starch to glucose.

Category (5) contains the *pyrophosphatases*, most of which split off the terminal phosphate of an organic pyrophosphate.

(7-7) $\begin{matrix}&\text{O}^-&&\text{O}^-&&&\text{O}^-&&\text{O}^-\\&|&&|&&&|&&|\\-&\text{P}-\text{O}-&\text{P}-\text{O}^- + \text{H}_2\text{O} \rightarrow &-&\text{P}-\text{O}^- + \text{O}=&\text{P}-\text{O}^- + 2\text{H}^+\\&\|&&\|&&&\|&&|\\&\text{O}&&\text{O}&&&\text{O}&&\text{O}^-\end{matrix}$

Myosin *adenosine triphosphatase* (*ATPase*), which hydrolyzes adenosine triphosphate to the corresponding diphosphate, has a vital function in muscle contraction (Chapter 6).

Enzymes of category (6) decarboxylate carboxyl derivatives by a hydrolytic mechanism.

(7-8) $\qquad\qquad \text{R}-\text{CO}\overline{\text{O}} + \text{H}_2\text{O} \rightarrow \text{RH} + \text{CO}_2 + \text{O}\overline{\text{H}}$

The second principal class of enzymes, called *transferring enzymes*, catalyze the transfer of a chemical group between two different substrates or between a subtrate and a coenzyme. The groups which may be transferred include:

(1) hydrogen
(2) nitrogenous groups
(3) phosphate groups
(4) acyl groups
(5) glycosyl groups
(6) methyl groups

Since our primary interest here is the relation of enzyme structure and function, our specific examples will be drawn from the class of hydrolytic enzymes, since the few cases for which definitive information is available come from this class. No detailed discussion of the transferring enzymes will be attempted here; individual cases important for the discussion of subsequent chapters will be described as they arise.

Active sites. Although all enzymes thus far identified have been shown to be proteins of reasonably high molecular weight ($> 10^4$), the available evidence indicates strongly that the enzymatic activity is not distributed uniformly over the protein surface, but is instead localized in certain specific areas of the enzyme. These active areas, or *sites*, appear to be of quite limited dimensions.

In addition to the disparate sizes of the enzyme and substrate in most cases, the evidence for the existence of active sites includes the following:

(1) Most enzymes are subject to inhibition upon binding various small molecules. In many cases the combination with only one molecule of inhibitor is sufficient to block activity.

(2) A number of cases are known in which a major fraction of the

enzyme may be removed by the action of proteolytic enzymes without destroying activity. Some examples are trypsin and myosin.

(3) Several enzymes, including trypsin and chymotrypsin, occur as inactive precursors which may be activated by localized alterations in the structure of a particular region of the molecule.

However, the active center appears in general to be complex in nature. In the few cases for which direct information is available, it appears to contain groups from different regions of the primary structure which must be held in juxtaposition by elements of the secondary and tertiary structure. Because loss of the latter will, in general, result in separation of the subelements of the active center, it is not surprising to find that disruption of the molecular organization by heat, pH, or chemical agents usually produces inactivation.

The enzyme-substrate complex. If the idea that enzyme action might occur by some kind of "action at a distance" can be rejected, it is necessary to postulate that enzymatic activity depends upon an intimate association of the substrate molecule with the active site of the enzyme. In other words, the substrate is bound by the active center to form an *enzyme-substrate* complex with a real, though fleeting, existence. This intermediate subsequently decomposes to form the products.

The existence of the enzyme-substrate complex was originally entirely hypothetical. However, in recent years it has been possible to demonstrate its presence directly in a number of cases by the use of special spectroscopic techniques (section 7-4).

Since the enzyme-substrate complex is ultimately broken down to regenerate free enzyme, it is clear that its formation cannot proceed by the establishment of stable chemical bonds between the substrate and the active site. The general means of attachment is probably by way of secondary forces, including van der Waals forces, electrostatic forces, dipole-dipole interactions, and in some cases the formation of a weak primary bond.

Since, with the exception of electrostatic forces, all the above are *short-range* interactions, which are attenuated rapidly with distance, it is to be expected that the dimensions and configuration of the substrate and the active site will be such as to maximize their mutual contact. There is a definite parallel with the case of antigen-antibody interaction, and the high degree of specificity shown by both processes probably has similar origins.

In a number of cases the conversion of the bound substrate is believed to be stepwise and to involve the formation of one or more intermediate species prior to the final release of the products and the regeneration of the enzyme.

The measurement of enzyme activity. The diverse character of the multitude of chemical reactions which are catalyzed by enzymes has required the development of a number of altogether different techniques to monitor them quantitatively. Usually the rate of appearance of products has been measured, rather than the rate of disappearance of substrate.

In many cases *spectrophotometric* methods have been utilized. If the substrate or the products, or both, absorb light in the visible or ultraviolet region, it is unlikely that the absorption spectra will be identical. It frequently happens that a wave-length exists at which a considerable change of absorbancy accompanies the conversion. This may be utilized to follow the progress of the reaction. Such is the case, for example, for the conversion of malate to fumarate by the enzyme *fumarase*. Changes in absorbancy have also been frequently used to follow reactions catalyzed by the numerous oxidizing enzymes.

If one of the products of the reaction is a gas, it is often convenient to measure the reaction rate by a *manometric* procedure. If due precautions are taken, the volume of gas liberated can serve as a sensitive indicator of the progress of the reaction. Thus the hydrolysis of urea to ammonia and CO_2 may be followed by this technique.

If the reaction results in the evolution or uptake of hydrogen ions it is often possible to monitor it by a *pH-stat* technique. The pH is maintained constant by the continuous addition of acid or alkali. The number of moles of acid or base consumed is a direct measure of the extent of the reaction. This approach has been used for many hydrolytic reactions including the hydrolysis of deoxyribonucleic acid by deoxyribonuclease and the hydrolysis of peptide bonds by peptidases.

In some cases *chromatographic* methods are useful. Because of the length of time required for development of a chromatogram, this method requires a sampling technique. Aliquots of the reaction mixture are withdrawn at intervals and subjected to chromatographic analysis. An example is the use of paper chromatography to follow the phosphorolysis of polyribonucleotides by *polyribonucleotide phosphorylase* (Chapter 10).

There are many cases where the physical properties of reactants and products do not differ sufficiently to provide a convenient means of monitoring the reaction. For such systems direct chemical analysis must be resorted to. The particular technique adopted will of course depend upon the nature of the products.

Since most enzymes are not available in purified form it is desirable to have some objective measure of their concentration and purity. It has been the practice to define an arbitrary system of *activity units* for each enzyme. A unit of activity is usually defined in terms of the number of moles of substrate converted in unit time under standard conditions. The number of enzyme activity units per unit weight of the enzyme preparation (usually 1 mg) is called the *specific activity*. The specific activity units which have been introduced for different enzymes vary widely.

In the case of enzymes available in high purity, it is sometimes convenient to introduce another parameter, the *turnover number*. This is equal to the number of moles of substrate converted per active site per minute. This has the advantage of giving a relatively clear idea of the rate of reaction of the enzyme.

Activators and inhibitors. Many enzymes can function only with the assistance of one or more small ions or molecules which appear to have an essential role in the catalytic process. In some cases, the effect of the activators is of a quantitative nature; in others, no activity whatsoever is observed in their absence. The activators may be grouped into the following categories:

(1) Nonspecific activators: These take no direct part in the chemical action. Their action consists in somehow bringing the enzyme itself into a catalytically active state. The most important activators of this class are metallic cations. Fifteen different cations have been found to activate one or more enzymes. The list includes K^+, Na^+, Rb^+, Mg^{++}, Cs^+, Zn^{++}, Ca^{++}, Cd^{++}, Cr^{++}, Mn^{++}, Fe^{++}, Co^{++}, Cu^{++}, Ni^{++}, and Al^{+++}.

The number of enzymes activated in this way is legion. Some examples are the activation by Mg^{++} of *polyribonucleotide phosphorylase, deoxyribonuclease*, and other enzymes acting on phosphorylated substrates (Chapters 9 and 10), the activation of *myosin adenosine triphosphatase* by Ca^{++}; and the activation of *glycyl-leucine dipeptidase* by Zn^{++}.

It is impossible to generalize as to the mode of binding of the metallic activator. In some cases, it is bound so tightly as to be virtually an integral part of the enzyme; in others, it combines reversibly and may be easily removed by dialysis.

(2) Specific activators or coenzymes: These are generally organic molecules of complex structure which intervene in the reaction itself, often as carriers of a particular chemical group, such as hydrogen or phosphate (Chapter 13).

Coenzymes are essential for the action of many transferring enzymes. In many cases they serve to link the action of two enzymes. Thus one enzyme transfers the group in question from a substrate to the coenzyme; a second transfers it to a second substrate. This is the pattern of action of many dehydrogenases (Chapter 13). In such cases the coenzyme may equally well be regarded as a substrate. This is the case for the di- and tri-phosphopyridine nucleotides, which serve as hydrogen carriers for various dehydrogenases (Chapter 13).

(3) Primers: These differ from the coenzymes in that they resemble the product chemically and in some cases are incorporated into it. Primers are required for a number of enzymes engaged in the assembly of polymeric nucleic acids. An example is the priming of polyribonucleotide phosphorylase by short ribonucleotide polymers (Chapter 10).

The primers are differentiated from ordinary substrates in that a single molecule of primer provides a nucleus for the conversion of many molecules of substrate. Primers are invariably similar in chemical nature to the product.

The velocity of many enzymatic reactions is reduced in the presence of particular ions or neutral molecules. These are known as *inhibitors* of the reaction in question. They generally act by combining with some element of the active center and thereby interfere with its catalytic function. Inhibition will be considered in detail in section 7-3.

7-2 ENZYME KINETICS

General remarks. Because of the basic catalytic property and function of enzymes in connection with chemical reactions, any fundamental study of their characteristics must be based upon quantitative measurements of the rate of the catalyzed reaction. What is known at the present time about the mechanism of enzymatic processes has been derived from studies of this kind.

The ultimate objective of such investigations is of course the direct relation of enzymatic activity to explicit structural features of the enzyme molecule. In no case has this goal been completely achieved, and its final realization in each case will have to await the availability of definitive information on all details of the molecular organization. (This is close to being the situation for hemoglobin and myoglobin).

However, a number of instances exist for which a direct correlation of structure and function is now in a fairly advanced state. This is true for several proteolytic enzymes, including trypsin, chymotrypsin, and ribonuclease; in other cases, important information about the nature of the active site has been obtained by kinetic analysis alone (for example, for such enzymes as acetylcholinesterase, urease, and catalase). Indeed, for those enzymes not yet available in a highly purified state, kinetic analysis is the *only* structural probe available.

The use of kinetics for this purpose generally hinges upon quantitative measurements of the dependence of the rate of the catalyzed reaction upon particular experimental parameters, such as pH, temperature, and the concentration of inhibitors. In order for information of this kind to be utilized intelligently, it is necessary to have a theoretical understanding of the more formal aspects of the kinetics of enzymatic reactions.

Enzyme-catalyzed reactions have special features which prevent the direct utilization of the standard equations of classical chemical kinetics and require the development of a new kinetics which takes explicit account of the role of the enzyme in the process (1, 2).

The Michaelis-Menten model. The first aspect of enzyme kinetics requiring attention is the dependence of rate upon the concentration of substrate. If the concentration of enzyme, as well as the other experimental parameters, is kept constant, it is generally found that in the limit of very low substrate concentrations there is a region in which the *velocity* of the reaction (the number of moles of product produced in unit time per unit volume) increases linearly with the substrate concentration (Fig. 7-1). At higher substrate concentrations the linearity is lost, the velocity attaining a plateau beyond which it does not increase with further addition of substrate. In some cases the velocity actually passes through a maximum with increasing substrate. This effect is known as *substrate inhibition*.

If the concentration of substrate is effectively maintained invariant by having it present in such excess that only a negligible fraction is transformed during the entire experiment, then the effects of varying the enzyme concentration may be examined separately. It is generally found that the velocity increases linearly with the enzyme concentration over a wide range of values for the latter (Fig. 7-1).

Fig. 7-1. Nature of dependence of reaction rate upon the concentrations of enzyme and of substrate. Part (a) represents constant enzyme concentration; (b) shows constant substrate concentration (1).

Instances where the relationship between velocity and enzyme level assumes an anomalous, nonlinear form are known and are usually traceable to the presence of either inhibitors or activators in the enzyme preparation.

We now turn to the problem of formulating a *quantitative* relationship between the concentrations of enzyme and substrate and the velocity of formation of the products. For simplicity it is desirable to limit the problem by restricting consideration to processes of the type involving formally only a *single* substrate:

(7-9) $\qquad S + \text{enzyme} \rightarrow \text{products} + \text{enzyme}$

At first glance it might appear that this restriction is very crippling, since only a few enzymatic reactions are of a strictly unimolecular character. The vast majority involve two reactants and are of the form:

(7-10) $\qquad S + X + \text{enzyme} \rightarrow \text{products} + \text{enzyme}$

However, in many cases one of the reactants is present in such excess that its concentration may be regarded as invariant for practical purposes and the process analyzed as if it were an unimolecular reaction. Such is the case for hydrolytic reactions when carried out in an exclusively aqueous medium.

(7-11) $\qquad S + H_2O + \text{enzyme} \rightarrow \text{products} + \text{enzyme}$

In this and similar instances the minute variations in the level of one of the reactants may be disregarded and the process analyzed formally as if it were of the type shown by equation 7-9.

Basic to the Michaelis-Menten treatment is the concept that the catalytic action of enzymes depends upon the formation by the enzyme and substrate molecules of a *complex species*, which has a definite, although transient, existence. This enzyme-substrate complex then decomposes to form the products. It is the rate of this final step which is the experimentally measured parameter.

(7-12) $\qquad E + S \rightleftarrows ES \rightarrow \text{products}$

The rate, or velocity, of the reaction has until now been discussed as if it were effectively constant for a given set of conditions. Clearly, this cannot be the case over the entire period of the reaction. If the above mechanism is correct then there must be an initial *transient* period during which the concentration of the enzyme-substrate complex is built up. In practice, this period is generally of very short duration (< 1 second) and requires elaborate experimental techniques for its detection.

If the reaction is allowed to proceed until a major fraction of the substrate is converted, its rate will of course slow down as equilibrium is approached (which is what occurs during reversible reaction). If the reaction is irreversible, its rate will decrease gradually to zero as the supply of substrate is exhausted.

The kinetic scheme to be discussed here presupposes that both of the above complications are avoided and thus is limited to that portion of the reaction where:

(1) The enzyme-substrate complex has attained its limiting concentration.

(2) Only a negligible fraction of the substrate has been converted, so that its concentration may be treated as constant during the course of the reaction.

Neither condition is very restrictive. Condition (1) is very difficult to avoid; condition (2) may be attained by maintaining the substrate in excess and confining measurements to the initial phase of the reaction.

The measured velocity (v) of the reaction is equal to the rate of decomposition of the enzyme-substrate complex.

$$(7\text{-}13) \qquad v = k_3[ES]$$

where k_3 is the first-order rate constant and $[ES]$ is the molar concentration of complex.

If v is to be constant, then $[ES]$ must be constant, or

$$(7\text{-}14) \qquad \frac{d[ES]}{dt} = 0$$

This will be the case if conditions (1) and (2) are met. It means that the loss of complex by conversion or dissociation must be exactly balanced by its formation through combination of enzyme and substrate. The rate of formation is given by:

$$(7\text{-}15) \qquad \left(\frac{d[ES]}{dt}\right)_f = k_1[E][S]$$

where k_1 is a rate constant. The rate of loss by *dissociation* into reactants and *decomposition* into products is given by:

$$(7\text{-}16) \qquad -\left(\frac{d[ES]}{dt}\right)_1 = k_2[ES] + k_3[ES]$$

where k_2 and k_3 are the rate constants for the dissociation ($ES \xrightarrow{k_2} E + S$) and decomposition ($ES \xrightarrow{k_3}$ products) reactions, respectively.

The change in complex concentration with time is equal to the sum of (7-15) and (7-16).

(7-17) $$\frac{d[ES]}{dt} = \left(\frac{d[ES]}{dt}\right)_f + \left(\frac{d[ES]}{dt}\right)_1$$
$$= k_1[E][S] - k_2[ES] - k_3[ES] = 0$$

Equation (7-17) may be solved for $[ES]$. Replacing $[E]$, the concentration of *free* enzyme, by $[E_t]$, the *total* concentration of enzyme, yields (as $[E_t] = [E] + [ES]$):

(7-18) $$[ES] = \frac{k_1[E_t][S]}{k_1[S] + k_2 + k_3}$$

From (7-13), the velocity is given by:

(7-19) $$v = k_3[ES]$$
$$= \frac{k_1 k_3 [E_t][S]}{k_1[S] + k_2 + k_3}$$

or

(7-20) $$\frac{v}{[E_t]} = k_3[S]/(K_m + [S])$$

where $$K_m = (k_2 + k_3)/k_1$$

Thus *two* kinetic constants are required to characteristic enzymatic processes of this kind. K_m is known as the *Michaelis constant* and occupies a central position in enzyme kinetics.

In practice, k_3, the *velocity constant*, and K_m are generally located by using (7-20) in inverted form:

(7-21) $$\frac{[E_t]}{v} = \frac{K_m}{k_3[S]} + \frac{1}{k_3}$$

Thus the intercept of a plot of $[E_t]/v$ as a function of $1/[S]$ yields the reciprocal of the velocity constant. From this and the slope, which is equal to K_m/k_3, the Michaelis constant may be computed (Fig. 7-2).

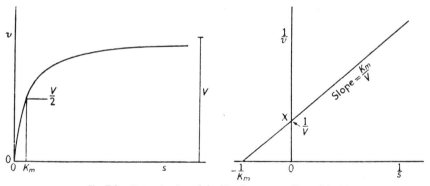

Fig. 7-2. Determination of the kinetic constants K_m and k_3 (1).

A word as to the significance of K_m is appropriate here. The Michaelis-Menten scheme postulates a *steady-state* concentration of ES. This is not in general equivalent to an *equilibrium* state. It becomes so if, and only if, k_3 is very small in comparison with k_2. In this case the concentration of the complex is determined solely by the balance of the association and dissociation reactions, and K_m reduces to:

$$(7\text{-}22) \qquad K_m = \frac{k_2}{k_1} \; (k_3 \ll k_2)$$

In practice, K_m is often loosely used as an index of the affinity of the substrate for the site. The smaller its magnitude, the stronger is the binding affinity of the site for the substrate.

The limiting, or *saturation*, rate of product formation, $v_{\max}$, or V, is equal to $k_3 [E_t]$. When the substrate concentration is such that

$$(7\text{-}23) \qquad v = \tfrac{1}{2} v_{\max} = \tfrac{1}{2} V$$
then $$[S] = K_m$$

Thus the nature of the dependence of velocity upon substrate concentration is determined by the two quantities $v_{\max}$ (or k_3) and K_m. If the purity of the enzyme preparation is such as to permit specification of $[E_t]$ in terms of molarity, then k_3 may be determined in absolute units.

Equation (7-19) predicts that the variation of v with $[S]$ will be of a hyperbolic form (Fig. 7-3). If, for convenience, the reduced quantities ϕ and σ are defined by

$$\phi = \frac{v}{v_{\max}}$$

and $$\sigma = \frac{[S]}{K_m}$$

then equation (7-20) may be rewritten as

$$(7\text{-}24) \qquad \phi = \frac{\sigma}{1 + \sigma}$$

A comparison of the predictions of equation (7-20) or (7-21) with actual behavior (Fig. 7-1) reveals that the Michaelis-Menten model is basically in accord with observation. Indeed this treatment, with occasional modifications, has provided an essential basis for the analysis of a host of very diverse enzymatic reactions.

In several instances, including the cases of chymotrypsin and acetylcholinesterase to be discussed later in this chapter, there is evidence that the conversion of the enzyme-substrate complex to products actually proceeds by two or more distinguishable steps:

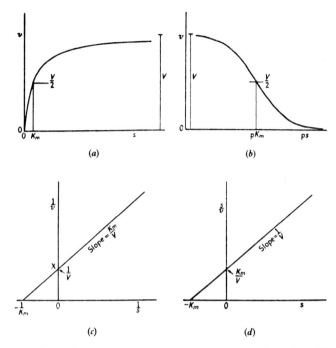

Fig. 7-3. Dependence of velocity upon substrate concentration predicted for Michaelis-Menten mechanism (1).

(7-25) $$ES_1 \underset{k_3}{\rightarrow} ES_2 \underset{k'_3}{\rightarrow} E + \text{products}$$

It can be shown that for this more complex case the kinetics are still formally of the Michaelis-Menten type. However, the significance of the parameters obtained is somewhat altered. Thus the apparent values of K_m and k_3 obtained by the use of equation (7-21) now have meanings defined as follows:

(7-26) $$\frac{1}{k_{3,\text{ apparent}}} = \frac{1}{k_3} + \frac{1}{k'_3}$$

$$K_{m,\text{ apparent}} = K_m / \left(1 + \frac{k_3}{k'_3}\right)$$

Thus the formal obeying of Michaelis-Menten kinetics does not necessarily mean that the mechanism of the reaction is accurately represented by equation (7-12).

Effect of inhibitors. A recurrent feature of enzyme-catalyzed processes is a sensitivity to the presence of added molecules or ions which decrease the velocity of the reaction. Such substances are said to be *inhibitors*. Inhibitors may be either *reversible* or *irreversible*.

Irreversible inhibitors require little discussion here. They act either by combining directly with an essential element of the active site, or else by altering the configuration of the enzyme and thereby indirectly disrupting the active center. An example of an irreversible inhibition is the inactivation of acetylcholinesterase by certain nerve gases, including diisopropylfluorophosphate (DFP). The degree of inactivation caused by an irreversible inhibitor depends only upon the *ratio* of the total concentrations of inhibitor and enzyme. It is independent of the absolute concentrations of enzyme, substrate, or inhibitor.

Reversible inhibitors, with which we shall be primarily concerned here, combine reversibly with the enzyme so that a dynamic equilibrium exists between free and bound inhibitor. Reversible inhibitors may be further classified into *competitive* and *noncompetitive* inhibitors.

Basically, competitive inhibitors act by combining with the enzyme at the same site as the substrate, so that substrate and inhibitor tend to *compete* for the site. A competitive inhibitor may be crowded off the site, and its inhibitory action reversed, by the presence of a sufficiently high concentration of substrate.

In the case of noncompetitive inhibition, the enzyme may combine simultaneously with both inhibitor and substrate. If the site of attachment of the inhibitor is such as to interfere with the *conversion* of the substrate but not with its *binding*, the inhibition will be of this type. This is rendered possible by the generally complex nature of the active center, so that the initial binding of the substrate and its subsequent reaction to form products are mediated by different elements of the site.

To anticipate, fully competitive inhibitors increase K_m, but do not affect k_3 or v_{max}. Noncompetitive inhibitors reduce k_3, but do not alter K_m.

The detailed kinetics of these cases are as follows:

(1) Competitive: In this case an equilibrium exists between bound and free inhibitor. The enzyme may bind either inhibitor or substrate, but not both.

(7-27)
$$E + I \underset{k_5}{\overset{k_4}{\rightleftarrows}} EI$$

where k_4, k_5 are rate constants. An equilibrium constant, K_i, may be defined for the *dissociation* of the enzyme-inhibitor complex.

(7-28)
$$K_i = \frac{k_5}{k_4} = \frac{[E][I]}{[EI]}$$

or
$$[EI] = [E][I]/K_i$$

Applying steady-state kinetics and assuming that the concentrations of ES and EI are constant, we have from equations (7-15), (7-16), and
(7-28) (since $[E_t] = [E] + [ES] + [EI]$):
(7-29) $$k_1[S]([E_t] - [ES] - [EI]) = (k_2 + k_3)[ES]$$
(7-30) $$k_4[I]([E_t] - [ES] - [EI]) = k_5[EI]$$

Solving for $[ES]$,

(7-31) $$[ES] = \frac{[E_t]}{1 + \frac{k_2 + k_3}{k_1[S]}\left(1 + \frac{k_4[I]}{k_5}\right)}$$

Since $v = k_3[ES]$, we have finally:

(7-32) $$v = \frac{k_3[E_t]}{1 + \frac{K_m}{[S]}\left(1 + \frac{[I]}{K_i}\right)}$$

or, as $v_{max} = k_3[E_t]$,

(7-33) $$\frac{v_{max}}{v} = 1 + \frac{K_m}{[S]}\left(1 + \frac{[I]}{K_i}\right)$$

In the case of fully competitive inhibition, the values of v_{max} and k_3, as obtained from the intercept of the (linear) plot of $1/v$ versus $1/[S]$, are independent of the concentration of inhibitor, $[I]$. However, instead of K_m, $K_m(1 + [I]/K_i)$ is obtained from the slope of $1/v$ as a function of $1/S$. Thus the *apparent* value of the Michaelis constant increases with the concentration of inhibitor (Fig. 7-4). In practice, this is a test for the presence of competitive inhibition.

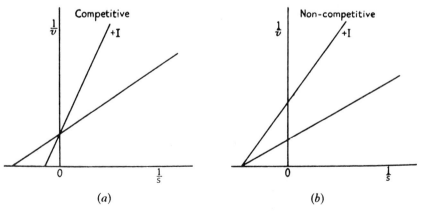

Fig. 7-4. Effect of competitive and noncompetitive inhibition upon kinetic behavior of an enzyme (1).

(2) Fully noncompetitive inhibition: The kinetic equations for this case are as follows:

(7-32) $\quad E + S \underset{k_2}{\overset{k_1}{\rightleftarrows}} ES \overset{k_3}{\rightarrow} \text{products} + E$

$\quad\quad\quad E + I \underset{k_5}{\overset{k_4}{\rightleftarrows}} EI$

$\quad\quad\quad EI + S \underset{k_2}{\overset{k_1}{\rightleftarrows}} EIS \overset{k_6}{\rightarrow} \text{products} + E + I$, where $k_6 \ll k_3$

$\quad\quad\quad ES + I \underset{k_5}{\overset{k_4}{\rightleftarrows}} EIS \overset{k_6}{\rightarrow} \text{products} + E + I$

The identity of the kinetic constants for the two pairs of reactions is a consequence of the fact that the binding of substrate and of inhibitor are mutually independent, occurring on different sites.

At this point it should be pointed out that completely noncompetitive inhibition is only possible *if the Michaelis constant is a true equilibrium constant* ($k_2 \gg k_3$). Otherwise, an alteration in k_3 cannot avoid changing K_m, in view of the definition of the latter quantity (equation 7-20).

The velocity of product formation is given by:

(7-35) $\quad\quad\quad v = k_3[ES] + k_6[EIS]$

By steps analogous to those of the preceding section, the rate equations may be solved for $[ES]$ and $[EIS]$, subject as before to the assumption that

$$\frac{d[ES]}{dt} = 0; \quad \frac{d[EIS]}{dt} = 0$$

Assuming that K_m is an equilibrium constant (equal to k_2/k_1),

(7-36) $\quad\quad\quad [ES] = [E][S]/K_m$

$\quad\quad\quad\quad\quad [EIS] = [EI][S]/K_m$

$\quad\quad\quad\quad\quad\quad\quad = [E][S][I]/K_m K_i$

Also $\quad\quad\quad [E] = [E_t] - [ES] - [EIS] - [EI]$

$$= \frac{[E_t]}{(1 + [S]/K_m)(1 + [I]/K_i)}$$

We obtain finally, using (7-35):

(7-37) $\quad\quad\quad \dfrac{v}{[E_t]} = \dfrac{k_3 + k_6 \dfrac{[I]}{K_i}}{\left(1 + \dfrac{K_m}{[S]}\right)\left(1 + \dfrac{[I]}{K_i}\right)}$

In this case the Michaelis constant is independent of $[I]$.

However, the apparent value of k_3 computed from the intercept by using equation (7-21) is now equal to $\left(k_3 + k_6 \dfrac{[I]}{K_i}\right) \Big/ \left(1 + \dfrac{[I]}{K_i}\right)$.

In practice, the existence of a noncompetitive inhibitor for a particular enzymatic reaction is evidence that its Michaelis constant is a true equilibrium constant.

The effect of pH. In general, the activity of enzymes is restricted to a limited range of pH and in the majority of cases a definite optimum pH is observed. The dependence of rate upon pH may arise from a change in the state of ionization of the substrate or of the enzyme itself. It is difficult to generalize about the former case, and the discussion here will be confined to the effects of pH upon the enzyme itself.

Such information as is available about the active centers of enzymes indicates that, in general, they are of a complex character and include more than one group. If the active center contains an ionizable site, it is to be expected that its kinetic parameters will depend upon its state of ionization.

The ionization of the active site may influence either K_m or $v_{\max}$, or both. The effect of pH upon $v_{\max}$ may, in practice, often be isolated by examining the pH profile of velocity in the presence of a saturating excess of substrate. Under these conditions, the limiting amount of substrate is bound, and variations in K_m are without influence.

Let us consider first the frequently encountered case of an enzyme whose rate passes through a maximum with pH and decreases at pH's acid or alkaline to the optimum range (Fig. 7-5). Behavior of this kind

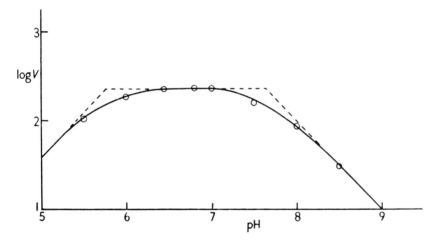

Fig. 7-5. The pH dependence of $v_{\max}$ for fumarase (1).

can be formally accounted for by postulating that the active center contains two ionizable groups, A and B. For activity to be present, A must be protonated and B must be unprotonated. All other combinations are inactive. If the enzyme is saturated with substrate,

(7-38)
$$ES_{AH^+,BH^+} \underset{H^+}{\overset{K_{SB}}{\rightleftarrows}} ES_{AH^+,B} \underset{OH^-}{\overset{K_{SAB}}{\rightleftarrows}} ES_{A,B}$$
$$\text{inactive} \qquad \text{active} \qquad \text{inactive}$$

$$K_{SB} = \frac{[H^+][ES_{AH^+,B}]}{[ES_{AH^+,BH^+}]}; \qquad K_{SAB} = \frac{[H^+][ES_{A,B}]}{[ES_{AH^+,B}]}$$

where K_{SB} and K_{SAB} are the two equilibrium constants for dissociation of hydrogen ion. The situation is capable of simple explanation provided that two further assumptions are made.

(1) pK_{SAB} is sufficiently larger than pK_{SB} (by 2 pH units or more) so that the two groups may be regarded as ionizing independently and, in particular, the species ES_{A,BH^+} is present in negligibly small concentration.

(2) The concentrations of the various ionized species are governed entirely by equilibrium considerations. In other words, the ionization reactions are sufficiently rapid so as not to be perturbed by the enzymatic reactions. It is possible to drop this restriction, but at the expense of obtaining expressions which are too complicated to be useful.

It is easily shown that in this case the fraction of enzyme in the active form is equal to $1 \Big/ \left(1 + \dfrac{[H^+]}{K_{SB}} + \dfrac{K_{SAB}}{[H^+]}\right)$.

Under conditions of substrate saturation, the velocity of reaction is given by:

(7-39)
$$v_{max} = \frac{k_3[E_t]}{1 + \dfrac{[H^+]}{K_{SB}} + \dfrac{K_{SAB}}{[H^+]}}$$

This equation predicts that the rate will pass through a maximum with pH. If, as was postulated, p$K_{SAB} \gg$ pK_{SB}, then in the region acid to the optimal zone the second term in the denominator of (7-39) may be neglected and the equation rewritten in reciprocal form as:

(7-40)
$$\frac{1}{v_{max}} = \frac{1 + \dfrac{[H^+]}{K_{SB}}}{k_3[E_t]}$$

Equation (7-40) predicts that a plot of $\dfrac{k_3[E_t]}{v_{max}}$ as a function of $[H^+]$

will be a straight line of slope $\dfrac{1}{K_{SB}}$. Similarly in the region alkaline to the optimal zone, equation (7-39) reduces to:

(7-41) $$\dfrac{1}{v_{\max}} = \dfrac{1 + \dfrac{K_{SAB}}{[\mathrm{H}^+]}}{k_3[E_t]}$$

Thus in the alkaline zone a plot of $\dfrac{k_3[E_t]}{v_{\max}}$ versus $\dfrac{1}{[\mathrm{H}^+]}$ yields a straight line of slope K_{SAB}.

In this manner the pK's of the ionizable groups in the active center which are involved in the actual *transformation* of bound substrate may be determined. This often provides a clue to their identity. Evidence of this kind has led to the proposal that a histidine group is included in the active centers of trypsin and chymotrypsin.

In general, changes in the state of ionization of the active site will of course affect the Michaelis constant as well as the maximum velocity. In order to describe the kinetic behavior at subsaturating concentrations of substrate, it is necessary to consider the complete ionization scheme.

(7-42) $$E_{A\mathrm{H}^+,B\mathrm{H}^+} \underset{\mathrm{H}^+}{\overset{K_B}{\rightleftarrows}} E_{A\mathrm{H}^+,B} \underset{\mathrm{OH}^-}{\overset{K_{AB}}{\rightleftarrows}} E_{A,B}$$
$$k_1 \updownarrow k_2$$
$$ES_{A\mathrm{H}^+,B\mathrm{H}^+} \overset{K_{SB}}{\rightleftarrows} ES_{A\mathrm{H}^+,B} \overset{K_{SAB}}{\rightleftarrows} ES_{A,B}$$
$$\downarrow k_3$$
$$E_{A\mathrm{H}^+,B} + \text{products}$$

The assumptions made in the case of saturating substrate concentration are retained. For this more general case the following expression is obtained by straightforward but tedious algebraic manipulation.

(7-43) $$v = \dfrac{k_3[E_t]}{1 + \dfrac{[\mathrm{H}^+]}{K_{SB}} + \dfrac{K_{SAB}}{[\mathrm{H}^+]}}{1 + \dfrac{Y}{[S]}\dfrac{k_2 + k_3}{k_1}}$$

where $$Y = \dfrac{\left(1 + \dfrac{[\mathrm{H}^+]}{K_B} + \dfrac{K_{AB}}{[\mathrm{H}^+]}\right)}{1 + \dfrac{[\mathrm{H}^+]}{K_{SB}} + \dfrac{K_{SAB}}{[\mathrm{H}^+]}}$$

In the limit of very high concentrations of $[S]$, equation 7-43 reduces to 7-39. If the usual procedures for the evaluation of K_m are applied, the value obtained is multiplied by Y. If pK_{AB} and pK_{SAB} are much greater than pK_B and pK_{SB}, respectively, then Y will be close to unity in the optimal pH range.

The simple treatment outlined above may of course be modified to account for variants of the above scheme, such as the case where two different ionized forms of the enzyme-substrate complex both decompose to form products, but at different rates, or the case where the substrate itself ionizes. Such special cases are however best considered individually as they arise.

Effect of temperature. The effect of temperature upon an enzymatic reaction is generally complex. It may reflect a reversible or irreversible change in the structure of the enzyme, a change in the state of ionization of the active center, or a change in the kinetic parameters K_m and k_3. In practice, it is not usually difficult to resolve these effects.

If measurements are confined to a range of temperature and pH in which neither the enzyme structure nor its state of ionization change significantly, then the kinetics at each temperature (T) can be analyzed in the usual way to yield the dependence of K_m and k_3 (or $v_{\max}$) upon T.

If the Michaelis constant is a true equilibrium constant, then its temperature dependence is similar in nature to that of any other equilibrium constant. In particular, we have

(7-44) $$\log K_m = -\frac{\Delta H}{RT} + \text{constant}$$

and $$\frac{d \log K_m}{dT} = \frac{\Delta H}{RT^2}$$

where R is the gas constant and ΔH is the over-all heat or *enthalpy* change accompanying the formation of the enzyme-substrate complex. This relation is known as the *Van't Hoff equation* and is of general applicability to equilibrium processes.

If K_m is not an equilibrium constant, the interpretation of its temperature dependence is no longer straightforward, representing a composite of the variations of k_1, k_2, and k_3. In this more general case, it is not usually possible to obtain useful information.

If the process is of the simple Michaelis-Menten type and involves only a single step in the decomposition of the enzyme-substrate complex to form products, then the variation of k_3 with T has a relatively simple interpretation.

(7-45) $$\log k_3 = -\frac{E^*}{RT} + \text{constant}$$

and $$\frac{d \log k_3}{dT} = \frac{E^*}{RT^2}$$

Here E^* is the classical Arrhenius activation *energy* and is equal to the net increment in *energy* accompanying the transition from the initial enzyme-substrate complex to an intermediate *activated state* prior to final decomposition into products.

(7-46) $$ES \rightarrow \underset{\substack{\text{activated}\\\text{state}}}{ES^*} \rightarrow E + \text{products}$$

7-3 STRUCTURE AND ACTIVITY OF TRYPSIN (3, 4, 5, 6)

General remarks. In this and subsequent sections we turn to the consideration of enzymes for which a definitive picture of the nature of the active site is beginning to be realized. Trypsin is a peptidase of molecular weight close to 24,000. It is synthesized and excreted by the pancreas in the form of an inactive precursor, trypsinogen. The amino acid composition of trypsin and its precursor has a preponderance of basic residues. As a consequence, its isoelectric point lies in the alkaline range, in the vicinity of pH 10.

The enzymatic specificity of trypsin has already been mentioned briefly in Chapter 3. It is more restrictive than those of most other peptidases (3, 5, 6). Peptide bonds which link the carboxyl group of a basic amino acid (arginine or lysine) to the α-amino group of another amino acid are hydrolyzed. The rate of hydrolysis is dependent upon the position and chemical environment of the peptide bond. In general, peptide bonds adjacent to an unmasked α-amino or α-carboxyl group are split relatively slowly. Also the proximity of an acidic side-chain appears to depress activity (Fig. 7-6). Substitution of the α-amino group of the basic amino acid enhances activity; substitution of the side-chain amino group abolishes it entirely.

The hydrolytic action of trypsin is not confined to peptide bonds. Ester and amide derivatives of the α-carboxyl group of lysine or arginine are split even more readily than peptide linkages (Fig. 7-6).

The activation of trypsin. The conversion of trypsinogen to trypsin is mediated by the action of any of several peptidases, including trypsin itself and *enterokinase*, an enzyme found in intestinal secretions. The latter appears to be the actual agent of physiological activation, for free trypsin cannot be detected in pancreas secretions. In both cases, side reactions occur which result in the production of some inert protein.

The molecular weights of trypsin and trypsinogen are so close as to be within the margin of experimental uncertainty of the physical techniques.

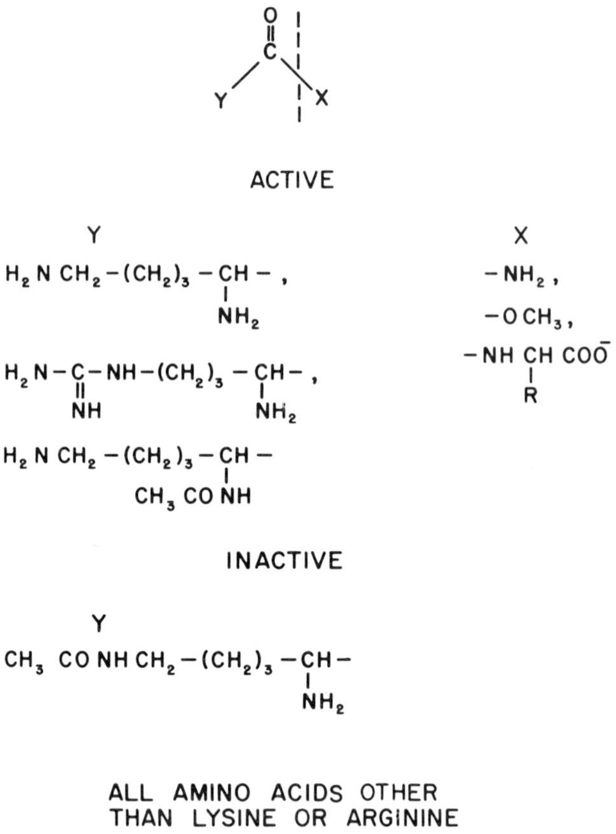

ACTIVE

$$\underset{Y}{\diagdown}\underset{}{\overset{\overset{O}{\underset{\|}{C}}}{\diagup}}\underset{X}{\diagdown}$$

ACTIVE

Y

$H_2N\ CH_2-(CH_2)_3-\underset{NH_2}{CH}-$,

$H_2N-\underset{\underset{NH}{\|}}{C}-NH-(CH_2)_3-\underset{NH_2}{CH}-$,

$H_2N\ CH_2-(CH_2)_3-\underset{CH_3\ CO\ NH}{CH}-$

X

$-NH_2$,

$-OCH_3$,

$-NH\ \underset{R}{CH}\ COO^-$

INACTIVE

Y

$CH_3\ CO\ NH\ CH_2-(CH_2)_3-\underset{NH_2}{CH}-$

ALL AMINO ACIDS OTHER THAN LYSINE OR ARGININE

Fig. 7-6. Enzymatic specificity of trypsin.

It is clear that the activation process cannot involve the amputation of a major fraction of the molecule.

The NH$_2$-terminal sequence of trypsinogen is Val (Asp)$_4$Lys Ileu Val Gly-. The NH$_2$-terminal residue of trypsin is isoleucine instead of valine. This suggests that the activation step may involve the hydrolytic splitting of the peptide bond between lysine and isoleucine.

A number of other facts support this hypothesis.

(1) A definite correlation exists between the rates of activation and splitting of a single peptide bond (as measured by hydrogen ion titration of the α-carboxyl produced).

(2) The hexapeptide Val (Asp)$_4$ Lys is the only peptide which can be detected chromatographically in the activation mixture.

(3) *Complete* acetylation of trysinogen prevents its activation by trypsin. The acetylation of the ε-amino group of lysine would of course

block the action of this enzyme on the Lys-Ileu bond. In contrast, trypsin itself can be acetylated without losing activity.

It is thus probable that the alterations in *primary* structure accompanying activation are confined to the splitting off of the NH_2-terminal hexapeptide Val $(Asp)_4$ Lys (Fig. 7-7).

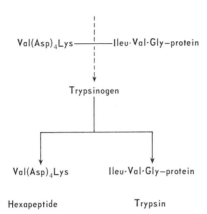

Fig. 7-7. The splitting of peptide bonds during the activation of trypsinogen (3).

In addition, an important change in the secondary structure of trypsinogen appears to accompany activation. The optical rotation of trypsinogen decreases sharply during activation, indicating that an important change in configuration occurs. The direction of the change is that commonly associated with an increase in α-helical content.

The active site of trypsin. The enzymatic activity of trypsin can be completely and stoichiometrically abolished by reaction with one molecule of diisopropylfluorophosphate (DFP). The one-to-one combining ratio of this irreversible inhibitor, together with the localized character of the peptide bond splitting accompanying activation, make it highly likely that trypsin contains only a single active site.

The DFP derivative of trypsin is stable enough to survive extensive acid or enzymatic hydrolysis of the protein. Examination of the products of limited hydrolysis has revealed that the phosphoryl group is attached to the serine group occurring in the sequence Gly Asp Ser Gly, which occurs in many peptidases. There is a definite implication that all, or part, of this sequence may be an element of the active site.

A third approach to the problem is to examine the pH dependence of the velocity of trypsin-catalyzed reactions. The activity of trypsin decreases at acid pH. The pH-activity curve can formally be accounted for in terms of the titration of a single group of pK 6.2. Since this is in the range for the imidazole group of histidine, it has been surmised that this group may also be an element of the active site.

The fragmentary information assembled above has yet to be fitted into a definitive picture of the active center of trypsin. One possible mechanism has been put forward by Neurath and Dixon (Fig. 7-8). The

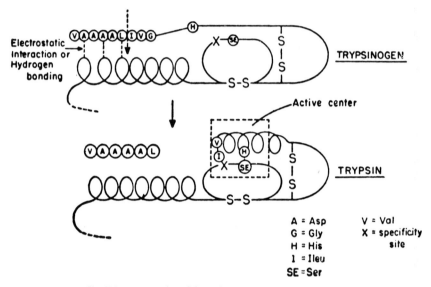

Fig. 7-8. Proposed model for the activation of trypsinogen (3).

presence of four adjacent negative charges in the $(Asp)_4$ segment of the NH_2-terminal peptide of trypsinogen undoubtedly keeps it in an extended, nonhelical configuration. When this electrostatic stress is relieved by splitting of the Lys-Ileu bond during activation, the remainder of the NH_2-terminal peptide may coil up further like a released spring, thereby bringing the essential groups into the appropriate positions. The conjectural nature of this model should, however, be recognized.

7-4 STRUCTURE AND ACTIVITY OF THE CHYMOTRYPSINS

General remarks. As in the case of trypsin, the various active forms of chymotrypsin are derived from an inactive precursor, *chymotrypsinogen,* which is produced and secreted by the pancreas. Two kinds of chymotrypsinogen, called A and B, are found in pancreas extract. The discussion here will be limited to chymotrypsinogen A and its derivatives, with which most of the existing studies have been concerned (7, 8, 9, 10).

The molecular weight of the basic molecular unit of the chymotrypsins is close to 25,000. Varying degrees of pH-dependent association occur in solution.

The enzymatic specificity of chymotrypsin is less restrictive than that of trypsin. Peptide, ester, and amide bonds involving the α-carboxyl group of an *aromatic* amino acid (tyrosine, phenylalanine, or tryptophan) are hydrolyzed. However, the requirement for an aromatic residue is not absolute, since bonds involving the α-carboxyl groups of asparagine, glutamine, leucine, and methionine are also attacked.

In analogy to the trypsin case, the proximity of a free α-amino group to the bond hydrolyzed has a depressing effect on the rate. This can be removed by acyl substitution of the amino group.

Chymotrypsin also catalyzes the hydrolysis of substances which contain no amino acid residues. One of these nonbiological substrates, p-nitrophenyl acetate (NPA), has figured prominently in studies of the mechanism of action of chymotrypsin.

The activation of chymotrypsinogen. Like trypsinogen, chymotrypsinogen may be activated by the hydrolytic scission of particular peptide bonds. Trypsin has been generally used in experimental studies of this process.

The original method of activation (*slow activation*) involved the use of relatively low levels of trypsin and a lengthy (48-hour) period of incubation. The primary product of this procedure was the variant α-*chymotrypsin*, which has the advantage of being more readily crystallizable than the other active forms. This is the classical chymotrypsin and is often referred to as simply "chymotrypsin."

More recently it has been found that other variants of chymotrypsin, of similar specificity but higher activity, could be obtained by incubation for shorter times in the presence of higher levels of trypsin. The consecutive products of stepwise activation have different mobilities, so that the process may readily be followed electrophoretically. This is the *fast activation* studied by Neurath and co-workers (Fig. 7-9).

Chymotrypsinogen has a single NH_2-terminal group, cystine, and a single COOH-terminal group, asparagine. The stepwise changes accompanying fast activation, as reflected by changes in the terminal groups, are as follows.

(1) Chymotrypsinogen is converted into an active form of different mobility called π-chymotrypsin. While no peptide is split off, one new NH_2-terminal (isoleucine) and one new COOH-terminal group (arginine) are produced. Only one peptide bond is hydrolyzed. The failure of any

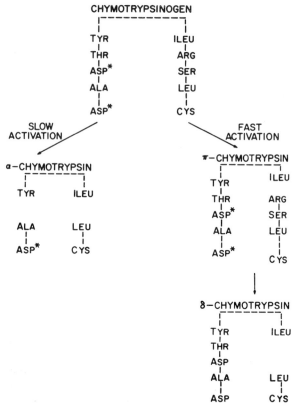

Fig. 7-9. Hydrolytic steps occurring in the activation of chymotrypsinogen (7). ASP* stands for asparagine.

free peptide to appear suggests that the bond hydrolyzed is part of a closed loop formed by cystine between two different parts of the same polypeptide.

(2) A second peptide bond is hydrolyzed to form a second variant, δ-*chymotrypsin*. Here the rapid activation process comes to a halt with the formation of this relatively stable end product. A Ser-Leu bond is split with the loss of one mole each of serine and arginine and the production of a new NH_2-terminal residue, leucine (Fig. 7-9).

From these results it is clear that a sequence Ileu Arg Ser Leu exists in chymotrypsinogen. During fast activation this sequence is first split open between Ileu and Arg to form π-chymotrypsin. Then the Ser Arg dipeptide is removed to form δ-chymotrypsin (Fig. 7-9).

During slow activation two further processes have time to occur.

Chymotrypsinogen is slowly attacked by chymotrypsin and δ-chymotrypsin slowly autolyzes. The process is complex and still incompletely understood. However, the formation of α-chymotrypsin appears to involve, in addition to the changes in the Ileu Arg Ser Leu sequence, the rupture of peptide bonds in a second Tyr Thr Asp-NH₂ Ala sequence (Fig. 7-9). A Thr Asp-NH₂ dipeptide is liberated from the interior of this sequence to produce a new NH₂-terminal alanine and a new COOH-terminal tyrosine. α-Chymotrypsin has a specificity similar to that of π- and δ-chymotrypsin, but its activity is significantly lower.

Two other distinguishable chymotrypsins, β- and γ-*chymotrypsin*, are formed in relatively small quantities during the slow activation process. The detailed mechanism of their formation and the structural points of difference from α-chymotrypsin are still obscure.

It is remarkable that full enzymatic activity is attained by the rupture of a single peptide bond in the case of π-chymotrypsin. The idea that the active center is preformed in the precursor and is simply exposed and rendered available during activation cannot be excluded entirely. However, the alternative explanation that activation involves the actual assembly of the active center from previously scattered elements appears more likely at present.

The attractiveness of this proposal is enhanced by the important changes in the secondary and tertiary structure which accompany activation. Thus, a significant decrease in optical rotation occurs. This structural change appears to be triggered in some incompletely understood way by the scission of the crucial Ileu Arg bond.

The activation of chymotrypsinogen does not result in major changes in the size and shape of the molecule. Both the zymogen and its activation products are compact globular proteins of low asymmetry, as judged from viscosity and frictional ratio measurements. Neither parameter changes significantly during activation.

The active site. Like trypsin, the chymotrypsins are completely and irreversibly inhibited by combination with a single molecule of DFP. This is consistent with, and suggests that, only one active center per molecule is present.

Examination of the products of limited hydrolysis of the DFP derivative of α-chymotrypsin has revealed that the organophosphoryl group is attached to a *serine* residue which forms part of the sequence Gly Asp Ser Gly. The parallel with trypsin is remarkable and suggests that a structural principle of general significance has been uncovered.

Again in parallel to the trypsin case, there is definite evidence implicating the imidazole group of histidine in the active center. The pH dependence of activity is consistent with this view, as is the fact that the destruction of one residue of histidine by photo-oxidation abolishes activity.

Much of the detailed information available about the active center of chymotrypsin has been derived from kinetic studies with a rather atypical substrate, p-nitrophenyl acetate (NPA), which is hydrolyzed to p-nitrophenol and acetate. This substrate has the advantage of being readily adaptable to spectrophotometric methods, because of its strong ultraviolet absorption.

Sturtevant and co-workers have made use of a special spectrophotometric technique to follow the initial transient phase of the hydrolysis of NPA by chymotrypsin. The method depends upon the use of a rapid mixing chamber (the "stopped-flow" chamber) for the enzyme and substrate and the monitoring of fast changes in absorbancy through the use of an oscilloscope. In this manner measurements can be extended to very short reaction times of the order of 0.01 to 0.02 seconds (8).

In this manner the *transient* period of the reaction, which precedes the attainment of the steady state, has been examined. Since the transient phase is over within about 5 seconds in the case of NPA, the more conventional methods of following the reaction cannot detect it.

This kind of approach has revealed that at least three steps are involved in the reaction:

(1) $E + S \leftrightarrows ES_1$

(2) $ES_1 \rightarrow ES_2 + P_1$

(3) $ES_2 \rightarrow E + P_2$

Step (1), which is very rapid, corresponds to the initial formation of the Michaelis-Menten complex. The conversion of the substrate occurs in two steps. The first of these, step (2), results in the formation of an acylated derivative (ES_2) of the enzyme and the liberation of one of the products, nitrophenol. The final step corresponds to the release of the acyl group and its transfer to a water molecule to form the other product, acetate. This is accompanied by the regeneration of free enzyme. It can be shown that the steady state kinetics predicted for this mechanism are still of the Michaelis-Menten type.

The site of attachment of the acetyl group is still uncertain. There is some evidence that it may be attached to the hydroxyl of serine.

Current thinking regards the active center as undoubtedly complex and probably involving both serine and histidine. Since these two residues do not form part of the same primary sequence, their correct mutual position must be maintained by virtue of the tertiary structure of the enzyme. As mentioned above, the active configuration may be formed by the structural rearrangement which accompanies the activation of chymotrypsinogen.

Sturtevant has proposed a general mechanism for chymotrypsin-catalyzed reactions (Fig. 7-10). This postulates that the imidazole and serine hydroxyl are initially hydrogen-bonded (8). This feature was introduced to account for the fact that the histidine pK appears to shift upon formation of acetyl-chymotrypsin. According to Sturtevant's model, acetylation occurs at the serine hydroxyl. The release of the nitrophenol group is accompanied by the transfer of a proton to the imidazole. The latter then hydrogen bonds to a water molecule which is then correctly positioned to attack the acetyl group (Fig. 7-10).

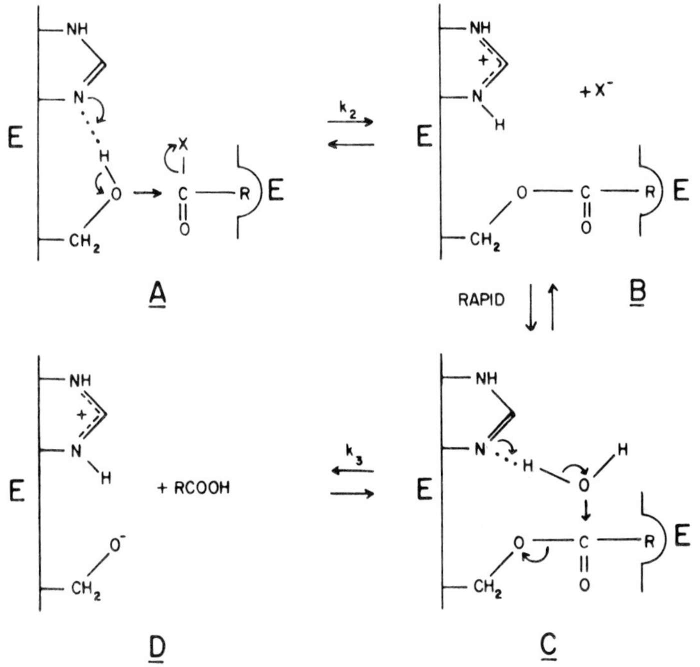

Fig. 7-10. Proposed mechanism for hydrolysis catalyzed by chymotrpsin (8). The arrows represent electron shifts. A represents formation of the initial enzyme-substrate complex (ES_1). B represents the formation of the acyl derivative (ES_2). C and D show the final step of deacetylation and regeneration of free enzyme.

This and competing mechanisms leave open the question of enzyme specificity. The entire subject is still in a state of rapid evolution at the time of writing, and a final decision must clearly be deferred.

7-5 ACETYLCHOLINESTERASE

General remarks. This enzyme has been identified in nerve tissue of all species thus far investigated. It catalyzes the hydrolysis of acetylcholine to acetate and choline according to:

(7-47)
$$(CH_3)_3\overset{+}{N}C_2H_4O-\underset{\underset{O}{\|}}{C}-CH_3 + H_2O \leftrightarrows (CH_3)_3\overset{+}{N}C_2H_4OH + CH_3CO\overset{-}{O} + \overset{+}{H}$$

While it is generally agreed that acetylcholine has an important role in nerve function, there is still considerable controversy as to the nature of its function. The more widely held view tends to limit its function to a restricted class of nerves, for which it is thought to be involved in neurohumoral transmission. A minority viewpoint assigns it a more general function as an intracellular agent influencing both transmission and conduction in all kinds of nerves.

Our present interest in this enzyme will be centered about its enzymatic properties for their own sake. Virtually all the information which is available as to the mechanism of action and the nature of the active site of acetylcholinesterase has been obtained from kinetic studies alone. This is necessarily the case because of the nonavailability of this enzyme in purified form. Acetylcholinesterase is an outstanding example of the power and versatility of this approach (11).

The impure character of the preparations obtained thus far has precluded any really definitive studies of its molecular properties. Rough estimates of the molecular weight have placed it in excess of 10^6.

Nature of the active site. Acetylcholinesterase is reversibly inhibited by a number of compounds which are analogous in structure to acetylcholine (Fig. 7-11). It is natural to suppose that they owe their inhibitory power to this structural similarity, which endows them with the capacity to be bound at the active site in place of the substrate.

All the potent inhibitors of this class share two structural features. They possess a cationic site separated by a nonpolar region from an ester or hydroxyl group. The absence of either site results in a profound depression of inhibitory activity. For example, a comparison of isoamyl alcohol and dimethylamino ethanol shows the latter to be a 30-fold stronger inhibitor (Fig. 7-11). The two molecules are structurally similar except for the absence of a positive charge in the case of isoamyl alcohol.

Similarly, neostigmine (Fig. 7-11), whose state of ionization is independent of pH, inhibits equally well at all pH's between 6 and 10. In

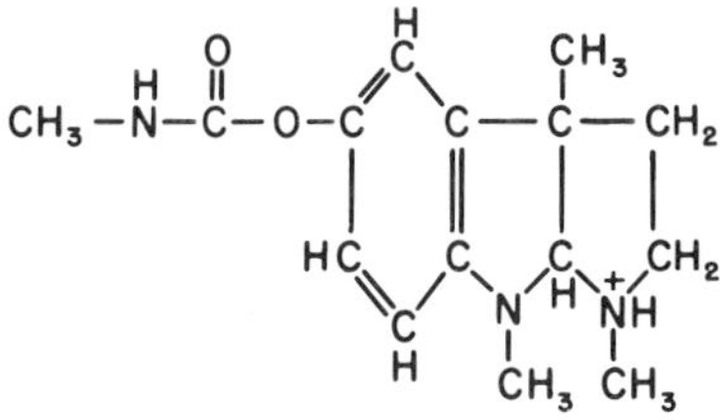

PHYSOSTIGMINE

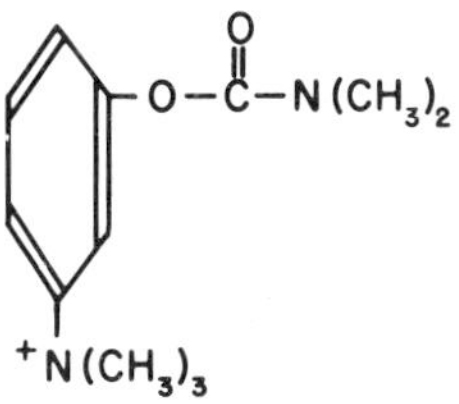

NEOSTIGMINE

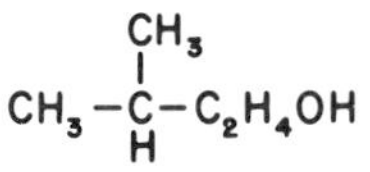

ISOAMYL ALCOHOL   DIMETHYLAMINO ETHANOL

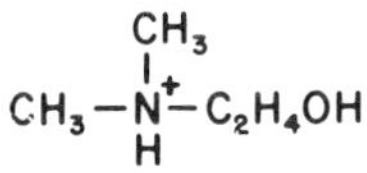

Fig. 7-11. Some competitive inhibitors of acetylcholinesterase.

contrast, physostigmine, with a pK of 8.1, shows a pronounced drop in inhibitory power at pH's above 8. Because the positive charge of the nitrogen is lost under alkaline conditions, this result indicates that the cationic form of physostigmine is more strongly bound by the active center than its conjugate.

Parallel experiments have been made with substrates. Dimethylaminoethyl acetate has a Michaelis constant which is smaller by a factor of 8 than isoamyl acetate (Fig. 7-12). If, as is likely, the Michaelis constant may be regarded as an equilibrium constant for dissociation in this case, isoamyl acetate is much less strongly bound than its charged structural analog.

Experiments of this kind have led to the proposal of a double character for the active site of acetylcholinesterase (Fig. 7-13). One subsite, which is anionic in character, is primarily concerned with specificity and serves as the point of attachment for the cationic site of the substrate or inhibitor. The other subsite, the *esteratic* site, is directly involved in the hydrolytic process.

However, the forces involved in binding at the anionic site do not appear to be exclusively electrostatic in character. Thus, a comparison of the inhibitory power of ammonium ion (NH^+_4) and the various methylamines ($CH_3NH_3^+$, $[CH_3]_2NH_2^+$, $[CH_3]_3N^+H$, and $[CH_3]_4N^+$) has shown that the inhibition constant increases steadily with increasing methyl substitution. This suggests that van der Waals interactions between nonpolar groups in the anionic site and the substrate or inhibitor make an important contribution to the strength of binding.

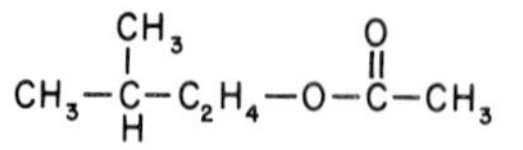

$$CH_3-\underset{H}{\overset{CH_3}{\underset{|}{C}}}-C_2H_4-O-\overset{O}{\underset{||}{C}}-CH_3$$

ISOAMYL ACETATE

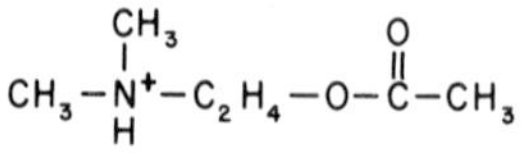

$$CH_3-\underset{H}{\overset{CH_3}{\underset{|}{N^+}}}-C_2H_4-O-\overset{O}{\underset{||}{C}}-CH_3$$

DIMETHYLAMINOETHYL ACETATE

Fig. 7-12. Two substrates of acetylcholinesterase.

The esteratic subsite appears to make a relatively slight contribution to the actual *binding* of substrate. Substrates which lack a cationic site and hence cannot interact strongly with the anionic subsite have Michaelis constants which are larger by orders of magnitude than their analogs which possess a positively charged site. This effect has already been noted in the case of isoamyl acetate ($K_m = 8 \times 10^{-3}$) and dimethylaminoethyl acetate ($K_m = 1 \times 10^{-3}$). Ethyl acetate, whose dimensions would preclude van der Waals, as well as electrostatic, interactions with the anionic site, has a Michaelis constant of 0.5.

The maximum velocity of hydrolysis (v_{max}) decreases at both acid and alkaline pH. The pH dependence of rate has been interpreted in

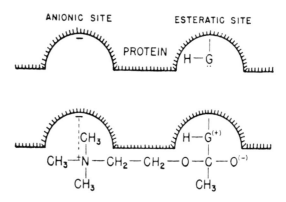

Fig. 7-13. Proposed nature of the active site of acetylcholinesterase (11).

terms of the presence in the esteratic subsite of an acidic and a basic group, whose pK's are 6.5 and 10.5, respectively.

(7-48) $$EH_2 \underset{\text{inactive}}{\overset{\overset{+}{H}}{\longleftarrow}} EH \underset{\text{active}}{\overset{\overset{-}{OH}}{\underset{}{\rightleftarrows}}} \underset{\text{inactive}}{E}$$

Acetylcholinesterase is irreversibly inhibited by DFP. This offers an interesting parallel to the cases of the peptidases trypsin and chymotrypsin and reinforces the idea that the active centers of all hydrolytic enzymes may have certain structural features in common.

Kinetics of the hydrolysis. Wilson has proposed that the hydrolysis of acetylcholine may proceed by way of the formation of an acetyl enzyme (11). The scheme is formally as follows:

(7-49) $$EH + S \underset{k_2}{\overset{k_1}{\rightleftarrows}} EH \cdot S \overset{k_3}{\rightarrow} ES' + ROH$$

$$ES' + H_2O \overset{k_4}{\rightarrow} EH + CH_3CO\overset{-}{O} + \overset{+}{H}$$

where ES' = acetyl enzyme

The equation for the rate of product formation is of the Michaelis-Menten type and is in fact formally equivalent to 7-21.

The evidence for the presence of an acetyl enzyme intermediate is indirect. This mechanism explains the fact that acetylcholinesterase also catalyzes the reverse reaction, namely, the synthesis of acetylcholine from choline and acetate. It also correctly predicts that the enzyme will catalyze the isotopic exchange of oxygen between acetate and water. If correct it would offer a suggestive parallel with chymotrypsin.

The kinetic behavior of the acetylcholine-acetylcholinesterase system is of the simple Michaelis-Menten type only at low substrate concentration. At higher substrate concentrations a decline in rate appears which can be explained by postulating the formation of a disubstrate complex which is inert.

(7-50) $$E + S \rightleftarrows ES$$

$$ES + S \rightleftarrows ES_2 \text{ (inactive)}$$

The ES_2 complex might, for example, involve the binding of two molecules of acetylcholine to the two subsites of the binary active center. The resultant strain might interfere with the proper alignment of the bound substrate at the esteratic site and thereby inhibit hydrolysis.

Wilson has proposed a mechanism for the hydrolytic process, which is shown in Fig. 7-14. The process is believed to occur entirely at the

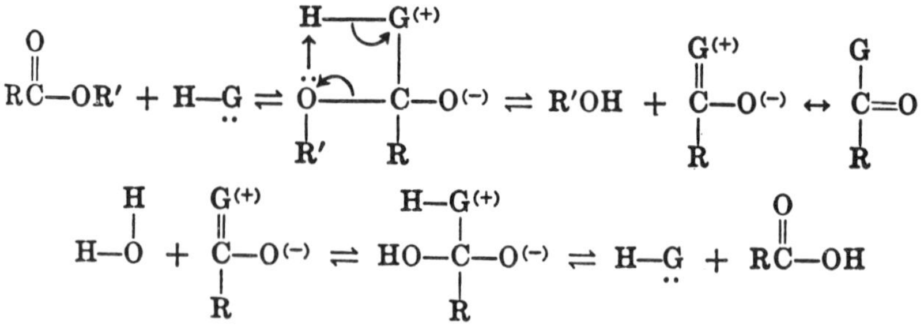

$$\text{RC}\overset{\text{O}}{\overset{\|}{-}}\text{OR}' + \text{H}-\text{G} \rightleftharpoons \overset{\text{H}-\text{G}^{(+)}}{\underset{\text{R}'}{\overset{\uparrow}{\text{O}}\cdots\overset{\downarrow}{\text{C}}-\text{O}^{(-)}}} \rightleftharpoons \text{R}'\text{OH} + \overset{\text{G}^{(+)}}{\underset{\text{R}}{\overset{\|}{\text{C}}-\text{O}^{(-)}}} \leftrightarrow \overset{\text{G}}{\underset{\text{R}}{\text{C}=\text{O}}}$$

$$\overset{\text{H}}{\underset{}{\text{H}-\text{O}}} + \overset{\text{G}^{(+)}}{\underset{\text{R}}{\overset{\|}{\text{C}}-\text{O}^{(-)}}} \rightleftharpoons \overset{\text{H}-\text{G}^{(+)}}{\underset{\text{R}}{\text{HO}-\overset{|}{\text{C}}-\text{O}^{(-)}}} \rightleftharpoons \text{H}-\text{G} + \text{RC}\overset{\text{O}}{\overset{\|}{-}}\text{OH}$$

Fig. 7-14. Proposed mechanism for hydrolysis of substrate by acetylcholinesterase (11). "G" represents the enzyme. The arrows stand for electron shifts.

esteratic site. It is rather analogous to that proposed for chymotrypsin in that the formation of an acetyl enzyme intermediate is postulated.

7-6 PANCREATIC RIBONUCLEASE

General remarks. The ribonucleases are a family of enzymes which catalyze the hydrolysis of particular phosphodiester linkages in ribonucleic acid. The discussion here will be confined to *pancreatic* ribonuclease, for which there is available extensive structural information.

The specificity and biological role of ribonuclease will be discussed in Chapters 8 and 10. Our interest for the present will be centered about the interrelationship of over-all structure and activity, for which there is more information than in the case of any other enzyme.

The molecular weight of ribonuclease, as determined from both sedimentation-diffusion measurements and from amino acid analysis, is 13,700. The hydrodynamic properties are those of a globular protein of low asymmetry. Thus the intrinsic viscosity and the frictional ratio do not exceed the values predicted for a spherical molecule of the same molecular weight by an amount greater than could be easily accounted for by hydration (Chapter 4). The best model for the over-all shape of ribonuclease does not deviate greatly from spherical symmetry.

Pancreatic ribonuclease hydrolyses phosphodiester bonds between two pyrimidine nucleotides or between a pyrimidine and a purine nucleotide (Chapters 8 and 10 and Fig. 10-3). A discussion of its specificity will be postponed to Chapter 10.

Primary structure and enzymatic activity. Ribonuclease consists of a *single* polypeptide chain which is internally cross-linked by four cystine residues. This follows from the failure of any drop in molecular weight to accompany the oxidation of all four cystines to cysteic acid and from the occurrence of only a single NH$_2$-terminal group, lysine, and only one COOH-terminal group, valine (12).

The determination of the primary structure of ribonuclease has been accomplished by methods basically similar to those which earlier were successful in the case of insulin (Chapter 3). After oxidation of all disulfide bridges, ribonuclease may be cleaved into a series of linear peptides by the action of trypsin, or some other enzyme of restricted specificity. The small fragments may be separated chromatographically and their individual composition determined by acid hydrolysis, as was described in Chapter 3.

The action of a second enzyme of different specificity produces a new set of peptides which overlap the first set. Consideration of the overlaps permits the arrangement of the fragments in a linear order.

By combining this kind of approach with direct sequence determination upon the small fragments produced by controlled acid hydrolysis, it has been possible to piece together the amino acid sequence of ribonuclease. The disulfide bridges have been located by hydrolyzing the unoxidized protein enzymatically to produce cystine—containing peptides. These are isolated, and split by performic acid oxidation into two cysteic acid peptides each. These can be located within the amino acid sequence (Fig. 3-2).

In this manner a provisional primary sequence has been determined for pancreatic ribonuclease (Fig. 7-15). While it is possible that minor alterations may be required in this structure, its basic correctness seems assured.

Ribonuclease may be subjected to considerable hydrolytic scisson without loss of activity. Thus two to five peptide bonds may be split by trypsin without inactivation. Presumably the disulfide bridges, together with the tertiary structure, are sufficient to maintain the elements of the active center in proximity, so as to permit activity. However, the removal by pepsin of the four COOH-terminal residues inactivates the enzyme.

More interestingly, Richards has shown that the cleavage of the Ser-Ala bond between residues 20 and 21 (numbering from the NH$_2$-terminal end) by the hydrolytic enzyme *subtilisin* results in no inactivation if carried out at low temperatures (13).

The NH$_2$-terminal "tail," which contains no cystine, may then be separated from the rest of the molecule by chemical fractionation. The two separated fragments are inactive. *However, if they are recombined, activity is regained.*

Two important conclusions follow from this result.

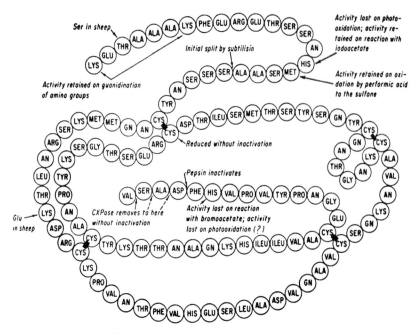

Fig. 7-15. Provisional primary structure of ribonuclease (12).

(1) The NH$_2$-terminal peptide, which contains 20 amino acids, either contains an essential element of the active site or has an essential function in maintaining the native configuration of ribonuclease (or both).

(2) The NH$_2$-terminal "tail" is integrated into the tertiary structure of ribonuclease, from which it may be detached, *reversibly*.

There is some evidence that histidine is essential for ribonuclease activity. Thus the photo-oxidation of two histidines inactivates the enzyme.

Configuration and activity. Application of optical rotatory dispersion and other criteria (Chapter 5) have led to estimates of 15 to 40% for the α-helical content of ribonuclease. In the presence of 8 M urea, a well-known reagent for breaking secondary bonds, important increases in intrinsic viscosity and specific rotation occur, suggesting that a major part of the native configuration of ribonuclease is lost under these conditions.

Nevertheless, ribonuclease retains its full activity in this solvent. Apparently, either much of the secondary and tertiary structure is unnecessary for activity, or alternatively the substrate may serve somehow to regenerate the native configuration.

Elimination of the cystine bridges by oxidation or reduction results in major changes in the molecular configuration of ribonuclease. The intrinsic viscosity rises by a factor of three, and the optical rotatory properties revert to those of an unorganized structure.

It thus appears that the cystine cross-links are essential for the stability of the α-helical regions. This stabilization is probably indirect, reflecting the shielding of the short α-helices from the solvent by the intact tertiary structure, to which the disulfide bridges are essential. The loss of the disulfide-stabilized secondary and tertiary structure permits a considerable expansion of the originally compact molecule. The general properties of the reduced, or oxidized, form of ribonuclease resemble in many respects those of an unorganized polypeptide.

Loss of the cystine cross-links produces total inactivation of the enzyme. In the case of reduced ribonuclease, activity may be regained by reoxidation through exposure to atmospheric oxygen. The reformation of —S—S— bridges is accompanied by recovery of the physical, as well as the enzymatic, properties of the native molecule.

This result provides a most dramatic example of the resiliency of the molecular configuration of this enzyme. Apparently, certain features of the primary structure of ribonuclease endow it with sufficient "memory" to permit reformation of the secondary and tertiary structure after a drastic disorganization.

GENERAL REFERENCES

1. *Enzymes*, M. Dixon and E. Webb, Academic Press, New York (1958).

SPECIFIC REFERENCES

Enzyme kinetics

2. "Enzyme Kinetics," J. Heason, S. Bernhard, S. Friess, D. Botts, and M. Morales in *The Enzymes*, P. Boyer, H. Lardy, and K. Myrback (eds.), vol. 1, p. 49, Academic Press, New York (1960).

Trypsin

3. "Trypsin," P. Desnuelle in *The Enzymes*, P. Boyer, H. Lardy, and K. Myrback (eds.), vol. 4, p. 119, Academic Press, New York (1960).
4. H. Gutfreund, *Trans. Faraday Soc.* **51,** 441 (1955).
5. H. Neurath and G. Schwert, *Chem. Rev.* **46,** 69 (1950).
6. N. Green and H. Neurath in *The Proteins*, H. Neurath and K. Bailey (eds.), vol. IIb, p. 1057, Academic Press, New York (1954).

Chymotrypsin

7. P. Desnuelle in *The Enzymes*, P. Boyer, H. Lardy, and K. Myrback (eds.), vol. 4, p. 93, Academic Press, New York (1960).
8. "The Mechanism of Action of Chymotrypsin," J. Sturtevant in *Protein Structure and Function*, Brookhaven Symposia in Biology, no. 13, p. 151 (1960).
9. F. Bettelheim and H. Neurath, *J. Biol. Chem.* **212,** 241 (1955).
10. M. Rovery, M. Poilroux, A. Curnier, and P. Desnuelle, *Biochim. et Biophys. Acta,* **565** (1955).

Acetylcholinesterase

11. I. Wilson in *The Enzymes*, P. Boyer, H. Lardy, and K. Myrback (eds.), vol. 4, p. 501, Academic Press, New York (1960).

Ribonuclease

12. C. Anfinsen and F. White in *The Enzymes*, P. Boyer, H. Lardy, and K. Myrback (eds.), vol. 5, p. 95, Academic Press, New York (1961).
13. F. Richards and P. Vithayathel in *Protein Structure and Function*, Brookhaven Symposia in Biology, no. 13, p. 115 (1960).

8

The Nucleotides

8-1 GENERAL REMARKS

The nucleotides are the basic chemical subunits of the nucleic acids, the structural relationship of the two being analogous to that of amino acids and proteins (1, 2, 3). The nucleotides found in nucleic acids are much fewer in number than the α-amino acids. Only four are of common occurrence for each of the two types of nucleic acid.

Each nucleotide consists of three elements. These are (1) a heterocyclic ring containing nitrogen and referred to as the *base;* (2) a five-carbon sugar, or *pentose;* and (3) a phosphate group.

The bases of common occurrence in nucleic acids are five in number. They are of two kinds—*purines* and *pyrimidines* (Fig. 8-1). The pyrimidines are 6-membered rings, while the purines consist of fused 5- and 6-membered rings. It is customary to designate the ring positions by the numbers 1 through 9 for the purines and 1 through 6 for the pyrimidines, as is indicated in Fig. 8-1.

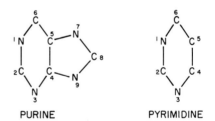

Fig. 8-1. Ring structure of purines and pyrimidines. The numbers indicate the conventional system of designating ring positions.

The chemical individuality of the bases is conferred by the nature of the substituents in the C_6, N_1, and C_2 positions (Fig. 8-2). These substituents are either amino or keto groups (Fig. 8-2).

The two principal purine bases are adenine and guanine. The pyrimidines are uracil, cytosine, and thymine, which differs from uracil only in having a methyl substituent in the C_5 position (Fig. 8-2).

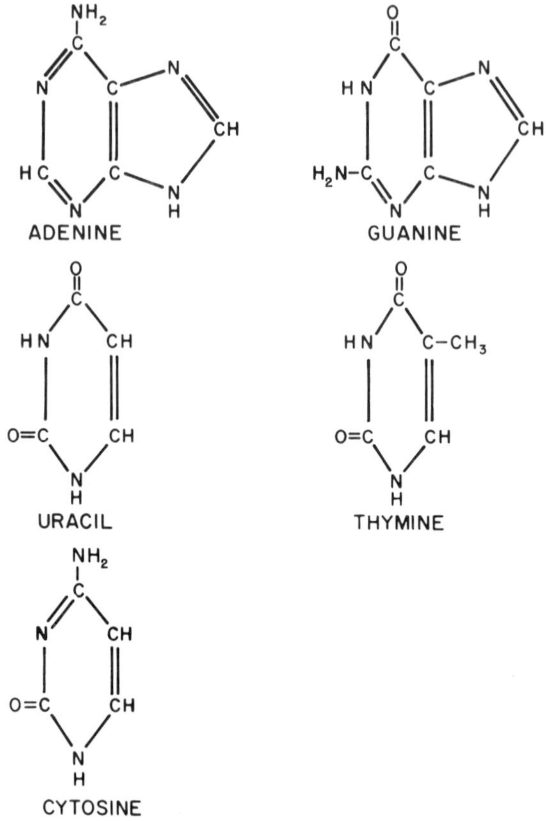

Fig. 8-2. Structures of the free bases.

There are two types of nucleic acid—ribonucleic acid (RNA) and deoxyribonucleic acid (DNA). Corresponding to these are the two nucleotide series, which differ in the nature of their pentose sugar. In the *ribonucleotide* series the sugar is *ribose* (Figs. 8-3 and 8-4). In the *deoxyribonucleotide* series the sugar is *deoxyribose*, which differs from ribose in that the $C_{2'}$ hydroxyl of the latter is replaced by a hydrogen (Fig. 8-5). (It is customary to designate positions in the pentose portion of nucleotides as 1' through 5' [Figs. 8-3, 8-4, and 8-5].) The bases commonly occurring in RNA are adenine, uracil, cytosine, and guanine. In DNA, uracil is replaced by thymine.

The bases are joined to the pentoses by N—C *glycosidic* bonds (Fig. 8-3). For the purines, the glycosidic bond is between the $C_{1'}$ position of the pentose and the N_9 position of the base. For the pyrimidines, the linkage joins the $C_{1'}$ and N_3 positions.

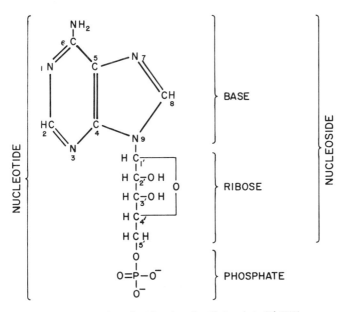

Fig. 8-3. The ribonucleotide adenosine-5'-phosphate (5'-AMP).

The nucleotides are esters of phosphoric acid (H_3PO_4) with one (or more) of the hydroxyls of the pentose. In the ribonucleotide series there are three positions available ($C_{2'}$, $C_{3'}$, and $C_{5'}$), and in the deoxyribonucleotide series, two ($C_{3'}$ and $C_{5'}$). Thus a number of isomeric nucleotides are possible, and examples of most of these have been identified. Systems containing only the base plus the pentose without a phosphate, are referred to as *nucleosides*.

Phosphoric acid is trifunctional and can form up to three ester bonds. While triply esterified phosphate does not appear to occur in nucleic acids, doubly esterified phosphate does occur and in fact provides the mode of linkage of the nucleotides to form the polymeric nucleic acids. The bond formed by a doubly esterified phosphate between

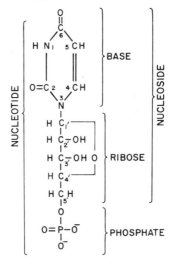

Fig. 8-4. The ribonucleotide uridine-5'-phosphate (5'-UMP).

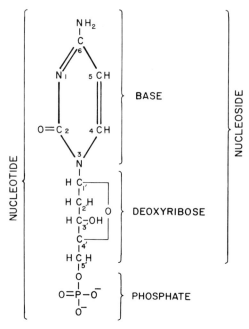

Fig. 8-5. The deoxyribonucleotide deoxycytidine-5'-phosphate (5'-dCMP).

the sugars of two different nucleotides is called the *phosphodiester* bond (Fig. 8-6).

The names and abbreviations of the principal nucleosides and nucleotides are cited in Table 8-1.

Nucleotides involving a double cyclic ester of a single phosphate are also known. The nucleotide in which the 2' and 3' hydroxyls of adenosine are esterified by a single phosphate is referred to as adenosine-2':3'-phosphate, abbreviated as 2':3'-AMP.

Finally, esters of pyrophosphoric acid are designated as adenosine-5'-diphosphate (5'-ADP or ppA), and so on. The corresponding triphosphate derivatives are referred to as adenosine-5'-triphosphate (5'-ATP, pppA), and so forth.

8-2 THE NUCLEOTIDE BASES

The purines. It is possible to write alternative tautomeric structures for both adenine and guanine (4, 5). Thus adenine could conceivably exist wholly, or in part, as an imino (=NH) derivative (Fig. 8-7). Similarly, guanine could, in principle, exist in an imino or an enol form (Fig. 8-7).

In actuality, neither possibility appears to be realized. X-ray diffraction studies have indicated that the C_6—O bond length (1.20 Å) of guanine is considerably shortened from the normal value (1.37 Å) for *single* bonds of this type. This is characteristic of, and provides strong evidence for, a predominantly double bond character for this linkage, as would be the case for a keto (C=O) group.

With respect to adenine, strong evidence for the existence of the external nitrogen as a primary amino group (—NH_2) has been obtained by the use of infra-red spectroscopy. The absorption of light in the infra-red region (> 7000 Å or 0.7μ) provides a sensitive and selective means of examining the state of particular bonds or groups. The different arrangement of double bonds for the amino and imino forms of adenine should lead to very different infra-red absorption spectra.

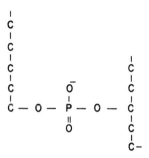

Fig. 8-6. The phosphodiester linkage.

The C_6-dimethylamino derivative of adenosine cannot tautomerize, for it lacks an available hydrogen. It thus should be analogous in structure to the amino form of adenosine and should have a similar absorption spectrum. This derivative has been synthesized and shown to have an infra-red spectrum very similar to that of adenosine itself, except for such changes as would be expected to accompany the replacement of the two amino hydrogens by methyl groups. This indicates that the ring structure and the nature of the C_6—N bond are the same for both compounds and hence that adenine is a C_6-amino purine.

The evidence in the case of the external nitrogen at the C_2 position of guanine is less conclusive. However, parallel studies upon other purine derivatives leave little doubt that it too is in the amino form and that guanine is a C_6-keto, C_2-amino purine (Figs. 8-2 and 8-7).

X-ray diffraction studies upon guanine and adenine have shown that both bases have a planar configuration within experimental error. The bond distances and angles of these two bases are shown in Figs. 8-8 and 8-9.

The question of the predominant tautomeric form of the bases is of considerable importance with regard to the hydrogen bonding which stabilizes the secondary structure of the nucleic acids (Chapters 9 and 10). The amino, but not the imino, structure of adenine allows the N_1 nitrogen to act as a hydrogen bond acceptor. The keto, amino form of

TABLE 8-1. NAMES AND ABBREVIATIONS OF THE PRINCIPAL NUCLEOTIDES AND NUCLEOSIDES

Name	Abbreviation
Adenine:	—
adenosine	A
2'-adenylic acid / adenosine-2'-phosphate / 2'-adenosine monophosphate	2'-AMP
3'-adenylic acid / adenosine-3'-phosphate / 3'-adenosine monophosphate	3'-AMP; Ap
5'-adenylic acid / adenosine-5'-phosphate / 5'-adenosine monophosphate	5'-AMP; pA
adenosine-5'-diphosphate	5'-ADP; ppA
adenosine-5'-triphosphate	5'-ATP; pppA
Uracil:	
uridine	U
2'-uridylic acid / uridine-2'-phosphate / 2'-uridine monophosphate	2'-UMP
3'-uridylic acid, etc.	3'-UMP; Up
5'-uridylic acid, etc.	5'-UMP; pU
uridine-5'-diphosphate	5'-UDP; ppU
uridine-5'-triphosphate	5'-UTP; pppU
Cytosine:	
cytidine	C
2'-cytidylic acid / cytidine-2'-phosphate / 2'-cytidine monophosphate	2'-CMP
3'-cytidylic acid, etc.	3'-CMP; Cp
5'-cytidylic acid, etc.	5'-CMP; pC
cytidine-5'-diphosphate	5'-CDP; ppC
cytidine-5'-triphosphate	5'-CTP; pppC
Guanine:	
guanosine	G
2'-guanylic acid / guanosine-2'-phosphate / 2'-guanosine-monophosphate	2'-GMP
3'-guanylic acid, etc.	3'-GMP; Gp
5'-guanylic acid, etc.	5'-GMP; pG
guanosine-5'-diphosphate	5'-GDP; ppG
guanosine-5'-triphosphate	5'-GTP; pppG

THE NUCLEOTIDES

TABLE 8-1. (Continued)

Name	Abbreviation
Deoxyribonucleotide series:	
deoxyadenosine	dA
3'-deoxyadenylic acid ⎫	
deoxyadenosine-3'-phosphate ⎬	3'-dAMP
3'-deoxyadenosine monophosphate ⎭	
5'-deoxyadenylic acid, etc.	5'-dAMP
deoxyadenosine-5'-diphosphate	5'-dADP
deoxyadenosine-5'-triphosphate	5'-dATP
deoxycytidine	dC
3'-deoxycytidylic acid, etc.	3'-dCMP
5'-deoxycytidylic acid, etc.	5'-dCMP
deoxycytidine-5'-diphosphate	5'-dCDP
deoxycytidine-5'-triphosphate	5'-dCTP
thymidine	T
3'-thymidylic acid, etc.	3'-TMP
5'-thymidylic acid, etc.	5'-TMP
thymidine-5'-diphosphate	5'-TDP
thymidine-5'-triphosphate	5'-TTP
deoxyguanosine	dG
3'-deoxyguanylic acid, etc.	3'-dGMP
5'-deoxyguanylic acid, etc.	5'-dGMP
deoxyguanosine-5'-diphosphate	5'-dGDP
deoxyguanosine-5'-triphosphate	5'-dGTP

guanine possesses a hydrogen bond donor in its N_1 nitrogen and an acceptor in its C_6 carbonyl. Both would be abolished if the external oxygen at the C_6 position were in the hydroxyl form.

The pyrimidines. As in the case of the purine bases, the hypothetical possibility of alternative tautomeric forms exists for cytosine and uracil (5). Thus cytosine could, in principle, exist partially or wholly as an imino derivative (Fig. 8-10), while a hydroxyl form would be a possibility for uracil or thymine (Fig. 8-10).

As in the case of the purines, these possibilities have been ruled out by the use of infra-red spectroscopy. Thus the dimethylamino derivative of cytidine has been prepared (Fig. 8-10) and shown to have an infrared spectrum similar to that of cytidine. This derivative cannot tauto-

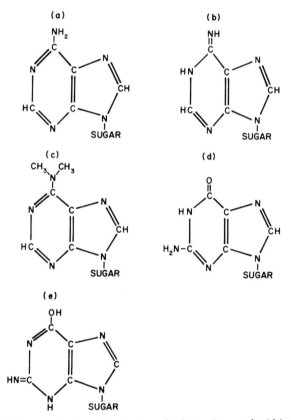

Fig. 8-7. Part (a) represents the amino structure of adenine (in a nucleoside); (b) the alternative imino structure of adenine (in a nucleoside); (c) the C_6-dimethylaminoadenosine; (d) the keto, amino form of guanine (in a nucleoside); and (e) the alternative hydroxyl, imino form of guanine.

merize, since it has no available hydrogen, so its structure should be analogous to that of the amino form of cytidine. In particular, the conjugation of the C_2 oxygen ($C_6=N_1-C_2=O$) should be retained, while the imino form would leave the C_2 oxygen unconjugated ($C_6-N_1-C_2=O$). Thus major differences in absorption spectra in the carbonyl (6μ) region would be expected for the two cases.

The ethyl derivative of the hypothetical C_6-enol form of uridine has been synthesized and shown to have an infra-red spectrum in the carbonyl region very different from that of uridine itself. Because the ring structure of the former would be equivalent to that of the hypothetical C_6-enol form of uridine, this provides evidence counter to the presence of the enol species (Fig. 8-10). In contrast, the C_1-methyl derivative of uridine has been shown to have an infra-red spectrum similar to that of uridine (Fig. 8-10). Since the ring structure of this derivative should

THE NUCLEOTIDES 255

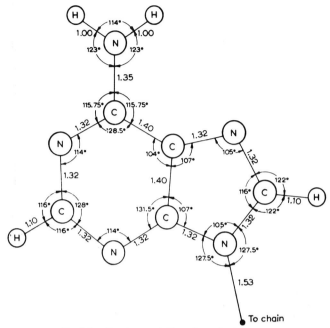

Fig. 8-8. Structure and dimensions of adenine.

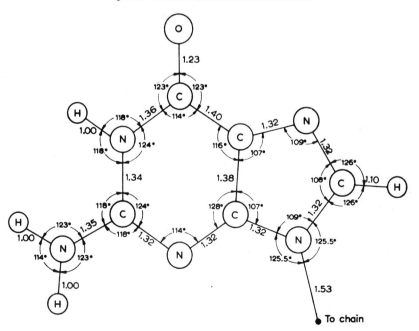

Fig. 8-9. Structure and dimensions of guanine.

be equivalent to that of the diketo form of uracil, it is very probable that this is the prevalent form.

In summary, the prevailing evidence indicates that cytosine may be assigned an amino, keto structure and that uracil and thymine are diketo derivatives.

X-ray diffraction studies have indicated that the pyrimidine rings have a planar configuration (Figs. 8-11 and 8-12).

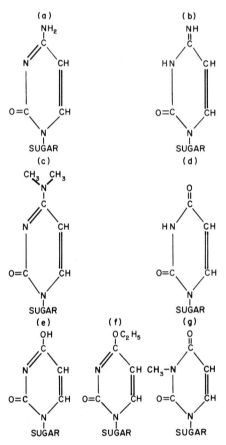

Fig. 8-10. Part (a) represents the amino form of cytosine (in a nucleoside); (b) the alternative imino form of cytosine; (c) the C_6-dimethylamino cytosine; (d) the diketo form of uracil (in a nucleoside); (e) the alternative hydroxyl from of uracil; (f) the ethyl derivative of hydroxyl form of uracil; and (g) the N_1-methyl uracil.

The less common bases. Bases other than the five discussed earlier are occasionally found in nucleic acids. Indeed, in one case, that of the DNA of several bacterial viruses, cytosine is replaced entirely by 5-hydroxymethyl cytosine. This differs from cytosine in that the C_5 hydrogen is replaced by a hydroxymethyl ($-CH_2OH$) group.

Other purine and pyrimidine bases found in small quantities in nucleic acids include 5-methyl cytosine, 6-N-methyl adenine, 6-N-dimethyl adenine, 2-N-methyl guanine, 1-methyl guanine, and 2-methyl adenine. By growing bacteria in media containing derivatives of other "unnatural" bases, a variety of such bases have been incorporated into their nucleic acids. These include 5-fluorouracil, 5-bromouracil, 5-chlorouracil, and thiouracil.

8-3 THE NUCLEOSIDES

The pentoses. The pentose sugar of the ribonucleosides has been identified as D-ribose (Fig. 8-13). Upon stepwise oxidation, it is converted first to optically active D-ribonic acid and then to optically inactive trihydroxyglutaric acid (Fig. 8-13).

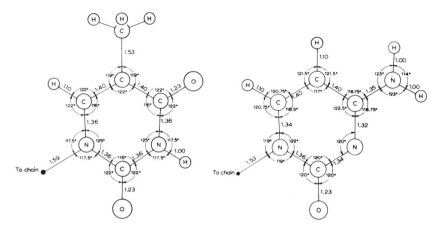

Fig. 8-11. Structure and dimensions of thymine.

Fig. 8-12. Structure and dimensions of cytosine.

In the free form, both ribose and deoxyribose behave as aldehyde derivatives. However, both sugars occur in nucleic acids and nucleosides as derivatives of alternative tautomeric forms in which the $C_{1'}$ carbonyl is in the hydroxyl state and a five-membered ring is formed between the $C_{1'}$ and $C_{4'}$ positions (Fig. 8-14). These forms are referred to as *ribofuranose* and *deoxyribofuranose* (6).

In ribofuranose derivatives the $C_{1'}$ carbon becomes asymmetric, so that there are two possible configurations. These are called the α- and β-con-

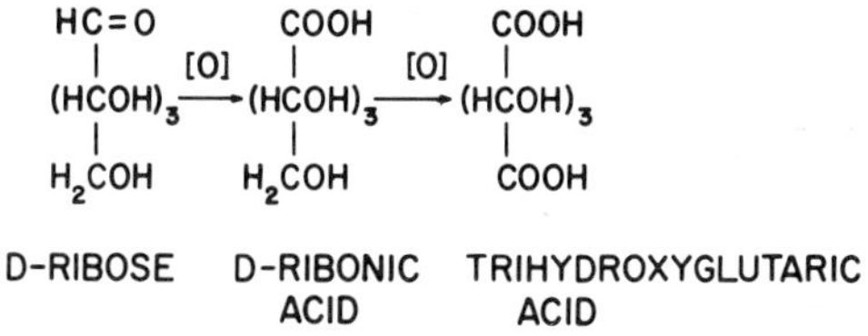

$$\begin{array}{ccc} \text{HC=O} & \text{COOH} & \text{COOH} \\ | & | & | \\ \text{(HCOH)}_3 \xrightarrow{[O]} & \text{(HCOH)}_3 \xrightarrow{[O]} & \text{(HCOH)}_3 \\ | & | & | \\ \text{H}_2\text{COH} & \text{H}_2\text{COH} & \text{COOH} \end{array}$$

D-RIBOSE   D-RIBONIC   TRIHYDROXYGLUTARIC
             ACID   ACID

Fig. 8-13. D-ribose and its oxidation products. D-ribose is shown in the aldehyde form.

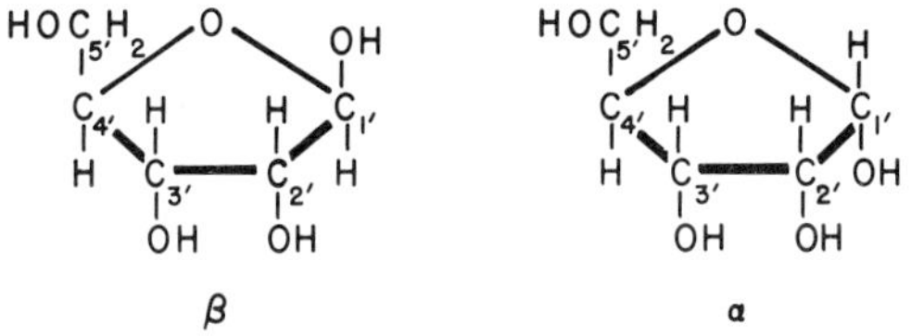

Fig. 8-14. Stereochemical formulas for the α- and β-forms of D-ribofuranose, showing the configuration about the $C_{1'}$ carbon.

figurations (Fig. 8-15). The nucleosides are formed by the elimination of water between the $C_{1'}$ hydroxyl of the ribofuranose form of ribose and an N—H group in the ring of the base. The resultant N—C bond is called a *glycosidic* linkage.

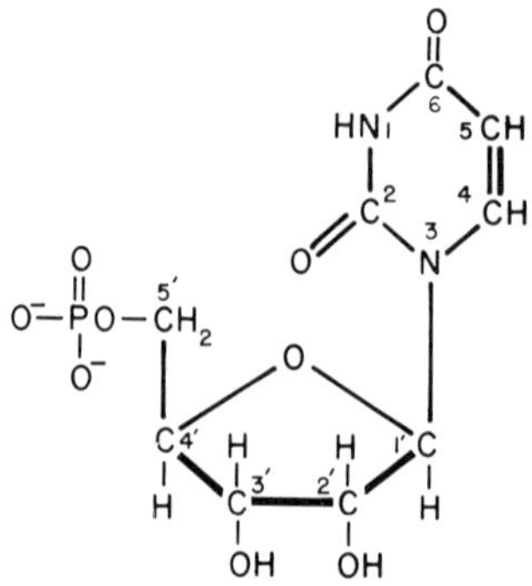

Fig. 8-15. Stereochemical formula for uridine-5'-phosphate, showing the configuration about the carbon atoms of the ribose.

Positions of the glycosidic linkage. The purine nucleosides are easily hydrolyzed by acid to free base and sugar. Because C—C bonds are normally resistant to hydrolysis, this was indicative that the base-sugar linkage was of the N—C type. By a comparison of the properties of adenosine and guanosine with those of synthetic analogs of known structure, it has been shown that the glycosidic linkage is to the N_9 position of the base (Fig. 8-3).

Similar studies for the pyrimidine nucleosides have placed the glycosidic linkage at the N_3 position of the pyrimidine base (Fig. 8-4).

Configuration of the glycosidic linkage. Because of the existence of spatial isomers of the ribofuranose (or deoxyribofuranose) ring, a question arises as to the configuration of the glycosidic linkage of the nucleosides; that is, whether they are derivatives of the α- or β-form of the pentose. A complicated series of conversions has shown that for all nucleosides of both the ribo- and deoxyribo-series the glycosidic linkage is in the β configuration. Figure 8-15 shows a "stereochemical" formula for uridine-5'-phosphate.

Interconversions of the nucleosides. Adenosine may be deaminated by treatment with nitrous acid to produce *inosine*, whose base, *hypoxanthine*, is a C_6-keto purine derivative (Fig. 8-16). While inosine is not found in natural nucleic acids, it is known to occur in biological systems.

Similarly, guanosine may be deaminated by nitrous acid to yield *xanthosine*, whose base, *xanthine*, is a diketo purine (Fig. 8-16).

The action of nitrous acid converts cytidine to uridine (Fig. 8-16). This reaction provides confirmatory evidence for the accepted structures of these compounds.

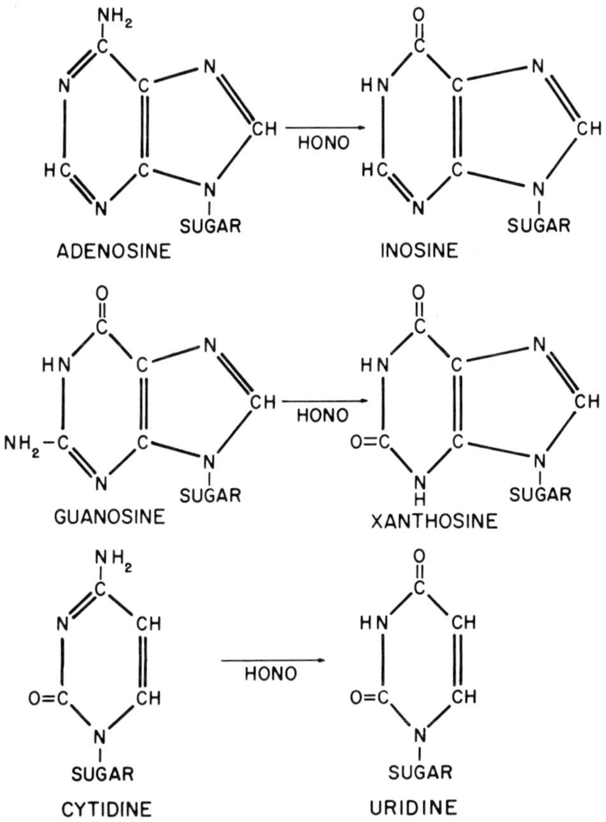

Fig. 8-16. Deamination of the bases. In each case the product is shown in the keto form, although the reaction may proceed by way of a transient hydroxyl intermediate.

8-4 THE NUCLEOTIDES

The nucleoside monophosphates. The ribonucleotide series has three possible positions for the phosphate ester. These are the ribose 2', 3', and 5' positions. Nucleotides of all three types have been isolated and identified (7).

Exhaustive hydrolysis of RNA by the enzyme *spleen phosphodiesterase* yields the 3'-nucleotides exclusively. However, alkaline hydrolysis results in a mixture of the 2' and 3' isomers. This has been shown to be a consequence of the mechanism of alkaline hydrolysis, which proceeds by way of a 2':3' cyclic phosphate (Fig. 8-17). Further hydrolysis of this intermediate results in the observed mixture of 2' and 3' monophosphates.

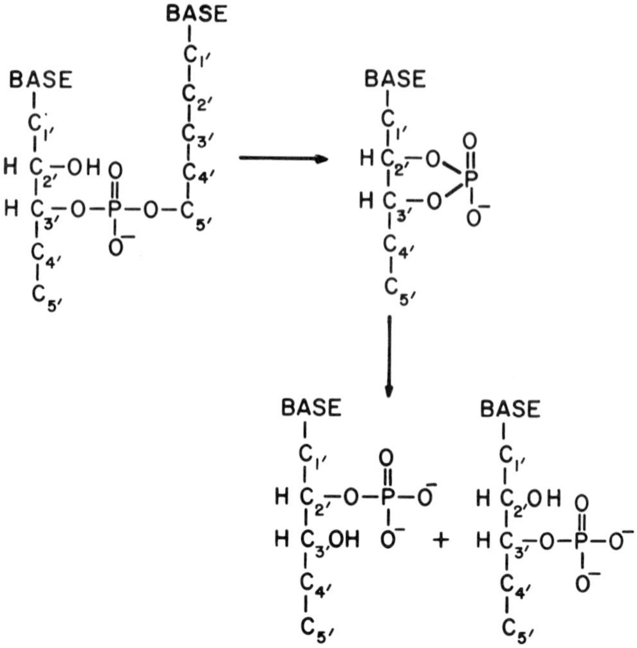

Fig. 8-17. Alkaline hydrolysis of the 3'-5' phosphodiester linkage of RNA.

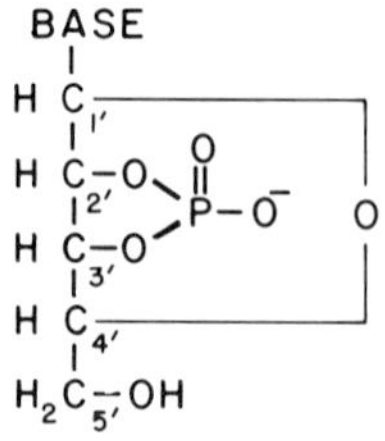

NUCLEOSIDE – 2':3'–PHOSPHATE

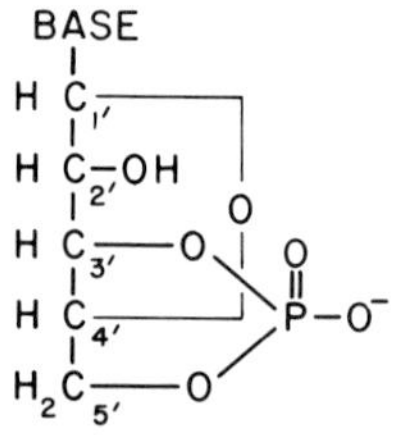

NUCLEOSIDE – 3':5'–PHOSPHATE

Fig. 8-18. Structures of two cyclic nucleotides.

The cyclic nucleotides have been prepared synthetically and are valuable as intermediates for the synthesis of nucleotide derivatives. Both 2':3' and 3':5' derivatives are known (Fig. 8-18).

In the deoxyribonucleotide series, only the 3' and 5' positions are available for esterification. The cyclic deoxyribonucleoside-3':5'-phosphates have been prepared synthetically.

The nucleoside di- and triphosphates. The naturally occurring diphosphates and triphosphates are esterified at the 5'-position of the pentose. Adenosine-5'-diphosphate (ADP) and adenosine-5'-triphosphate (ATP) are of ubiquitous occurrence in biological systems and have roles of central importance (Fig. 8-19).

The ribonucleoside-5'-diphosphates are the substrates for the enzyme

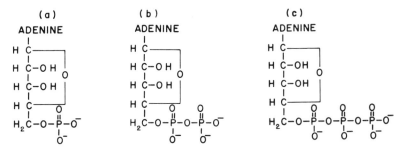

Fig. 8-19. The 5'-adenine nucleotides. Part (a) represents the adenosine-5'-phosphate (5'-AMP); (b) the adenosine-5'-diphosphate (5'-ADP); and (c) the adenosine-5'-triphosphate (5'-ATP).

polynucleotide phosphorylase, which catalyzes their polymerization to polyribonucleotides. The deoxyribonucleoside-5'-triphosphates are the substrates for the enzyme *DNA polymerase* which is responsible for the biosynthesis of DNA. The action of these enzymes will be discussed in Chapters 10 and 9, respectively.

8-5 PROPERTIES OF THE NUCLEOTIDES

Ionization. Each of the bases occurring in the ribo- and deoxyribonucleotides has one ionizable site, except for guanine, which has two. The bases of adenosine and cytidine are uncharged at neutral and alkaline pH, but bind a proton at acid pH to acquire a cationic charge. The bases of thymidine and uridine are uncharged at neutral and acid pH, but dissociate a hydrogen ion at alkaline pH to become negatively charged (Fig. 8-20). The base of guanosine is uncharged at neutral pH and both binds a proton at acid pH and dissociates one at alkaline pH.

Phosphoric acid is tribasic and ionizes in three steps of widely separated pK. These are:

(8-1) primary $H_3PO_4 \rightleftarrows H^+ + H_2PO_4^-$ (pK < 2)

secondary $H_2PO_4^- \rightleftarrows H^+ + HPO_4^=$ (pK $\cong$ 6)

tertiary $HPO_4^= \rightleftarrows H^+ + PO_4^{\equiv}$ (pK $\cong$ 9)

The mononucleotides are formed by the establishment of a phosphomonoester linkage between the pentose and phosphoric acid. As a con-

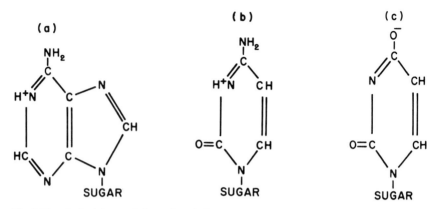

Fig. 8-20. Ionized forms of the nucleoside bases. Part (a) represents the acid form of adenine (in adenosine); (b) the acid form of cytosine (in cytidine); and (c) the alkaline form of uracil (in uridine).

sequence, the tertiary phosphate ionization is lost, leaving only the primary and secondary ionizations. The primary ionization occurs at so low a pH as to correspond for practical purposes to that of a strong acid. Thus the phosphate group of the nucleotides carries a single negative charge at acid pH's and a double negative charge at alkaline pH's (pH > 6).

$$(8\text{-}2) \qquad -\text{C}_{5'}-\text{O}-\overset{\overset{\text{O}}{\|}}{\underset{\underset{\text{O}^-}{|}}{\text{P}}}-\bar{\text{O}} + \text{H}^+ \rightleftarrows -\text{C}_{5'}-\text{O}-\overset{\overset{\text{O}}{\|}}{\underset{\underset{\text{O}^-}{|}}{\text{P}}}-\text{OH}$$

Thymidylic acid and uridylic acid are anionic over the entire pH range, while cytidylic acid and the purine nucleotides are anions at alkaline pH and zwitterions at acid pH.

There are a total of three dissociation constants for guanylic acid at pH's above 2 and two for each of the other nucleotides. The problem of analyzing the ionization of the nucleotides is simplified by the separation of the ionizable sites, which is much greater than in the case of the α-amino acids. The influence of the charged phosphate group upon the ionization of the base is relatively slight and, to a good approximation, they may be regarded as ionizing independently. The base pK is altered by only a few tenths of a pH unit upon going from the nucleoside to the nucleotide.

In this case the dissociation constant for a given site does not depend greatly upon the state of ionization of the other site. Since the pK's of the base and phosphate are separated by over 2 pH units in each case, there is little inaccuracy involved in associating each dissociation constant with the ionization of a particular site. The pK's of the ionizable sites of the nucleotides have been tabulated on this basis (Table 8-2).

TABLE 8-2. pK VALUES FOR RIBONUCLEOTIDES

Nucleotide (3')	Primary Phosphate	Secondary Phosphate	Base
AMP	0.9	6.0	3.7
UMP	1.0	5.9	9.4
CMP	0.8	6.0	4.2
GMP	0.7	6.0	2.4, 9.3

Combination of nucleotides to form a polynucleotide results in the elimination of the secondary phosphate ionizations except for that arising from the terminal phosphate, which is not involved in a phosphodiester linkage.

Loci of ionization. The sites of proton binding by adenine, cytosine, and guanine are by no means obvious and have required the use of X-ray diffraction for their identification. In the cases of adenine and cytosine the most probable positions for the bound protons are the N_1 nitrogens (Fig. 8-20). In the case of guanine the position is less certain.

The ionization of the pyrimidine bases results in the loss of the N_1 hydrogen. The negative charge is localized in either the C_2 or C_6 external oxygen (Fig. 8-20).

Ulraviolet absorbancy. The sugar and phosphate groups do not absorb light in the visible or near-ultraviolet wave-lengths (> 220 mμ). However, the presence of the bases endows the nucleotides with very intense absorption in the near ultraviolet. The position of maximum absorption is in the range 260-280 mμ, depending upon the base.

The ultraviolet absorption spectra of the nucleotides are shown in Fig. 8-21. The molar absorbancy is dependent upon the state of ionization of the base.

GENERAL REFERENCES

1. E. Chargaff and J. Davidson, *The Nucleic Acids*, Academic Press, New York (1955).
2. R. Steiner and R. Beers, *Polynucleotides*, Elsevier, Amsterdam (1961).
3. D. Jordan, *The Chemistry of Nucleic Acids*, Butterworth, London (1960).

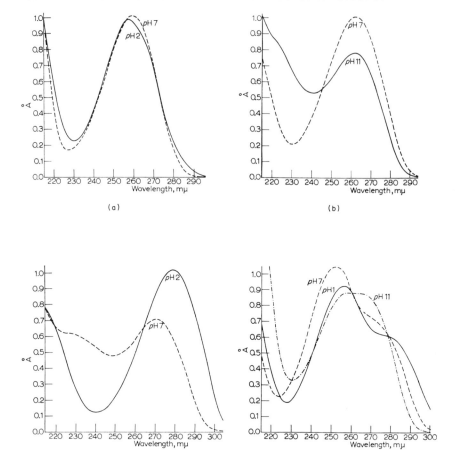

Fig. 8-21. The molar absorbancies, at several pH's, of the 5'-ribonucleotides. The values are almost identical for the corresponding 2'- and 3'-ribonucleotides and for the deoxyribonucleotides.
  (a) 5'-AMP. The moral absorbancy at 259 mμ (pH 7) is 15.5 × 10^3.
  (b) 5'-UMP. The molar absorbancy at 262 mμ (pH 7 is 10.0 × 10^3.
  (c) 5'-CMP. The molar absorbancy at 271 mμ (pH 7) is 9.0 × 10^3.
  (d) 5'-GMP. The molar absorbancy at 252 mμ (pH 7) is 13.7 × 10^3.

SPECIFIC REFERENCES

The purine and pyrimidine bases

4. A. Bendich in reference 1, vol. 1, p. 81.
5. Reference 2, chap. 2.

Ribose and deoxyribose

6. W. G. Overend and M. Stacey in reference 1, vol. 1, p. 9.

The nucleotides

7. J. Baddiley in reference 1, vol. 1, p. 137.

PROBLEMS

1. Why is the problem of determining nucleotide sequence in the ribonucleic acids more difficult than that of determining amino acid sequence in the proteins?

2. Compute a complete hydrogen ion titration curve for an adenylic acid, using the constants of Table 8-2.

3. Suppose that one is confronted with the problem of determining the composition of a mixture of 5'-AMP and 5'-UMP in unknown proportion. How might this be accomplished by measurement of ultraviolet absorption alone?

4. How might the molecular weight of a ribonucleic acid of low molecular weight ($<$ 10 nucleotide units) be determined from its hydrogen ion titration curve?

9

The Deoxyribonucleic Acids

9-1 GENERAL REMARKS

Deoxyribonucleic acid (DNA) is found in all biological systems, with the exception of certain viruses. In microorganisms of the simplest kind, such as the bacteriophages (Chapter 11), it often accounts for a major fraction of the organism, occurring as a central core encased in protein.

In higher multicellular organisms DNA is confined to the cell nuclei, where it is present as a complex with various basic proteins, including protamines and histones. The distribution of DNA is by no means uniform, being localized in the chromosomes (Chapter 1). The presence of DNA, which accounts for about half of the chromosomal mass, endows the chromosomes with distinctive staining properties, including a high affinity for basic dyes.

The evidence which led to general recognition of the central genetic role of the chromosomes has been briefly summarized in Chapter 1. A compelling and constantly expanding body of evidence indicates that DNA is the genetically active component of chromosomes and functions as the basic carrier of genetic information.

The earliest evidence leading to this concept was indirect. The cellular distribution of DNA shows a quantitative dependence upon the function of the cells. Ordinary somatic cells (Chapter 1), which are not directly concerned with reproduction, generally contain twice the amount of DNA present in the haploid sperm cells. This is just the distribution which would be anticipated for a genetic determinant, for the latter cells contain only single copies of genetic information instead of the two copies present in diploid somatic cells. Moreover, many chemical and physical agents known to alter DNA are mutagenic, suggesting that a modification in its structure is reflected by an alteration of its genetic message.

The discovery of transforming principle (section 9-9) and the studies upon the replication of the bacterial viruses (Chapter 11) strongly reinforced the central hypothesis and permitted its extension to organisms of the simplest kind. The discussion of the genetic role of DNA contained in this book will be largely centered about systems of this type, which are amenable to description at a molecular level.

The currently accepted picture of the biological function of the nucleic acids may be summarized as follows:

(1) DNA is the fundamental carrier of genetic information for all living systems (with the exception of certain viruses) and, in particular, functions as the ultimate determinant of the primary structure of proteins. Each classical gene may be identified with a sequence of nucleotides within a DNA molecule. The amino acid sequence of any natural polypeptide chain is uniquely related to the linear sequence of nucleotides within a DNA molecule. The genetic message, which is translated into the specific primary structure of a protein, is stored in the corresponding DNA as a specific order of nucleotides, much as a telegraph message in Morse code is specified by a particular sequence of dots and dashes.

(2) DNA does not directly guide the synthesis of proteins. The physiological function of the gene depends upon a process called *transcription*, where DNA molecules direct the synthesis of RNA, in which the original genetic message is preserved. The specific deoxyribonucleotide sequence of DNA is thereby transcribed into a specific ribonucleotide sequence of RNA, which serves as a secondary carrier of genetic information.

(3) The RNA transcript subsequently directs the assembly of amino acids into the protein molecule.

According to this model DNA can participate in two distinct chemical processes, which are catalyzed by different enzymes. In the first of these, called *replication*, the purely genetic function of DNA is performed. The parent DNA molecule serves to guide the polymerization of deoxyribonucleotides to form an identical copy or *replica* of its nucleotide sequence. The second process is the transcription already described.

The next three chapters will be primarily concerned with the justification and amplification of this basic scheme. However, a detailed account of these more biological aspects should be prefaced by an account of the chemical nature and molecular organization of DNA.

9-2 THE PRIMARY STRUCTURE OF DNA

Chemical nature. DNA is a polymer of deoxyribose nucleotides, to which it may be hydrolyzed quantitatively by certain enzymes. The only nucleotides normally present in greater than trace quantities are deoxyadenylic acid, deoxyguanylic acid, deoxycytidylic acid, and thymidylic acid (1, 2, 3, 4, 5).

Physical studies upon DNA solutions have indicated that the molecular weight is generally very high—10^6 or greater, depending upon the source. Hydrogen ion titration data upon carefully prepared DNA have failed to detect any significant *secondary* phosphate ionization. This indicates that singly esterified phosphate groups are not present, except presumably for a single group at the terminus of each polynucleotide chain. Because of the high molecular weight of DNA, such terminal phosphomonoester groups would be too few to be detected by hydrogen ion titration.

The absence of secondary phosphate ionization can only be explained by the involvement of at least two of the three functional groups of each phosphate in some form of chemical linkage. Since DNA can be hydrolyzed to nucleotides by enzymes whose specificity is restricted to the hydrolysis of phosphate esters, it follows that these linkages must be of the ester type and must be involved in the internucleotide linkages (4, 5).

The internucleotide linkages. Since deoxyribonucleoside monophosphates are the sole products of the exhaustive hydrolysis of DNA by enzymes specific for phosphate esters, the *only* internucleotide linkages present must be of the phosphodiester type. Since no 2'-hydroxyls are present, the only positions possible for phosphodiester bonds are between the 3' and 5' positions of adjacent deoxyriboses or alternating between 3'-3' and 5'-5' positions.

The problem was ultimately solved by examination of the products of exhaustive hydrolysis of DNA by the combined action of the enzymes *pancreatic deoxyribonuclease* and *snake venom phosphodiesterase*. These enzymes catalyze the hydrolysis of esters of nucleoside-5'-phosphates, which are cleaved so as to leave the phosphate in the 5'-position. Their combined action converts DNA quantitatively to a mixture of deoxyribonucleoside-5'-phosphates (Fig. 9-1), which can be separated and identified by chromatographic analysis. This evidence conclusively eliminated the possibility of alternating 3'-3' and 5'-5' bonds and established the primary structure of DNA as a linear chain of deoxyribonucleotides united by 3'-5' phosphodiester linkages (5).

The absence of secondary phosphate ionization indicates that branching is probably not present to an important degree. In DNA such branching could only occur by a triply esterified phosphate. Since one phosphate is present for each base, this would leave a singly esterified phosphate at each branch terminus, which should possess a secondary phosphate ioni-

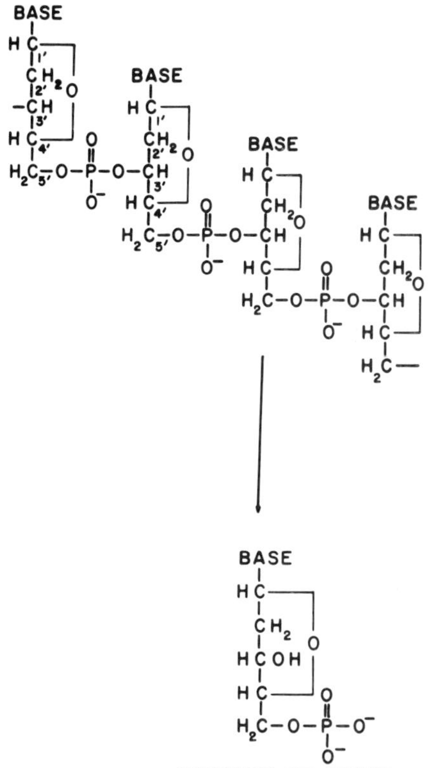

Fig. 9-1. Hydrolysis of the phosphodiester linkages of DNA by snake venom phosphodiesterase.

zation. Further evidence against the presence of branches has been obtained from electron microscopic observation, which has shown DNA to have a linear structure.

Base composition and sequence. The base compositions of DNA from a wide variety of biological sources exhibit a strikingly consistent pattern, to which there are almost no exceptions (6). This may be summarized as follows:

$$\text{adenine} = \text{thymine}$$
$$\text{guanine} = \text{cytosine}$$
$$\text{adenine} + \text{guanine} = \text{thymine} + \text{cytosine}$$
$$\text{adenine} + \text{cytosine} = \text{guanine} + \text{thymine}$$

In the DNA of several bacterial viruses (Chapter 11), cytosine is replaced by 5-hydroxymethyl cytosine.

The almost universal character of the above relationship suggests that it is directly related to the molecular organization of DNA. As will be discussed in the next section, this is indeed the case.

Species variations in the base contents of natural DNA's are correlated with parallel variations in several physical parameters. One of these is the buoyant density (7), which shows a systematic linear increase with increasing fraction of guanine plus cytosine (Fig. 9-2). The density

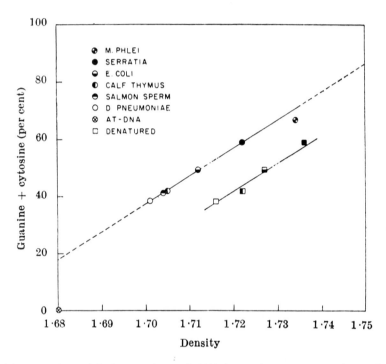

Fig. 9-2. Dependence of the buoyant density of DNA's from various sources upon the guanine + cytosine content (7).

differences are too small to be measured readily by the standard methods, but can easily be detected by the CsCl banding technique (section 9-7), which can also detect *heterogeneity* in density.

The DNA from relatively simple organisms, including the bacterial viruses and numerous species of bacteria, shows little evidence of any distribution of densities (7). This is consistent with the idea that DNA from these sources does not consist of molecules differing greatly in base composition, in harmony with the currently prevalent belief that DNA from most microorganisms consists of only a few, or a single, molecular species. However, rather surprisingly, DNA's from different bacterial sources often show a wide divergence of composition and density.

In contrast, DNA from mammalian sources, such as calf thymus, generally displays an important degree of heterogeneity of density, suggesting that it represents a mixture of molecules of different compositions. This conclusion is supported by the finding that calf thymus DNA may be fractionated into species of varying base content (6).

While accurate data are available for the over-all base compositions of DNA's from many sources, the task of nucleotide *sequence* determination has scarcely been started. Sequence determination for natural DNA is a much more formidable undertaking than the analogous polypeptide case, because of the very high molecular weights encountered and the ambiguities arising from the presence of only four different subunits.

In the case of the *biosynthetic* DNA produced by the action of the DNA-polymerase of *E. coli* (section 9-8), Kornberg and co-workers have succeeded in determining the relative frequency of occurrence of the various possible *nearest neighbor* dinucleotide sequences, for which there are a total of 16 possibilities. For example, in the sequence -TATCC-, the nearest neighbor dinucleotides are TA, AT, TC, and CC. The technique used was similar to that described in section 10-6 for the parallel case of biosynthetic RNA.

No very obvious pattern has emerged from such studies. However, the important generalization can be made that the frequencies of occurrence show wide deviations from those expected for a purely *random* nucleotide sequence. For example, since the adenine and thymine contents of DNA are equivalent, it would be expected that the pairs AA and AT would occur with equal frequency, if the nucleotide sequences were determined by purely statistical factors. This is not generally the case, discrepancies of up to a factor of two having been observed.

This non-randomness of base sequence is in accord with expectations aroused by the biological function of DNA. To be sure, the determinations cited above were made with *biosynthetic* DNA (section 9-8). However, the production of biosynthetic DNA by DNA-polymerase (section 9-8) requires the presence of a natural DNA primer, whose base sequence is believed to be preserved in the product. Thus the distribution of dinucleotides for the primer and the product should be equivalent.

Hydrolysis of DNA. DNA is relatively resistant to alkaline hydrolysis. This is a consequence of the absence of 2'-hydroxyl groups, which precludes hydrolysis of the phosphodiester linkages by way of the formation of the 2':3' cyclic ester (5), as occurs in the case of RNA.

The enzyme *pancreatic deoxyribonuclease* (often referred to as deoxyribonuclease I or simply deoxyribonuclease) converts DNA to a mixture of oligonucleotides terminating in a nucleoside-5′-phosphate. This enzyme and *snake venom phosphodiesterase* together reduce DNA to a mixture of 5′-nucleotides. DNA is rather refractory to the action of venom phosphodiesterase alone, probably because of the nature of its secondary structure.

Spleen deoxyribonuclease (deoxyribonuclease II) hydrolyzes DNA to oligonucleotides terminating in a nucleoside-3′-phosphate. Another phosphodiesterase has been isolated from the bacterial species *E. coli*. This likewise hydrolyzes DNA to nucleoside-3′-phosphates. However, its action is confined to *single-stranded* DNA (section 9-7). Ordinary double-stranded DNA (section 9-3) is not attacked.

9-3 THE SECONDARY STRUCTURE OF DNA

The Watson-Crick model. The currently accepted picture of the molecular organization of DNA is essentially that proposed in 1953 by Watson and Crick (8). This structural proposal, which was put forward well in advance of the detailed physical studies which subsequently confirmed and refined it, represents probably the most inspired single hypothesis in the history of molecular biology. Its dramatic success in rendering coherent an enormous body of information invites comparison with such advances as Bohr's theory of atomic spectra.

The basic features of the Watson-Crick model are as follows:

(1) The molecular configuration of DNA is highly ordered and helical. However, the helical form is entirely different from those discussed in Chapter 5 for polypeptides. The DNA molecule consists of *two* helically wound polynucleotide chains rather than a single chain (Fig. 9-3).

(2) In contrast to the polypeptide α-helix, the DNA helix is stabilized by hydrogen bonding between the bases of the two *different* strands. Two types of hydrogen-bonded base pairs are present: adenine-thymine and guanine-cytosine (Fig. 9-4). In the former case two hydrogen bonds are present. These are between the C_6 amino group of adenine and the C_6 carbonyl of thymine and between the N_1 nitrogen of adenine and the N_1 nitrogen of thymine (Fig. 9-4). In the guanine-cytosine pair three hydrogen bonds are present, which link the C_6 amino of cytosine and the C_6 carbonyl of guanine, the two N_1 nitrogens, and the C_2 amino of guanine and the C_2 carbonyl of cytosine (Fig. 9-4).

(3) The bases form the core of the double helix, with the sugar-phosphate backbones on the periphery. The two (planar) bases of each pair

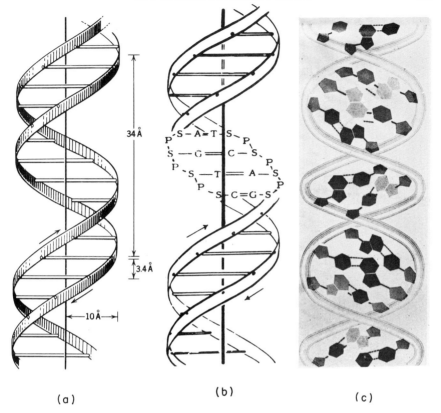

Fig. 9-3. (a) General form and dimensions of the doubly stranded helical structure of DNA. (b) Schematic representation of DNA. S, P, A, T, G, and C = sugar, phosphorus, adenine, thymine, guanine, and cytosine, respectively. (c) Location of the bases in the double helix.

lie in the same plane and are roughly perpendicular to the axis of the helix. The consecutive base pairs are thus mutually parallel.

This structure provides a logical explanation for the characteristic features of the base compositions of natural DNA's and circumvents many steric difficulties encountered by structural proposals based on models other than the helical duplex.

Thus the equivalence of adenine to thymine and of guanine to cytosine has an obvious explanation in terms of this secondary structure. In view of this equivalence, the helical content of DNA would be expected to be very high, as is indeed the case.

Since each base pair consists of a purine plus a pyrimidine, their dimensions are very similar. In this manner steric problems arising from the unequal sizes of the purines and pyrimidines are avoided, as is the

274   THE CHEMICAL FOUNDATIONS OF MOLECULAR BIOLOGY

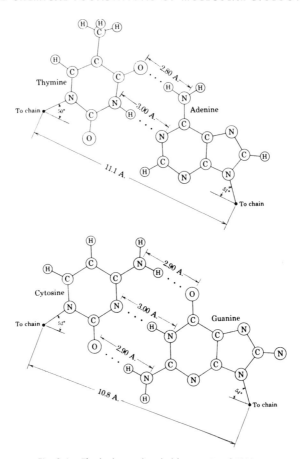

Fig. 9-4. The hydrogen-bonded base pairs of DNA.

need for any periodicity of base sequence along the individual polynucleotide strands.

An important feature of the Watson-Crick model is that the nucleotide sequences of the two strands are *complementary*. Thus the partner of each strand has the sequence and composition predicted for the replacement of each base of the first strand by the other member of its hydrogen-bond pair. This has an important bearing upon possible mechanisms for the biological replication of DNA, as will be discussed in Chapter 11.

In principle, such a doubly stranded structure could tolerate the existence of occasional breaks in either of the strands, provided that two breaks do not occur in equivalent positions in the two. However, this possibility does not appear to be realized in native DNA, and there is no evidence for any significant interruption of the chains.

It will be recalled that three hydrogen bonds stabilize the guanine-cytosine pair while only two join the adenine-thymine pair. Thus it would be expected that the former pair would be more stable. This appears indeed to be the case and is reflected by a dependence of the thermal stability of DNA's from different sources upon their base composition.

Details of the DNA structure. Subsequent work has consistently confirmed and reinforced the Watson-Crick model. The task of refining the model so as to bring the predicted X-ray diffraction pattern into quantitative agreement with observation has been a challenging one and has required many years of exacting work (9, 10, 11).

X-ray diffraction studies upon fibers of DNA have revealed that several variants of the fundamental helical duplex exist, which differ structurally to a minor extent. The relative stability of these is governed by the humidity and by the nature of the cation.

All forms share the following features:

(1) The two strands are *anti-parallel;* that is, the terminal phosphates of the two strands are at opposite ends of the molecule. As a consequence, rotation by 180° does not alter the appearance of the molecule.

(2) The planar purine and pyrimidine bases of each strand are aligned parallel to each other and roughly perpendicular to the fiber axis. The bases are thus "stacked" in a parallel array (Fig. 9-5).

(3) The components of each base pair lie in the same plane.

At low relative humidities (< 70%) the sodium salt of DNA crystallizes in the A form. This is very highly regular and crystalline and gives an X-ray diffraction pattern with over 100 independent reflections. The bases are not strictly perpendicular to the fiber axis, being tilted by about 25° to the normal to that axis.

At higher relative humidities a transition occurs to the B form, whose degree of crystallinity is less than for the A form. In this case the base pairs are normal to the helical axis. The spacing between adjacent (parallel) base pairs is 3.4 Å. There are ten residues per turn of the helix. The phosphates on the periphery of the molecule are at a radius of 9Å and are separated by 7 Å along a given chain. The ribose ring is planar and is inclined at an angle to the helical axis.

The B form of DNA appears to be the prevalent structure in solution and in the living cell. Its molecular coordinates appear to be similar for DNA's from a wide variety of sources. A representative X-ray diffraction pattern is shown in Fig. 9-6.

A third variant, the C form, arises for the lithium salt of DNA at relative humidities of 44% or less. This differs from the B form in that the base pairs are moved about 2 Å away from the helical axis and

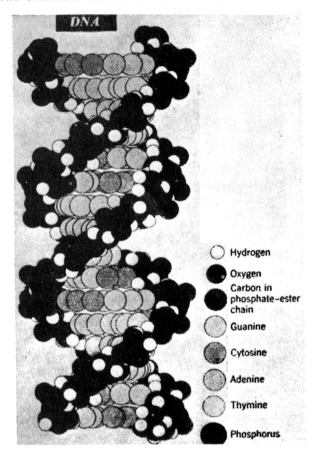

Fig. 9-5. Molecular model of the B form of DNA (10).

tilted by about 5°. The pitch of the helix is altered so that there are about 9 instead of 10 residues per turn.

A number of reasonable predictions can be made about the over-all physical properties of the DNA double helix.

(1) The structure is essentially completely rigid. Any appreciable bending would require an intolerable distortion of bond angles or distances. Thus a *perfect* double helix would have the configuration of a thin rigid rod.

(2) Since there is no evidence for any periodicity of base sequence along the individual polynucleotide strands, the maintenance of the integrity of the DNA structure depends upon the two strands being in complete register. Any extensive disruption of the complementary pairing would not be expected to be rapidly reversible.

9-4 SOLUTION PROPERTIES OF DNA

Size and shape. A complicating factor in the physical characterization of DNA from many sources is its heterodisperse character, a given preparation consisting of a collection of molecules differing in molecular size and base composition. This molecular inhomogeneity is absent only in the case of DNA from the simplest organisms, as the bacterial viruses, and becomes more pronounced for DNA from mammals and other complex living systems.

A satisfactory procedure has yet to be developed for the fractionation of DNA with respect to molecular weight. Hence the physical parameters which have been reported for the mammalian DNA's, which have been used in most investigations, represent only average quantities.

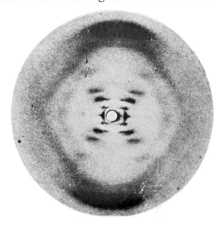

Fig. 9-6. An X-ray diffraction pattern for fibers of the B form of DNA.

A second problem is the tendency of DNA molecules, particularly those of high molecular weight, to be degraded mechanically by shearing stress, which is difficult to avoid in the usual preparative procedures. As a consequence, most of the molecular weights which have been reported tend to be smaller than those of the intact molecules present in the biological system.

Nevertheless, a definite pattern does emerge from the existing information as to molecular weight and dimensions. The following generalizations can be made (12, 13, 14):

(1) The average molecular weight of natural DNA is invariably high. In the case of the DNA from the thymus gland of the calf, for which the largest number of determinations are available, most estimates from light scattering have ranged between 6×10^6 and 10×10^6. Because of the complications mentioned earlier these numbers must be regarded as lower limits, and the average molecular weight of the material present in the living cells may be very much higher. In the case of the DNA from T-2 bacteriophage, a value close to 1.3×10^8 has been reported.

(2) The molecular dimension, as measured by light-scattering, is also large, but by no means as large as would be predicted for a *completely* extended double helix of the type described in section 9-3. For a *perfect*

double helix of the Watson-Crick form, the expected length (in angstroms) is equal to 3.4 times the number of nucleotides in each strand, as the separation of adjacent base pairs along the helical axis is 3.4 Å. The radius of gyration is equal to $1/\sqrt{12}$ times this length.

On this basis it would be predicted that a DNA sample of molecular weight 6×10^6 would have a radius of gyration close to 9000 Å. The experimental values from light scattering are one-third to one-quarter of this. This pattern is followed consistently by DNA samples of widely varying molecular weight.

While a high order of accuracy cannot be assigned to the existing light scattering data, it can probably be safely concluded that the extension of DNA in solution does not approach its limiting value.

The above should be qualified by mentioning that electron microscope photographs of DNA generally show very extended filaments, which approach a rodlike configuration (Fig. 9-7). However, such photographs

Fig. 9-7. (a) Schematic version of the configuration of DNA in solution at left. (b) Electron microscope photograph of DNA at right.

represent drying patterns and could easily be influenced by a change in elongation during the drying process.

Since any important degree of torsional bending of the Watson-Crick structure is not feasible, the existing flexibility can only be explained on the basis of local deviations from that structure, such as the occasional occurrence of a mismatched base pair. These might provide amorphous "hinge points" which could endow the molecule as a whole with an appreciable degree of flexibility.

Whatever the explanation, the best model for the physical state of native DNA in solution appears to be a system of "stiff" coils, coiled up to about one-third their maximum extension.

Hypochromism. It is customary to express the ultraviolet absorbancy of DNA in terms of the extinction coefficient per mole of nucleotide, which is equal to the extinction per mole of phosphorus and is often denoted by

ϵ_p. At neutral pH and room temperature, ϵ_p at 260 mμ is smaller by almost 40% than the value predicted for a non-interacting mixture of nucleotides of the same composition. This decrease in absorbancy is referred to as the *hypochromism* of DNA.

The degree of hypochromism is a sensitive function of the physical state of DNA and has often been used as an index of the departure of DNA from its native doubly stranded helical state. The theory of hypochromism is not as yet in a completely satisfactory state. However, there is general agreement that the effect arises from the mutual interaction of the electron systems of the bases. This interaction is dependent upon the configuration of the bases and is most important when they are stacked in a parallel array, as is the case for the helical duplex structure of native DNA. In this state the large degree of overlap of the planar bases and their small separation serve to maximize the interactions responsible for hypochromism. In general, any departure from the completely ordered helical conformation is reflected by a loss of hypochromism and hence by an increase in absorbancy (14).

Optical rotation. As in the polypeptide case (Chapter 5), the optical activity of DNA is composite in origin and includes, in addition to the contribution from the three asymmetric centers of deoxyribose, a second major component arising from the asymmetric nature of the (right-handed) double helix itself. The specific rotation at 589 mμ ($[\alpha]_D$), predicted for a mixture of nucleosides in the same ratio as for a typical DNA, is close to zero. However, the value of $[\alpha]_D$ for native DNA is generally large and positive—usually about 150°.

Thus, as in the case of the α-helical polypeptides, the magnitude of the specific rotation of DNA provides a valuable index of the state of its molecular organization. In general, any disruption of the helical structure of DNA will be reflected by a fall in the magnitude of its specific rotation. In this manner optical rotation can supplement ultraviolet hypochromism as a means of monitoring any loss of the native structure of DNA.

9-5 THE ATYPICAL DNA OF $\phi \times 174$ VIRUS

An account of the molecular properties of DNA would be incomplete without mention of one example which violates most of the generalizations of sections 9-3 and 9-4. The DNA of $\phi \times 174$ virus was isolated by Sinsheimer (15), who found it to have a molecular weight close to 1.7 ×

10^6. From the known size and DNA content of the virus, this would allow only one DNA molecule per virus particle.

The base composition of this DNA was found to show a definite deviation from the Watson-Crick pattern. Thus the relative base contents are adenine, 1.0; thymine, 1.3; guanine, 1.0; and cytosine, 0.8.

The degree of hypochromism was much less than is usually observed for native DNA. Moreover, the temperature profile of absorbancy showed a relatively gradual increase instead of the abrupt transition characteristic of DNA's from other sources (section 9-6). The dimensions of the molecule, as estimated from light-scattering, were indicative of a much higher degree of coiling at moderate ionic strengths (0.1 M NaCl) than is usual for DNA. Moreover, the molecule displayed a considerable degree of flexibility, expanding at low ionic strengths in a manner similar to RNA (section 10-4). From these results it was clear that this molecule has neither the high degree of order nor the rigidity characteristic of the intact helical duplex.

The above evidence, together with the nature of the kinetics of its enzymatic degradation, led Sinsheimer to the conclusion that this DNA consists of single polynucleotide strands whose molecular organization is incomplete and probably resembles that of RNA of comparable molecular weight (Chapter 10).

9-6 THE DENATURATION OF DNA (16, 17, 18, 19, 20)

Thermal denaturation. Ultraviolet hypochromism provides the simplest and most widely utilized means of observing the denaturation of DNA. The ultraviolet absorbancy of the individual nucleotides, or bases, shows no dependence upon temperature. Hence any thermal dependence of absorbancy observed in the case of DNA itself must arise from changes in the degree of hypochromism which result from alterations in its structure (14, 18).

At neutral pH and in the presence of 0.1 M KCl the absorbancy of DNA solutions is likewise constant up to a critical temperature range which depends upon the base composition and which is usually between 80° and 100°C. A rather sharp increase in absorbancy at 260 mμ, amounting to 20% or more of the original value, occurs in the critical range (Fig. 9-8). The transition is remarkably abrupt, going to completion over a temperature interval of 5°-10°. The critical temperature zone depends upon the ionic strength and decreases with decreasing ionic strength.

The shape of the reverse curve obtained upon cooling from a temperature above the transition zone is dependent upon the rate of cooling. If

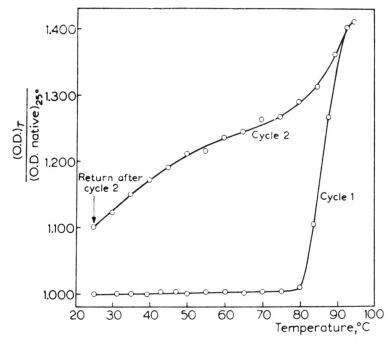

Fig. 9-8a. Thermal denaturation of calf thymus DNA, as monitored by ultraviolet absorbancy at 260 mµ (0.15 M NaCl, pH 7). The second cycle represents the thermal profile of a solution after exposure to the highest temperature, followed by rapid cooling to room temperature (14).

the solution is cooled *rapidly* back to room temperature, the initial thermal profile is not retraced. A major degree of hysteresis occurs between the heating and cooling curves, and only a partial recovery of the initial absorbancy is observed. If the cooled solution is reheated, a new forward curve is obtained (Fig. 9-8).

The rise in absorbancy (or drop in hypochromism) occurring in the critical temperature range is paralleled by important changes in other physical properties, including a major drop in viscosity and in specific rotation. Light scattering measurements upon DNA solutions before and after exposure to temperatures above the critical zone indicate a drop in radius of gyration of threefold or more (16).

All the phenomena cited above can be fitted into a definite physical picture. At elevated temperatures the organized helical structure of DNA is disrupted. The resultant loss of the ordered array of bases serves to reduce the degree of hypochromism. Once the structural rigidity conferred by the helical organization is lost, the polynucleotide strands assume a randomly coiled state. This is accompanied by a drop in viscosity.

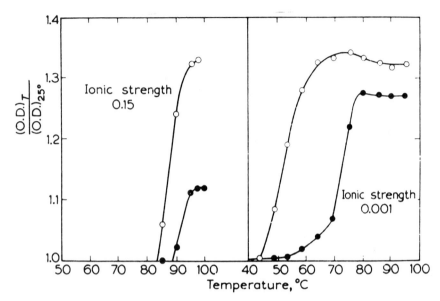

Fig. 9-8b. A comparison of the ultraviolet absorbancy of calf thymus DNA when measured directly at an elevated temperature (open circles) and when measured after cooling to room temperature (filled circles) (14).

As was discussed in Chapter 8, the presence of a third hydrogen bond between the guanine-cytosine pair would be expected to endow this pair with an enhanced stability as compared with the adenine-thymine pair. This is reflected by an increase in the temperature of denaturation with increasing content of the former bases (Fig. 9-9).

The question of whether an actual separation into individual strands occurs upon denaturation remained unsettled for a long time. The issue is obscured by the occurrence of both nonspecific aggregation and the scission of phosphodiester bonds. However, in recent years clear-cut evidence for an actual dissociation into single strands has been obtained by several approaches (section 9-9), indicating that there are no bonds of a refractory nature between the two strands and that there are no intrinsic kinetic barriers to their unwinding or disentanglement.*

Denaturation of DNA at extremes of pH. The groups involved in the hydrogen bonding of both the adenine-thymine and the guanine-cytosine pairs include ionizable sites (section 8-5). Changes in the state of ionization

* It should be mentioned that the occurrence of complete strand separation is still not universally accepted. For a contrary view, see L. Cavalieri and B. Rosenberg, *Biophys. J.*, 1, 317 (1961).

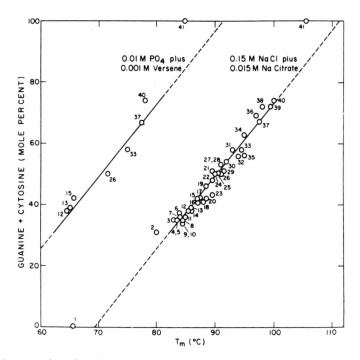

Fig. 9-9. Dependent of "melting point" (midpoint of thermal transition) of DNA's from various sources upon their guanine-cytosine content (22).

of the bases at extremes of pH would be expected to influence the stability of the DNA molecule.

Thus the binding at acid pH's of a proton by the N_1 nitrogen of either adenine or cytosine would eliminate this site as a hydrogen-bond acceptor and hence serve to break one of the two (or three) hydrogen bonds stabilizing the corresponding base pair. Moreover, the electrostatic repulsion of similarly charged bases packed into the core of the double helix should contribute to the weakening of this structure. Hence it is to be expected that the stability of DNA will be reduced at acid pH's in the zone of titration of the bases.

A similar situation prevails at alkaline pH. The loss of a hydrogen ion from the N_1 positions of thymine or guanine results in a loss of the capacity of these groups to serve as hydrogen-bond donors, thereby eliminating the corresponding hydrogen bonds. As in the case of acid pH's, the acquisition of charge by the bases serves to further destabilize the double helix.

Thus it is not surprising to find that DNA undergoes denaturation at extremes of pH. The first direct evidence for this process was the demonstration of irreversibility in the hydrogen ion titration curve of DNA. At 25°C a pronounced hysteresis occurs between the forward and reverse branches of the titration curve between neutral pH and pH 2 or pH 13 (Fig. 9-10). The implication of this finding was that a structural transition occurs at extremes of pH, as a consequence of which the effective pK's of the bases are altered (17).

The appearance of irreversibility in the hydrogen ion titration curve was found to be accompanied by a dramatic transition in the physical properties of DNA. Thus the viscosity drops by an order of magnitude (Fig. 9-11). Light scattering measurements reveal a major decrease in the radius of gyration (Table 9-1).

The above observations can be fitted into a logical pattern. The combined effects of electrostatic stress and partial rupture of the hydrogen bonding so weaken the DNA helical duplex as to render it unable to withstand the thermal motion occurring at 25°. Once the rigidity conferred by the helical structure is lost, the molecule collapses into randomly coiled polynucleotide chains. The over-all characteristics of the process are very analogous to those of the helix-coil transition of polypeptides.

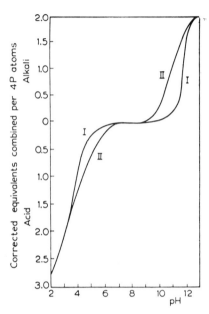

Fig. 9-10. Hydrogen ion titration curve of DNA (17). Curve I includes the acid and alkaline branches of the curve obtained upon starting from neutral pH. Curve II represents the acid and alkaline branches of the reverse curve obtained upon back titration from an acid or alkaline pH.

The degree of hydrogen ion binding by its bases which the DNA helix can tolerate is dependent upon the temperature and increases at lower temperatures. If titration is carried out at $-0.5°$ (in 1 M KCl), the

TABLE 9-1. COMPARISON OF THE PROPERTIES OF NATIVE AND DENATURED CALF THYMUS DNA (16)

	Molecular Weight ($\times 10^{-6}$)	Radius of Gyration (Å)	Intrinsic Viscosity
Native	7.7	3000	72
Denatured	—	1000	4

molecular organization of DNA can survive exposure to pH's as low as 2 (2).

DNA denaturation as a helix-coil transition (21). The denaturation of DNA at elevated temperatures or extremes of pH is in many respects very analogous to the helix-coil transition already described for the polypeptide case (Chapter 5). Both processes involve the transition from an ordered helical state to a disordered, randomly coiled configuration. In both cases an increase in energy resulting from the rupture of hydrogen bonds is compensated for by an increase in entropy accompanying a gain in "randomness."

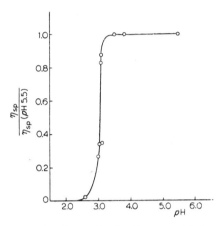

Fig. 9-11. Change in viscosity upon acid denaturation of DNA (19).

The abnormal sharpness of the thermal transition likewise reflects in both instances a resistance to the introduction of *new* breaks in the helical structure, as opposed to the enlargement of existing breaks. However, this has different origins in the two cases. In the case of α-helical polypeptides it is a consequence of the necessity for breaking three *consecutive* hydrogen bonds in order to liberate a residue from the helical configuration. In the DNA case the dominant factor appears to be the additional stabilization energy arising from the mutual interactions of adjacent "stacked" bases. This "nearest neighbor" interaction energy is maximized if the bases are arranged in the parallel, closely spaced array present in the DNA helix. The interaction energy originates from the mutual interaction of the electron systems of the bases and is of the type referred to as *van der Waals* energy of interaction.

The enlargement of a pre-existing random sequence by one nucleotide pair results in the loss of only one additional pair of base-base "nearest neighbor" interactions of this type. In contrast, creation of a *new* random sequence involves the loss of two such interactions. Thus there is a penalty imposed upon the initiation of new breaks in the helical structure. This is responsible for the sharpened character of the thermal transition, for reasons identical to those discussed in section 5-6.

Another difference between the two processes is that the random regions in the α-helix cases are *linear* polypeptide chains, while in the DNA case they are closed *loops* (Fig. 9-12).

Finally, because of the complementarity of the base compositions of the two strands, a reformation of the double helix requires that the two be completely in register. Thus it would be expected that the reversal of DNA denaturation would be difficult to achieve. In actuality the reversal requires a slow "annealing" process at intermediate temperatures, and is generally incomplete.

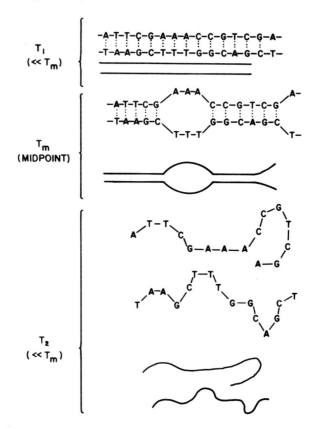

Fig. 9-12. Schematic model for thermal denaturation of DNA.

9-7 THE RENATURATION OF DENATURED DNA AND THE FORMATION OF MOLECULAR HYBRIDS

General remarks. Most of the earlier work upon the denaturation of DNA utilized experimental conditions such as to preclude the possibility of any major recovery of structure. The discovery of conditions under which a reunion of the strands and a reformation of the helical duplex become possible has occurred only recently (22-25).

With the recognition that denaturation and strand separation were not intrinsically irreversible, there naturally arose speculation that the DNA's from different, but closely related, species might be sufficiently similar in nucleotide sequence to permit the formation of true molecular hybrids upon renaturation of a mixture of the two denatured species.

This possibility has in fact been realized for DNA's from distinct strains of bacteria having a close genetic relationship. Indeed, the capacity for hybrid formation appears to offer a potential criterion for the classification of bacterial species (22).

Since the CsCl density gradient technique is of central importance to the studies to be discussed in this section, a description of them should be preceded by an account of this method.

Density gradient ultracentrifugation. If a solution of cesium chloride in high concentration ($8 \, M$) is subjected to centrifugation at very high speeds (60,000 rpm), a redistribution of the solute occurs under the influence of the centrifugal field so that the concentration increases toward the bottom of the cell. Thus a gradient of both concentration and density within the ultracentrifugal cell will be produced (26).

If a second solute of high molecular weight is present, then the speed and direction of its migration in the centrifugal field will depend upon the *local* density in a particular region of the cell. If the range of solvent densities *spans* the density of the biopolymer then there will exist a particular radial distance at which the density of the latter is equal to that of the medium. At this point no migration of the polymer will occur. *Sedimentation* of the polymer from the less dense region toward the top of the cell and *flotation* from the more dense zone at the cell bottom will occur. Thus the material collects in a narrow band at the point at which the densities of solute and solvent are equal.

If two components of different density are present then two bands will appear. In practice, density gradient centrifugation is an extremely sensitive means of detecting the presence of several components differing slightly in density. In particular, two molecular species differing only in the extent of isotopic labeling can easily be resolved by this technique.

The concentrations at which density gradient sedimentation is usually carried out ($< .1$ gm/l.) are much too low to permit the use of the schlieren technique discussed in Chapter 4. The distribution of DNA at equilibrium has usually been measured by its absorption of ultraviolet light. With an appropriate optical system the degree of darkening of an ultraviolet-sensitive photographic plate by light passing through a particular region of the ultracentrifuge cell is a sensitive measure of the reduction of light transmisison by absorption by solute in that region. The darkening of the plate may be measured quantitatively by means of a photoelectric densitometer. In this manner the distribution of polynucleotide in the cell may be determined (Fig. 9-13).

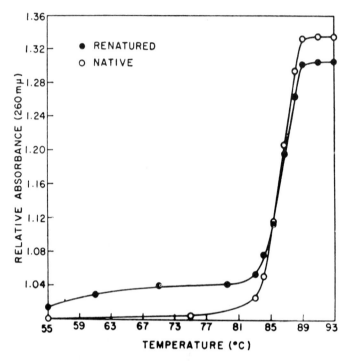

Fig. 9-13a. Thermal stability of the native and renatured forms of the DNA of *Mycoplasma gallisepticum* in 0.3 M NaCl, as monitored by absorbancy (22).

Factors influencing renaturation. The thermal denaturation of DNA, as examined by a number of physical criteria, is accompanied by a loss of the organized secondary structure and a splitting of the helical duplex into separate polynucleotide strands. In view of the highly specific base sequence of each strand and the requirement that the bases of the two very long strands be in complete register for the helical form, it would intuitively appear that the reversal of denaturation would be difficult to achieve. It was in fact widely believed for a long time that reformation of the original conformation from the completely denatured state would encounter insuperable kinetic difficulties.

Nevertheless, it has been found in recent years that by a proper choice of the conditions of cooling, it is possible to renature a major fraction of thermally denatured DNA. This has been observed thus far only with bacterial DNA and not for mammalian. The reason for this appears to reside in the more heterogeneous character of mammalian DNA, which consists of a large number of different molecules of varying base sequence. In such a system the vast majority of collisional contacts which a particular DNA strand undergoes will be with *noncomplementary* strands and cannot result in reformation of the double helix.

In contrast, bacterial DNA consists of only a few different molecular species (possibly of only one). Thus a much larger fraction of the collisional contacts of the separated strands of denatured DNA will be between complementary strands and can provide the basis for reunion and renaturation.

The crucial experimental parameter for renaturation is the *rate* of cooling. If thermally denatured bacterial DNA is cooled *rapidly* to room temperature, very little renaturation is observed. While the hypochromism increases, this appears to be due to a more or less random intramolecular reformation of hydrogen bonds to yield a product which does not contain helical regions extending over a major portion of its length. If the solution is reheated, the ultraviolet absorbancy increases gradually,

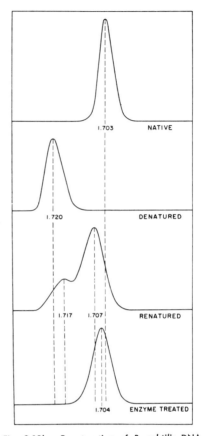

Fig. 9-13b. Renaturation of *B. subtilis* DNA, as observed by CsCl banding (22). The ordinate is proportional to the concentration of DNA at a particular level within the ultracentrifugal cell. The abscissa is proportional to the density, which increases from right to left. The figures represent the densities at the midpoints of the peaks. The second curve from the top shows the density distribution for a (completely denatured) sample after heating at 100° C followed by rapid cooling. The third curve from the top shows the density distribution produced by slow cooling. A residual shoulder indicates that some denatured material is present. The bottom curve shows the result of treatment of the renatured material by *E. coli* phosphodiesterase.

with no sign of the abrupt transition characteristic of native DNA (14).

The degree of recovery of the original hypochromism is dependent upon its exact thermal history. Slow cooling appears to be essential for the attainment of a hypochromism approaching that of the original material. If the solution is held at a constant temperature somewhat below the melting point, a fall in absorbancy at 260 mµ with time may be observed. This is a direct measure of the renaturation process, whose rate depends upon the composition of the DNA and the temperature. For pneumococcal DNA in 0.3 M NaCl, the optimum temperature is about 65°. As the temperature is further lowered, the rate of recovery falls rapidly, becoming nil at room temperature.

By prolonged exposure to temperatures at which recovery is rapid, a renatured product may be obtained whose hypochromism at 25° is only slightly less than that of the original native material. A second forward heating curve shows an abrupt increase in absorbancy at a temperature close to that characteristic of native DNA (Fig. 9-13).

Other criteria are suggestive of a major recovery of the original helical conformation. The optical recovery is paralleled by a regaining of biological activity in the case of *transforming DNA* (section 9-10).

The density of DNA, as determined by CsCl banding, increases upon denaturation. If cooling is carried out rapidly, virtually all the material is present in the denatured band (Fig. 9-13). Slow cooling produces a band of density close to that of the native material. Treatment with the phosphodiesterase of *E. coli*, which attacks single strand DNA specifically, eliminates the denatured band (Fig. 9-13).

The reformation of native structure at intermediate temperatures is often referred to as an *annealing* process. The annealing temperature must be high enough so that hydrogen bonds may be broken and reformed at a relatively rapid rate. Under these conditions the strands will possess a considerable lateral mobility, which permits the exploration of competitive configurations, until that of maximum stability is selected. In this way, irregularities and areas of mismatching may be progressively eliminated. Rapid cooling to room temperature tends to freeze imperfections into the structure.

Molecular hybridization. If a mixture of two native DNA preparations from the same species, one of which is heavily labeled with N^{15}, is examined by the CsCl banding technique, two distinct bands corresponding to the two densities will be observed (Fig. 9-14). If the mixture is first thermally denatured and then annealed carefully prior to examination, a total of five bands are observed. These correspond to, in order of in-

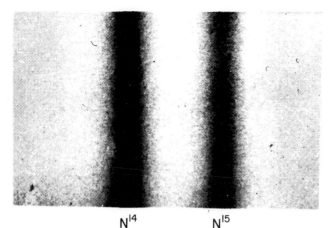

Fig. 9-14. CsCl banding of a mixture of normal and N^{15}-labeled DNA (30).

creasing density, renatured normal, denatured normal, *hybrid*, renatured labeled, and denatured labeled (Fig. 9-15).

Treatment with *E. coli* phosphodiesterase reduces the five peaks to three, corresponding to renatured normal, hybrid, and renatured labeled. The density of the hybrid is close to that expected for a molecule with one normal and one labeled strand.

Still more interestingly, if the same experiment is performed with DNA from two *different* but genetically *related* strains, such as *E. coli* strains B and K, hybrid formation has also been shown to occur. The implication of this result is that *extensive regions of similar nucleotide sequence are present in the two*.

This is not surprising for species having such a close taxonomic and genetic relationship. Marmur, Doty, and co-workers have examined for hybridization the DNA's from a wide range of bacterial species. From these studies the following generalizations can be made as to the requirements for hybrid formation:

(1) The over-all base composition of the two species must be similar. This is of course a minimum requirement to permit the extensive regions of base complementarity needed to stabilize the hybrid.

(2) The two species must be *genetically* related. The occurrence of genetic interchange, as by transformation (section 9-10) or by transduction (section 11-5), may be regarded as positive evidence for a close genetic relationship.

An example of the operation of these criteria is provided by the occurrence of hybridization between *E. coli* and *B. subtilis*, which satisfy both conditions. In contrast, *E. coli* does not hybridize with many species whose base compositions are similar but which are genetically unrelated. This high degree of specificity is a strong argument against the view that hybridization reflects simply a random physical aggregation.

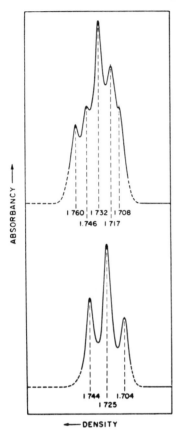

Fig. 9-15. Upper: CsCl banding pattern produced by a renaturation of a mixture of normal and N^{15}-labeled DNA (22). Lower: pattern of same mixture after treatment with *E. coli* phosphodiesterase (22).

The methods which have thus far been developed for the study of hybridization are not sufficiently refined to permit a differentiation of DNA's differing in base sequence only to a minor extent. Thus various mutant strains of *E. coli* have been shown to be indistinguishable by this criterion.

It is desirable to conclude this section with a few words of caution. While the occurrence of hybridization appears to be well established, the detailed molecular mechanism of the process is still imperfectly understood. While the discussion of the preceding pages has been developed in accord with the general picture that hybridization proceeds by complete strand separation followed by recombination, it would be premature to exclude the occurrence of alternative mechanisms in particular instances.

9-8 BIOSYNTHETIC DNA

Enzymatic synthesis. Kornberg and co-workers have isolated from *E. coli* an enzyme which has the property of catalyzing the polymerization of deoxyribonucleoside triphosphates. The isolation of the enzyme, *DNA-polymerase*, involves a lengthy fractionation of the extract of an ultrasonically ruptured suspension of *E. coli* (27-29). For an extensive synthesis of DNA to occur, the following constituents must be present in the reaction mixture, in addition to enzyme:

(1) all four deoxyribonucleoside triphosphates—dATP, TTP, dCTP, and dGTP;

(2) Mg^{++}; and

(3) *primer* DNA, which may be either native or denatured.

Pyrophosphate is split out by the reaction:

(9-1) $\quad\quad\begin{array}{l} n \text{ dATP} \\ n \text{ dCTP} \\ n \text{ dGTP} \\ n \text{ TTP} \end{array} \xrightarrow[\text{primer}]{\text{enzyme}} (\text{dAp, dCp, dGp, Tp})_n + pp$

If no primer DNA is present in the reaction mixture, there occurs after a long induction period an important degree of synthesis of a polymer containing the bases adenine and thymine only. This interesting material is called the *dAT copolymer*.

Structure of biosynthetic DNA. The analysis of biosynthetic DNA by the same procedures as have already been described for the case of natural DNA (section 9-2) has shown conclusively that its primary structure is based upon the same 3′-5′ phosphodiester linkage as the latter. There is no evidence for any appreciable branching.

The molecular weight of biosynthetic DNA is of course variable, but preparations have been obtained whose average molecular weight is of the order of several million. The general physical properties resemble those of natural DNA, and its conformation appears to be identical.

The Watson-Crick base ratios are preserved in biosynthetic DNA. Moreover, the base composition parallels closely that of the added primer. Thus it has been shown that, for primers of widely varying base composition, the base composition of the DNA synthesized is, within wide limits, independent of the ratios of deoxynucleoside triphosphates present in the reaction mixture and corresponds to the primer composition.

This correspondence appears to extend to the base *sequence* as well. While a rigorous proof, which is not accessible at present, would require a comparison of the *complete* nucleotide sequences of primer and product, the above picture can be assigned a high order of probability by virtue of the following results.

In an experiment already alluded to (section 9-2), it has been shown that the distribution of dinucleotide nearest neighbors in the biosynthetic product is non-random and unique. *If biosynthetic DNA is, in turn, used* as a primer, its unique distribution of dinucleotides is preserved in the new product. This result is consistent with, and suggests that, the primary structures of primer and product are essentially identical.

This has naturally led to the proposal that the mechanism of the guided biosynthesis of DNA by DNA-polymerase involves the linear assembly of the deoxyribonucleotide subunits upon the individual strands of the primer. This model represents the primer function as that of a *template*, the correct positioning of the nucleotides being attained by base interactions of the Watson-Crick type. Each strand of primer would thus direct the synthesis of its *complement*. For example, the sequence—ATTCCG—in the primer would produce the complementary sequence—TAAGGC—in the daughter strand (Fig. 9-16).

This model obviously requires that the synthetic process be accompanied or preceded by an unwinding of the primer helix. This view is reinforced by the finding that denatured DNA, as well as the single stranded DNA of $\phi \times 174$, are also efficient primers.

Incorporation of unnatural bases. The DNA-polymerase of *E. coli* acts only upon deoxynucleoside triphosphates. However, some latitude is permitted

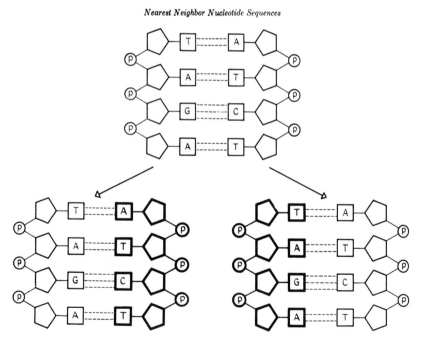

Fig. 9-16a. Proposed model for the biosynthesis of DNA by DNA-polymerase (28).

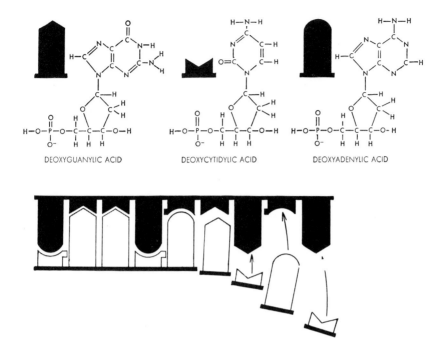

Fig. 9-16b. Schematic picture of the semiconservative model for the replication of DNA.

as to the nature of the base, *provided that its structure is such as to permit the Watson-Crick pairings.* A number of unnatural bases have been incorporated when supplied as the deoxynucleoside triphosphate. These share an undiminished capacity for hydrogen bonding of the Watson-Crick type. Some examples are 5-bromouracil, 5-bromocytosine, 5-methylcytosine, and hypoxanthine. The contents of purines and pyrimidines are equal in the product, just as in the product obtained using ordinary bases.

9-9 THE BIOLOGICAL REPLICATION OF DNA

Hypotheses as to DNA replication. The molecular mechanisms which have been proposed for the synthesis of DNA by biological systems fall into three main categories:

(1) *Dispersive.* Into this class fall all mechanisms which postulate a general breakup of the parent DNA and its more or less uniform distribution among the progeny. It is difficult to propose a model of this type which can plausibly allow the preservation of nucleotide sequence.

Moreover, the existing experimental evidence is uniformly inconsistent with this kind of scheme.

(2) *Conservative.* This represents the other extreme of replication models. Mechanisms of this class postulate that *both* the primary and the secondary structure of the parental DNA are conserved. Thus the replication process would involve neither the rupture of the individual strands nor the unwinding of the helical duplex.

The latter feature has endowed this model with some intrinsic attractiveness, and considerable effort has been expended in developing ingenious variants of it.

The most detailed of these proposes that the sequence of hydrogen-bonded base *pairs,* rather than individual bases, serves as the genetic determinant. According to this picture, particular base pairs might preferentially bind specific nucleotides and determine their sequential position in the daughter polynucleotide.

However, it is difficult to design explicit models for this process which are structurally and chemically plausible. While it would be premature to rule out the conservative mechanism entirely at present, it cannot be assigned a high order of probability.

The conservative model allows for no redistribution of parental DNA among the progeny. The original DNA is transferred as an intact unit or not at all.

(3) *Semiconservative.* Models of this category specify that the replication of DNA involves the unwinding of the double helix, but no rupture of the separated polynucleotide strands. Thus the *primary,* but not the *secondary,* structure of the parental DNA is conserved during the replication process.

According to this mechanism, the unwinding of the original strands would be accompanied or followed by the guided synthesis of the *complementary* strand of each of the two separated polynucleotide strands. The linear assembly, catalyzed by enzymes, of the nucleotides of the complementary sequence would be guided by their pairing with their Watson-Crick partners in the original strand. In this manner the proper nucleotide sequence would be automatically assured (Fig. 9-16). The DNA helix would thereby be regenerated, each of the strands of the original helical duplex uniting with its freshly synthesized complement to form a total of two new doubly stranded structures. The semiconservative model thus predicts that entire polynucleotide strands of the parental DNA will be passed on to the progeny in intact form. Each molecule of progeny DNA will receive either one-half or none of its nucleotides from the parental DNA.

This model has the obvious merit of providing a means for the preservation of nucleotide sequence which requires no new or untested principles. It is of course strongly reinforced by the discovery of DNA-polymerase. It is still probably too early to regard any of the above models as conclusively established for all biological systems. However, it can be stated that the available evidence definitely favors the semiconservative model for the systems which have been examined. The experiment of Meselson and Stahl is particularly compelling (30).

The Meselson-Stahl experiment. It is natural to raise the question of to what extent the process catalyzed by DNA-polymerase *in vitro* represents the actual mechanism of DNA synthesis by biological systems. While it is not yet possible to answer this question with finality, such a hypothesis is certainly very attractive.

Meselson and Stahl, in an elegant experiment, have provided a highly provocative clue as to the mechanism of DNA replication by *E. coli* (30). If successive generations of this microorganism are grown upon a nutritive medium in which nitrogen is present exclusively as the heavy N^{15} isotope, then the ordinary N^{14} isotope is progessively replaced by N^{15} until ultimately almost all the nitrogen atoms present in the *E. coli* belong to this isotopic species.

The DNA isolated from such a culture has a higher density than that obtained from ordinary cultures grown on a medium in which nitrogen is present as the N^{14} isotope. The difference in density is readily detectable by CsCl banding. A mixture of N^{15}-labeled and ordinary DNA forms two distinct bands (Fig. 9-14).

Meselson and Stahl subjected the N^{15}-labeled culture to an abrupt change of environment by diluting with a ten-fold excess of normal (N^{14}-containing) medium. Successive generations arising after the change in medium were sampled and their DNA isolated and examined by CsCl banding.

During the entire course of the experiment only three bands were observed whose densities corresponded to 100% labeled, 50% labeled, or unlabeled DNA. During the period between the change in environment and one generation time, the 100% labeled form was progressively depleted, with a concomitant increase in the concentration of the 50% labeled species. *At one generation time virtually all the material was present as the 50% labeled form.* At later times *only* 50% labeled DNA and the completely unlabeled species were present, the proportion of the latter progressively increasing with time (Fig. 9-17).

The following conclusions were drawn from these results:

(1) *The nitrogen of a DNA molecule is divided equally between two subunits which remain intact through many generations.* The 50%

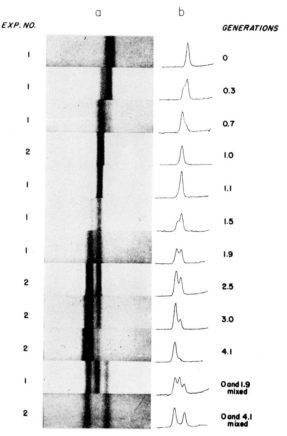

Fig. 9-17. Progressive change of CsCl banding pattern for the DNA of successive generations of E. coli after a change from an N^{-15} to an N^{-14} containing medium (30).

labeled form corresponds to DNA, each of whose molecules consists of one completely labeled and one wholly unlabeled subunit.

(2) *Each molecule of progeny DNA receives one parental subunit during replication.* Otherwise, the first generation would contain some unlabeled and some fully labeled DNA molecules, derived from progeny which received two or no parental subunits, respectively.

The only subunits of DNA present in equal quantities and likely to persist in intact form during replication are of course the two polynucleotide strands. The result of the Meselson-Stahl experiment is exactly what would be predicted if the replication of DNA were preceded, or accompanied, by an unwinding of the two strands, with each strand serving as a template for the synthesis of its complementary strand.

After each replication each strand of parental DNA remains paired with a newly synthesized strand (Fig. 9-18). In this way a doubling of the number of DNA molecules occurs at each replication (Fig. 9-18).

The bearing of this result upon the general question of the biological function of DNA will be discussed further in Chapter 11.

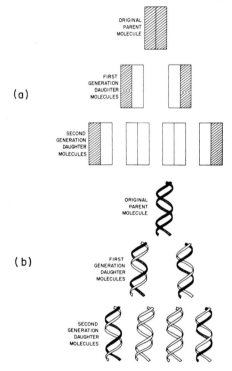

Fig. 9-18. Proposed model for the replication of the DNA of *E. coli* (30).

Thermal denaturation of hybrid DNA. If the 50% labeled DNA is subjected to thermal denaturation, followed by rapid cooling, prior to banding, *two* bands are observed, whose densities are close to those expected for denatured normal and denatured labeled DNA. This result, in combination with an observed halving of the molecular weight, leaves little doubt that the thermal denaturation of *E. coli* DNA produces an actual separation of the strands.

9-10 TRANSFORMING PRINCIPLE

Transformation of pneumococcus. The classical initial observations of a bacterial transformation were made by Avery, MacLeod, and McCarty with cultures of pneumococcus bacteria (31). This organism occurs as a number of distinguishable types which differ in the chemical nature of the external polysaccharide capsules in which the bacterial cells are encased. The exterior capsules are readily differentiated by immunological techniques.

In addition, the encapsulated strains may, as a result of spontaneous mutation, lose the capacity to form a capsule. Such unencapsulated variants may be obtained which show no sign of reversion to the capsulated form when cultured over many generations. The capsulated and unencapsulated strains are referred to as *smooth* (S) and *rough* (R), respectively, from the appearance of their cultures when grown upon an agar plate.

The DNA from an S strain can be isolated in purified form by standard procedures. When added to a culture of an R form under the proper conditions it has the property of inducing an alteration of the characteristics of the R strain so that it becomes encapsulated (Fig. 9-19).

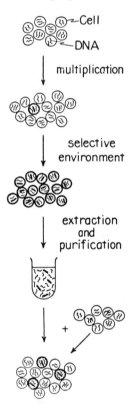

Fig. 9-19. Operations involved in bacterial transformation (32).

The importance of these findings was underlined by these particularly significant features:

(1) The polysaccharide capsules of the new S strain always have the same chemical characteristics as those of the parent S strain from which the transforming DNA was obtained.

(2) Transformation was accomplished with R strains which bred true, that is, which had no tendency to revert over many generations.

(3) The transformed culture would reproduce indefinitely as an S strain, becoming itself a source of transforming DNA.

These results indicated clearly that the transforming principle had performed two functions of a genetic character: it had induced at least one specific inheritable property (capsule synthesis), and it had instigated its own replication (31, 32, 33).

Capsule synthesis is only one of the many distinguishable characterties with which a particular pneumococcus strain can be endowed by transforming principle from another strain. The list of transferable properties includes resistance to specific drugs, the production of enzymes, and probably a host of other characteristics less readily recognizable.

Drug resistance has become the property most commonly utilized in detailed studies of bacterial transformations. The use of this property has obvious advantages in that the transformed species can be easily separated and identified by plating on a medium containing the drug to which resistance is developed. Streptomycin has been often used for this purpose.

Susceptibility to a transforming principle is not confined to pneumococcus bacteria. A number of other bacterial species have strains which can be interconverted in this way, including *H. influenzae*.

Deactivation of transforming principle. Conclusive proof that the activity of transforming principle arises exclusively from DNA, rather than from some contaminant, has been obtained from studies of its inactivation by enzymatic or physical means. In brief, treatment which is known to produce a loss of primary or secondary structure invariably results in a loss of transforming activity.

Thus hydrolytic degradation of transforming principle by deoxyribonuclease results in a total and irreversible loss of activity. In contrast, treatment with enzymes which catalyze the hydrolysis of polypeptides is without effect.

Similarly, the denaturation of transforming DNA at extremes of pH or at elevated temperatures is accompanied by inactivation. The loss of activity parallels the denaturation process, as followed by ultraviolet absorbancy.

The above evidence provides very compelling proof that the activity of transforming principle is indeed due to DNA and to DNA alone.

Details of the transformation process. Most quantitative studies upon bacterial transformations have utilized the property of drug resistance. In practice, DNA from a drug-resistant strain is added to a culture of a nonresistant strain. After an appropriate interval the transformed culture is diluted into a growth medium containing a sufficient concentration of the drug to kill or inhibit the growth of all cells except the transformants.

Extensive studies have been made by Hotchkiss upon the acquisition of resistance to streptomycin by pneumococci (33). Interpretation is simplified by limiting the time of exposure to transforming DNA to a short, clearly defined interval, as 5 minutes. This may be done by adding deoxyribonuclease after a suitable interval. This hydrolyzes DNA and eliminates its transforming ability.

Under these conditions, the extent of transformation is found to be dependent upon: (1) the length of time following exposure to transforming DNA; (2) the total concentration of transforming DNA; and (3) the composition of the DNA added, in particular the fraction of inert DNA in the mixture.

The time course of transformation in such experiments is of a two-step nature. There is an initial very rapid rise in the number of transformed cells beginning about one cell division period after exposure to DNA. This is followed by a period of several hours during which the number of transformants is static. After this interval there is a slow and exponential increase. This second phase appears to reflect the replication of transformed cells. The over-all extent of the transformation is fractional, not more than about 20% of the cells of the original culture being transformed in the limit.

The total number of cells transformed increases with the concentration of added DNA. A linear log-log relationship is obeyed at low DNA concentrations. At high concentrations a plateau is attained. The value of the limiting level of transformation approached is dependent upon the purity of the DNA preparation. The plateau value decreases with increasing fraction of inert DNA and cannot be increased by adding more of the same impure mixture. This points to a definite inhibiting action of nontransforming DNA.

Biological significance of transformation. It is evident that there are many parallels in function of transforming principle and the genes of classical genetics. Just as many genes can be assembled from diverse origins into a single cell, so several transforming abilities can be introduced independently into the DNA of a particular strain.

In general, if a particular transferable property, such as resistance to streptomycin, is acquired through a series of *stepwise* mutations, then the inception of this property in a transformed strain is likewise stepwise (32). The action of transforming principle from the maximally resistant end product of the process brings the newly transformed strain only to the first level of resistance (32). A second exposure to the same transforming principle brings it to the next highest level, and so forth.

It thus appears that the extracted DNA does not represent solely the limiting maximally resistant state of the donor strain, but includes a collection of determinants representing all the successive mutational experiences of this strain. Thus the transfer of genetic determinants appears to parallel their mutational production. If the acquisition is stepwise, the transfer is stepwise too.

The results with transforming principle provide another illustration of the manner in which alterations in inheritable characteristics of living cells are faithfully registered by alterations in their DNA. They thus furnish a valuable confirmation of the general scheme outlined in the earlier sections of this chapter.

The mechanism of action of transforming DNA within the cell is still uncertain. The process may resemble that of genetic recombination in bacteriophage (section 11-4).

Thermal denaturation of transforming DNA. Like DNA from other sources, transforming DNA undergoes denaturation at elevated temperatures and extremes of pH. As in the case of other DNA's, the process is probably accompanied by a separation of strands. In a pioneering investigation, Marmur and Lane have examined the effects of thermal denaturation followed by rapid cooling, upon the transforming ability of a pneumococcal DNA in which was incorporated the capacity to induce three separate transformations, which conferred resistance to three different antibiotics (25). This provided three different "markers," which could be assayed independently.

A major drop in transforming ability of all three types occurred in the region of thermal denaturation (as measured by changes in ultraviolet absorbancy). However, the inactivation profiles were appreciably different for the three, suggesting that the resistance markers were located in zones of different base composition.

If the transforming ability of samples which had been *rapidly* cooled to room temperature was measured as a function of the maximum temperature to which the DNA was exposed, it was found that activity fell to a low level at temperatures above the denaturation zone, but did not quite disappear. Thus a marginal activity (0.05%) survived autoclaving at 120° for 10 minutes, suggesting that some degree of recombination of strands occurred upon rapid cooling.

The *rate* of cooling and the total concentration of DNA proved to be the controlling factors as to degree of recovery. The extent of recovery was greater by orders of magnitude for slowly cooled solutions.

Both observations are consistent with, and suggest that, the molecular event associated with recovery of transforming ability is the reformation of the original doubly stranded helical structure of the DNA. The importance of slow cooling lies in its facilitation of the "annealing" process whereby irregularities in structure are corrected in a temperature range where the DNA strands possess considerable lateral mobility. Rapid cooling tends to freeze defects and areas of base mismatching into the structure.

Parallel studies using the CsCl gradient ultracentrifugation technique, as well as other physical methods, demonstrated that the recovery process is indeed correlated directly with the reunion of the separated strands of denatured transforming DNA and the reformation of the helical structure. Thus it is clear that the integrity of the DNA helical duplex is essential for transforming activity.

Moreover, the recovered activity per unit weight was found to be not reduced by the addition of denatured DNA from the *normal*, untransformed pneumococcus strain. This has been shown to arise from the formation of *hybrid* molecules, which retain transforming ability. This result indicates that both strands of transforming DNA need not be present in the same molecule to permit activity. Apparently a doubly stranded molecule, one of whose strands is normal, can still induce transformation. The stability of the hybrid indicates that the nucleotide sequences of normal and transforming DNA cannot be very different for most of their length, for no such hybridization occurs between unrelated DNA's. The failure of the specific activity to increase suggests that only one strand of the original transforming DNA is active. If both strands were active, it would be expected that each helical duplex would give rise, upon denaturation and renaturation, to two active hybrid molecules, with a consequent doubling of the specific activity.

There remains the question of why the integrity of the double helix is essential for transforming activity, in view of the currently accepted picture that the genetic information stored in DNA is determined by its base sequence alone. While this question cannot be answered with finality at present, there is some evidence that denatured DNA is not readily taken up by the bacterial cell and hence cannot exert its residual transforming activity (34). Alternatively, it may be more susceptible to the action of phosphodiesterases.

GENERAL REFERENCES

1. *The Nucleic Acids*, E. Chargaff and J. Davidson, Academic Press, New York (1955).
2. *Polynucleotides*, R. Steiner and R. Beers, Elsevier, Amsterdam (1961).
3. *The Chemistry of Nucleic Acids*, D. Jordan, Butterworth, London (1960).

SPECIFIC REFERENCES

Primary structure of DNA

4. D. Brown and A. Todd, *J. Chem. Soc.*, **52** (1952).
5. D. Brown and A. Todd in reference 1, vol. 1, p. 409.
6. E. Chargaff in reference 1, vol. 1, p. 307.

Density of DNA

7. N. Sueoka, J. Marmur, and P. Doty, *Nature*, **183**, 1429 (1959).

Secondary structure of DNA

8. J. Watson and F. Crick, *Nature*, **171**, 964 (1953).
9. A. Rich, *Rev. Mod. Phys.*, **31**, 191 (1959).
10. M. Feughelman, R. Langridge, W. Seeds, A. Stokes, H. Wilson, C. Hooper, M. Wilkins, R. Barclay, and L. Hamilton, *Nature*, **175**, 834 (1955).
11. R. Langridge, W. Seeds, H. Wilson, C. Hooper, M. Wilkins, and L. Hamilton, *J. Biophys. Biochem. Cytol.*, **3**, 767 (1957).

Solution properties of DNA

12. Reference 2, chap. 7.
13. C. Sadron in reference 1, vol. 3, p. 1.
14. P. Doty, H. Boedtker, J. Fresco, R. Haselkorn, and M. Litt, *Proc. Natl. Acad. Sci. U. S.*, **45**, 482 (1959).

DNA of $\emptyset \times 174$ virus

15. R. Sinsheimer, *J. Molec. Biol.*, **1**, 43 (1959).

Denaturation of DNA

16. E. Geiduschek and A. Holtzer, *Advances in Biological and Medical Physics*, vol. 6, p. 431, Academic Press, New York (1958).
17. D. Jordan in reference 1, vol. 1, p. 447.
18. Reference 2, chap. 7.
19. P. Doty, *J. Cell. and Comp. Phys.*, **49**, suppl. 1, 27 (1957).
20. J. Marmur and P. Doty, *Nature*, **183**, 1427 (1959).

Helix-coil transition of DNA

21. Reference 2, chap. 9.

Renaturation of DNA and formation of molecular hybrids

22. J. Marmur, C. Schildkraut, and P. Doty in *The Molecular Basis of Neoplasia*, p. 9, University of Texas Press, Austin (1962).

23. J. Marmur and P. Doty, *J. Molec. Biol.*, **3**, 585 (1961).
24. P. Doty, J. Marmur, J. Eigner, and C. Schildkraut, *Proc. Natl. Acad. Sci. U. S.*, **46**, 461 (1960).
25. J. Marmur and D. Lane, *Proc. Natl. Acad. Sci. U. S.*, **46**, 453 (1960).

Density gradient ultracentrifugation

26. M. Meselson, F. Stahl, and J. Vinograd, *Proc. Natl. Acad. Sci. U. S.*, **43**, 581 (1957).

Biosynthesis of DNA

27. I. Lehman, M. Bessman, E. Simms, and A. Kornberg, *J. Biol. Chem.*, **233**, 163 (1958).
28. J. Josse, A. Kaiser, and A. Kornberg, *J. Biol. Chem.*, **238**, 864 (1961).
29. I. Lehman, *Annals of the N. Y. Academy of Science*, **81**, 745 (1959).

Replication of DNA in E. Coli

30. M. Meselson and F. Stahl, *Proc. Natl. Acad. Sci. U. S.*, **44**, 671 (1958).

Transforming principle

31. O. Avery, C. MacLeod, and M. McCarty, *J. Exptl. Med.*, **79**, 137 (1944).
32. R. Hotchkiss in reference 1, chap. 27.
33. R. Hotchkiss in *The Chemical Basis of Heredity*, W. McElroy and B. Glass (eds.), p. 321, Johns Hopkins Press, Baltimore (1957).
34. L. Lerman and L. Tolmack, *Biochim. et Biophys. Acta*, **33**, 371 (1959).

PROBLEMS

1. It is known that a considerable number (up to 100 per molecule) of the phosphodiester bonds of DNA may be hydrolyzed by deoxyribonuclease with only a minor (20%) drop in molecular weight. Why is this?

2. What factors would be expected to influence the rate of separation of the two strands of denatured DNA?

10

The Ribonucleic Acids and the Biosynthesis of Proteins

10-1 GENERAL REMARKS

The biological function of RNA. We turn now to the later phases of the general mechanism which has been presented for the directed synthesis of proteins (section 9-1). It will be recalled that, according to this model, RNA functions as a secondary carrier of genetic information, serving as an intermediary in the process whereby the information stored as a specific deoxyribonucleotide sequence in DNA is translated into a specific amino acid sequence of a polypeptide (1-6).

The earliest evidence for the central role of RNA in protein biosynthesis was indirect. Cytological studies showed that the RNA contents of the cells of different tissues were correlated directly with their activity in protein formation (5). These and other studies (4, 5) led to the proposal of the general mechanism cited in section 9-1. Subsequent, more conclusive, investigations have consistently confirmed this hypothesis, which can now be regarded as well established.

To anticipate, RNA is a linear, single-stranded polymer of ribonucleotides, whose primary structure is analogous to that of the individual polynucleotide strands of DNA (sections 10-2 and 10-3). The discovery of the enzyme *RNA-polymerase* (section 10-6), which catalyzes the synthesis of an RNA molecule whose base sequence is complementary to that of a strand of DNA primer, provides a plausible means for the transfer of genetic information from DNA to RNA. The individual polynucleotide strands of DNA direct the formation of complementary strands of RNA, thereby producing a *transcript* of its coding sequence of nucleotides.

The formulation of a feasible mechanism whereby a specific *nucleotide* sequence might be translated into a specific *amino acid* sequence presents obvious difficulties which are not encountered for DNA or RNA replication. There is no evidence for any direct interaction of amino acids with nucleotide bases. The widely varying sizes of the amino acids make it difficult to conceive of a model for their linear assembly along a polynucleotide template, unless they undergo a preliminary attachment to some form of *adaptor*.

More serious still is the difficulty that there are 20 different amino acids and only four different bases. Thus, unlike the case of DNA replication, the transfer of genetic information from polynucleotide to polypeptide resembles the translation of a message from one language to another which is altogether different. The term "genetic coding" is often used to denote the control of amino acid sequence by nucleotide sequence.

It is clear that a one-to-one correspondence between nucleotides and amino acids is out of the question. The only way in which specificity of sequence can be attained is for each amino acid to be coded for by a *combination* of nucleotides. In particular, a triplet code has gained considerable support (section 10-7). This model supposes that a sequence of three consecutive nucleotides corresponds to each amino acid.

The fraction of cellular RNA which has an active directive function in protein synthesis is called *messenger RNA* (section 10-7). This acts as a *template* for the linear assembly of amino acids into an extended polypeptide chain. Combination of the amino acids with the messenger RNA template does not occur directly, but requires a preliminary combination of each amino acid with a molecule of *transfer RNA*. This form of RNA, which is of relatively low molecular weight, is believed to serve as an adaptor, which can recognize and become attached to specific patches of nucleotides in the nucleotide sequence of messenger RNA (Fig. 10-1), thereby positioning its amino acid for incorporation into polypeptide. Each amino acid corresponds to a particular type of transfer RNA, which is specific for it.

Cellular distribution of RNA. The distribution of RNA within cells is very uneven. While a small but definite amount occurs in the nucleus,

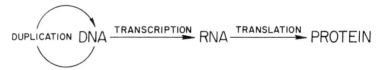

Fig. 10-1a. The flow of information from DNA to synthesized protein.

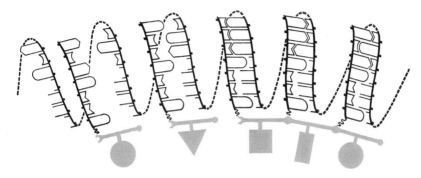

Fig. 10-1b. Highly schematic version of the assembly of amino acids, each attached to a transfer RNA adaptor (represented by a wavy line), on a molecule of messenger RNA to form a polypeptide. While messenger RNA is here shown to have a helical conformation, its actual configuration while serving as a template is unknown.

the bulk of cellular RNA is found in the cytoplasm. Ultracentrifugal examination of the RNA isolated from many biological sources has often revealed the presence of three components of sedimentation coefficients close to 4S, 16S, and 23S. The fraction of lowest sedimentation coefficient, called *soluble RNA*, is distributed throughout the cytoplasm without being localized in any preformed body. This fraction includes transfer RNA.

The greater part (85%) of cytoplasmic RNA is incorporated into discrete particulate entities called *ribosomes*. The 16S and 23S components are localized in these subcellular particles.

The ribosomes are morphologically well-defined elements which occur in the cytoplasm of the cells of a variety of mammalian, fungal, bacterial, and plant sources. They are readily observed by electron microscopic examination and appear as round or slightly asymmetric particles of average diameter 100 to 200 Å. In some cases they occur in a free state within the cytoplasm; in others, they appear to be attached to cell membranes. The ribosomes are particularly rich in RNA, which accounts for about half their mass. The balance of the particles consists primarily of protein, with a high concentration of bound Mg^{++} (7).

The ribosomes may be readily isolated by fractional centrifugation of tissue homogenates. Ultracentrifugal examination of purified preparations has generally revealed the presence of several discrete components. The usual range of sedimentation coefficient is from 20S to 100S, but dimerization can occur to produce larger particles. Ribosomes of different sizes can often be isolated from a single source. For example, particles of 30S, 50S, and 70S are found in *E. coli*. However, the 70S species appears to be an aggregate of one 30S and one 50S particle. Ribosomes of sedimentation coefficient 70 to 80S are found in preparations from a number of biological sources.

The Mg^{++} present in ribosomes appears to be an important stabilizing factor. In the case of *E. coli* ribosomes a dissociation of the 70S species to 30S and 50S particles occurs in media of low Mg^{++} concentration (10^{-4} M).

The RNA of ribosomal particles has been isolated and studied by physical methods. The molecular weights appear to be of the order of several hundred thousand. The physical properties of ribosomal RNA from different species are similar and resemble those of high molecular weight RNA from other sources, such as that of tobacco mosaic virus.

The ribosomes have been clearly shown to be an essential component of the protein-synthesizing systems of living cells (section 10-7). All the cell-free systems thus far investigated which display synthetic properties have included ribosomes as an essential component. It should however be noted that messenger RNA is believed to be quite distinct from ribosomal RNA and is not thought to have more than a fleeting association with ribosomes (section 10-7).

Viral RNA. Viruses may contain either DNA or RNA. However, each particular virus generally contains nucleic acid of only one type. Some examples of RNA viruses are influenza, turnip yellow mosaic, tomato bushy stunt, tobacco necrosis, and the particularly well-studied tobacco mosaic virus (TMV).

10-2 THE PRIMARY STRUCTURE OF RNA

Internucleotide linkages. The ribonucleic acids are polymers of ribonucleotides, to which they may be converted quantitatively by enzymatic or alkaline hydrolysis. There is no evidence for the presence of structural elements other than nucleotides.

As in the case of DNA, the earliest evidence as to the nature of the internucleotide linkages was derived from hydrogen ion titration curves. For carefully prepared RNA which has not undergone degradation during the process of isolation, the occurrence of secondary phosphate ionizations is essentially nil. This indicates that at least two of the three functional sites of each phosphate must be esterified and strongly suggests that the mode of internucleotide linkage must involve phosphodiester bonds. (There is of course a singly esterified phosphate at the terminus of each RNA molecule which undergoes a secondary phosphate ionization. These are however too few in number to make a detectable contribution to the hydrogen ion titration curve.)

Since a 2′-hydroxyl is present in each of the ribonucleotides, the problem of the assignment of the ribose positions bridged by the phosphodiester group was more difficult than in the case of DNA. It was necessary to choose between 2′-3′, 2′-5′, 3′-5′, and (alternating) 2′-2′ and 3′-3′ (or 5′-5′) linkages. As in the case of DNA, examination of the products of exhaustive hydrolysis by enzymes of known specificity was decisive in making the structural assignment (8, 9, 10).

The enzyme spleen phosphodiesterase converts RNA to a mixture of ribonucleoside-3′-phosphates. Since the action of this enzyme is known to be confined to esters of 3′-nucleotides, this result indicates that the phosphodiester linkages of RNA involve the 3′-positions of ribose. Also, the action of snake venom phosphodiesterase hydrolyzes RNA to ribonucleoside-5′-phosphates. The specificity of this enzyme is restricted to the hydrolysis of esters of nucleoside-5′-phosphates. Hence the phosphodiester bonds of RNA must involve the 5′-position of ribose, as well as the 3′-position (8, 9, 10, 11).

These results are virtually conclusive in establishing the mode of internucleotide linkage of RNA as exclusively 3′-5′. There is no evidence for any significant degree of branching. Hence the primary structure of RNA, like that of DNA, is that of linear polyribonucleotides, joined by 3′-5′ phosphodiester bridges (8, 9, 10, 11), as illustrated by Fig. 10-2.

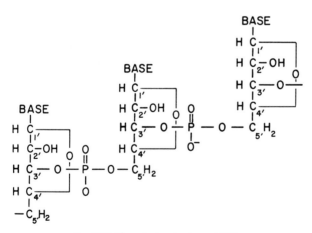

Fig. 10-2. The primary structure of RNA.

Hydrolysis by ribonuclease and by alkali. Ribonuclease may be regarded as a phosphodiesterase of somewhat restricted specificity. Esters of *pyrim-*

idine 3′-*ribo*nucleotides are attacked readily. The ultimate product of hydrolysis is a pyrimidine ribonucleoside-3′-phosphate. However, unlike spleen and venom phosphodiesterase, the mechanism of hydrolysis is biphasic (Fig. 10-3). The initial product is a nucleoside-2′:3′-cyclic phosphate. This is in turn hydrolyzed to a nucleoside-3′-phosphate.

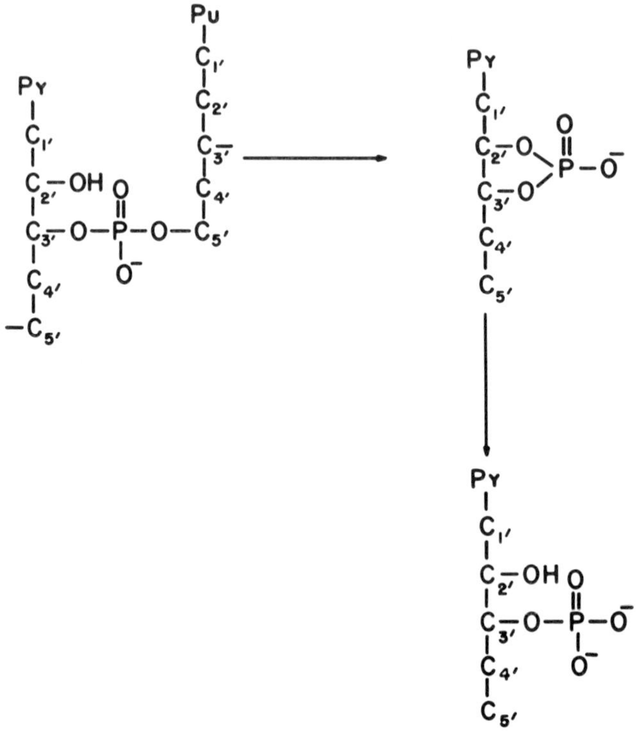

Fig. 10-3a. The bonds attacked during the hydrolysis of RNA by ribonuclease.

Esters of purine 3′-nucleotides are attacked at a rate which is very much slower. Thus the action of ribonuclease upon RNA converts it to a mixture of purine oligonucleotides terminating in a pyrimidine nucleoside-3′-phosphate, as well as pyrimidine 3′-mononucleotides.

RNA is much more susceptible to alkaline hydrolysis than is DNA. This has been shown to be a consequence of the presence of a free 2′-hydroxyl, which serves to labilize the 3′-5′ phosphodiester bonds (10).

As in the case of ribonuclease-catalyzed hydrolysis the reaction proceeds by way of a nucleoside-2′:3′-phosphate intermediate. However, in contrast to the hydrolysis by ribonuclease, the subsequent hydrolytic splitting of the cyclic phosphate is essentially random, so that a mixture of 2'- and 3'-nucleotides is obtained.

Base composition. Preparations of RNA from a particular source (apart from the RNA viruses) are generally quite heterogeneous and consist of

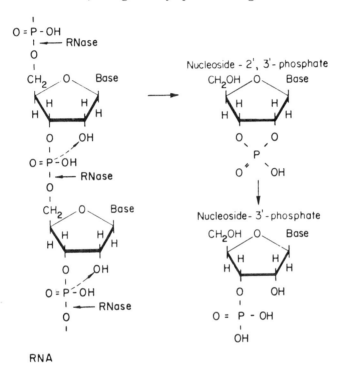

Fig. 10-3b. The complete mechanism for the hydrolysis of RNA by ribonuclease.

a collection of molecules of different base composition. This heterogeneity persists for RNA obtained from different elements of the same cell. Thus the base compositions of nuclear, ribosomal, and soluble RNA are usually quite different.

The base content of RNA does not exhibit the high degree of regularity characteristic of that of DNA (Table 10-1). The base ratios are nonintegral and vary in an irregular way with the source of the RNA. No general pattern analogous to that of DNA has been found for ribosomal, soluble, or viral RNA.

An important difference in base composition exists between soluble and ribosomal, or viral, RNA. Soluble RNA contains a relatively high proportion of "unusual" bases other than the commonly occurring four.

The anomalous bases occurring in quantities of several per cent or more include 5-ribosyl uracil, 5-methyl cytosine, 6-N-methyl adenine, 6-N-

TABLE 10-1. BASE COMPOSITIONS OF SEVERAL VIRAL RIBONUCLEIC ACIDS

	Relative Occurrence			
Source	Adenine	Uracil	Cytosine	Guanine
Tobacco mosaic virus	1.0	0.9	0.6	0.9
Tomato bushy stunt virus	1.0	0.9	0.7	1.0

dimethyl adenine, 1-methyl guanine, 2-N-dimethyl guanine, and 2-N-methyl guanine. The bearing of this difference in composition upon the biological function of soluble RNA remains to be assessed.

The problem of determining the nucleotide sequence of RNA is far more difficult than the analogous case of polypeptides, and relatively little progress has been made in this direction, except in the case of biosynthetic RNA (section 10-6).

In view of current ideas as to the mechanism of RNA biosynthesis (section 10-5), the question of the correspondence of the base compositions and sequence of the DNA and RNA from the same cell assumes considerable importance. The correlation in base *composition* is marginal at best and disappears entirely in the case of ribosomal RNA (35).

Nevertheless, there is compelling evidence that complementarity of sequence exists between ribosomal RNA and a *portion* of the polynucleotide strands of the corresponding DNA (35). Thus the formation of molecular hybrids between the ribosomal RNA of *E. coli* and thermally denatured DNA from the same source has been demonstrated by the CsCl banding technique (35). This is consistent with, and suggests that, ribosomal RNA represents a complementary transcript of a *limited* region of the corresponding DNA molecule. The lack of correlation in over-all base composition is thereby understandable.

10-3 THE SECONDARY STRUCTURE OF RNA

General remarks. Generalization as to the secondary structure of RNA is probably less justified than in the case of DNA, as the available information is much less extensive. Nevertheless, the existing studies do permit a number of conclusions about the configuration of RNA which are likely to survive the more detailed investigations now in progress.

A discussion of the solution properties of RNA should be prefaced by a description of the behavior of structureless polymers which contain similarly charged groups. Such charged polymeric molecules are referred to as *polyelectrolytes*.

In the absence of charges the configuration of polymer coils which are not subject to structural restraints (as the hydrogen bonding of the α-helix or the DNA helix) is governed by purely statistical considerations. This is the case for the randomly coiled polymers discussed in section 4-5. If similarly charged groups are present, their mutual electrostatic repulsion results in a deformation of this limiting randomly coiled state (Fig. 10-4). The magnitude of the electrostatic repulsive forces depends upon the concentration of added electrolyte. Roughly speaking, small ions of charge opposite to that of the polyelectrolyte tend to cluster preferentially about the charged groups of the latter. This diminishes their effective electrical field and reduces the mutual repulsion of the charged sites of the polyelectrolyte.

The configuration of polyelectrolytes represents a compromise between statistical or entropic (section 5-5) factors, which tend to return the molecule to a randomly coiled state, and the mutual repulsion of similar charges, which tends to increase the average separation of the charges, thereby minimizing electrostatic stress. The latter factor becomes less important with increasing electrolyte concentration, so that the polymer contracts (Fig. 10-4).

In general, the intrinsic viscosity of polymer coils increases with their spatial extension. Thus the viscosity of flexible polyelectrolytes increases with decreasing concentration of added electrolyte.

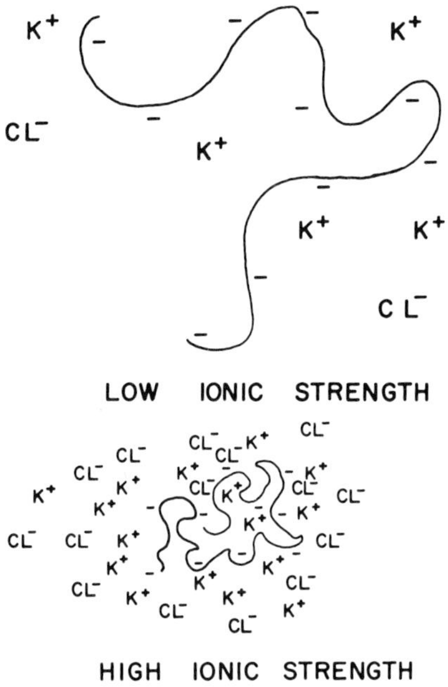

Fig. 10-4. Schematic version of the expansion of a polyelectrolyte under electrostatic stress.

Hydrodynamic and light scattering properties of viral and ribosomal RNA. The hydrodynamic properties of RNA vary with the source. What follows refers in particular to the RNA's of high molecular weight which have been the subject of most of the physical studies made thus far, such as

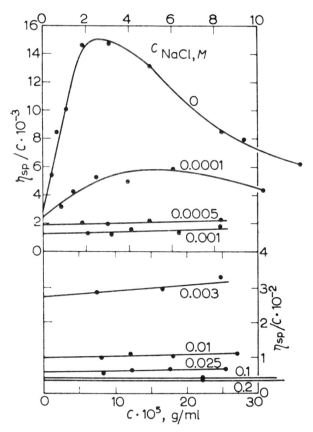

Fig. 10-5. Dependence upon electrolyte concentration of the viscosity of solutions of the RNA from Escherichia Coli.

the RNA of tobacco mosaic virus (TMV) and ribosomal RNA. Transfer RNA will be discussed separately.

In general, the viscosity of these RNA solutions at neutral pH decreases with increasing concentration of added electrolyte (Fig. 10-5). At an ionic strength of the order of 0.1 or higher the intrinsic viscosity of RNA is very much less than that of DNA under comparable conditions (Table 10-2).

The dependence of the intrinsic viscosity of viral and ribosomal RNA upon electrolyte concentration is indicative of an appreciable degree of flexibility which endows RNA to some extent with the properties of an unorganized polyelectrolyte. In contrast, the intrinsic viscosity of native DNA is independent of ionic strength.

Light scattering measurements indicate that, at moderate ionic

TABLE 10-2. A Comparison of Some Physical Properties
of DNA and an RNA of High Molecular Weight (2)

Material	Molecular Weight	Intrinsic Viscosity*	Radius of Gyration*
DNA (calf thymus)	6.0×10^6	72	2700 Å
RNA (TMV)	2.0×10^6	0.7	375 Å

* pH 7, ionic strength 0.1.

strengths, the radius of gyration (R_G) of the RNA of TMV is very much less than that of DNA, even if allowance is made for any difference in molecular weights (Table 10-2). Thus the ratios R_G/M and R_G^2/M are much smaller in the case of RNA.

In summary, all the physical data upon RNA in solution point to a configuration which is much more compact and flexible than the extended filaments characteristic of DNA. It is not safe to extend the generalization further than this.

Hypochromism and optical rotation. The flexibility and compactness indicated by the hydrodynamic and light scattering properties of RNA indicate clearly that the molecular organization is less complete than in the case of DNA and leave open the question of the extent and nature of the secondary structure. However, evidence of another kind shows that an important degree of secondary structure is present (13-16).

The degree of hypochromism of RNA varies with its source and base composition, as well as with such experimental parameters as ionic strength, pH, and temperature. At neutral pH, room temperature, and an ionic strength of 0.1 or more, the degree of hypochromism at 260 mμ is usually large, ranging between 30 and 40%. Moreover, the hypochromism is temperature-dependent and decreases at elevated temperatures (Fig. 10-6).

Since the presence of an appreciable degree of hypochromism is invariably a consequence of the incorporation of purine nucleotides into a polynucleotide, the presence of hypochromism cannot in itself be regarded as conclusive evidence for the existence of an organized secondary structure. However, the *thermal dependence* of hypochromism is very difficult to explain on any basis other than the thermal disruption of *some* structure (13, 14).

The specific rotation of RNA differs widely from that expected for a mixture of free nucleotides of the same composition. Such a mixture would have a specific rotation at 589 mμ ($[\alpha]_D$) close to zero. In contrast $[\alpha]_D$ for most preparations of RNA is 180° to 200°. Furthermore,

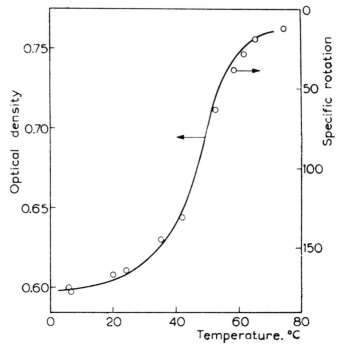

Fig. 10-6. Thermal denaturation of the RNA of tobacco mosaic virus, as monitored by ultraviolet absorbancy and optical rotation (14).

the optical rotation, like the ultraviolet absorbancy, is strongly temperature-dependent (13, 14).

The limiting value of $[\alpha_D]$ for a completely helical polyribonucleotide appears to be close to 300° (14). The difference between this figure and that for DNA (150°) is probably a consequence of the different nature of the pentose in the two cases.

Indeed, both the optical rotation and ultraviolet absorbancy data are definitely consistent with the existence of an important degree of molecular organization. Since the only structures which have been convincingly proposed for polynucleotides are helical in nature and since both the hypochromism and the optical rotation of RNA are consistent with this kind of structure, it is *a priori* rather likely that the molecular organization of RNA is based upon a helix.

This conclusion can be reconciled with the molecular compactness and flexibility indicated by the viscosity and light-scattering data cited earlier, if it is postulated that the molecular organization is incomplete and that the helical regions are separated by unorganized or "amorphous" zones. Such an intermittently helical system could be folded into a relatively compact configuration in a manner rather reminiscent of the globular proteins (Chapter 5). The observed expansion of RNA at low ionic strengths probably reflects both the flexibility conferred by these amorphous regions and the disruption of the helical zones by the enhanced electrostatic stress under these conditions.

Unfortunately, the basic theory for both hypochromism and optical rotation is insufficiently developed to permit quantitative values for helical content to be cited with confidence. The existing estimates vary widely, depending upon the criterion and the origin of the RNA. Figures ranging from 50% to 90% have been cited for the helical contents of RNA of high molecular weight. There is some evidence that the helical content of soluble RNA may be somewhat higher.

X-ray diffraction. The lower degree of molecular organization of RNA, as compared with DNA, has rendered it impossible to obtain X-ray diffraction patterns of comparable quality. The information which has been obtained from RNA fiber diagrams is suggestive of the presence of helical regions of the same type as the helical duplex of DNA, but relatively limited in extent. These are presumably stabilized by hydrogen-bonded base pairs of the Watson-Crick type (Fig. 9-4), with uracil replacing thymine.

Over-all picture of RNA. While the spatial organization of RNA cannot be specified in as much detail as that of DNA, the available information does permit the following conclusions (13, 14, 16):

(1) There is no evidence for the existence of more than one polyribonucleotide strand in the natural RNA's thus far examined and, in contrast to the case of DNA, no fall in molecular weight has been found to accompany denaturation, provided that the hydrolytic scission of phosphodiester bonds is avoided.

(2) Hence the helical regions of natural RNA arise from the hairpin-like bending of the single strand back upon itself (Fig. 10-7). The

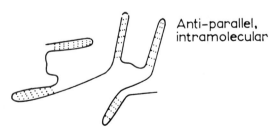

Fig. 10-7a. Schematic picture of one version (4a) of the over-all structure of RNA.

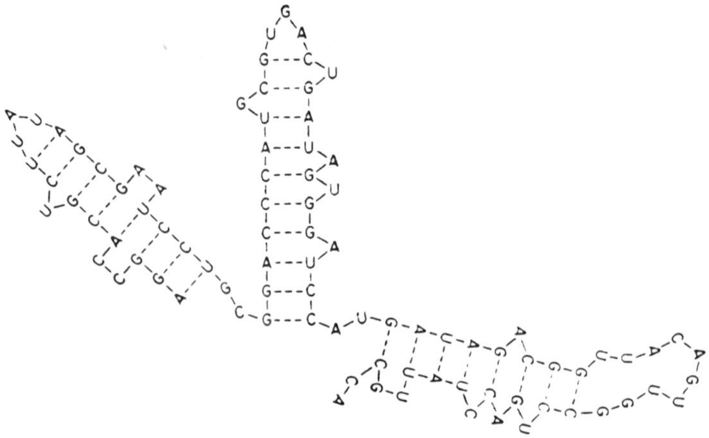

Fig. 10-7b. Nature of the external loops postulated in model (4b) for the structure of RNA (16).

helical segments are thus necessarily anti-parallel, as is the case for the DNA helix.

(3) The helical regions are stabilized by A . . . U and C . . . G interactions similar to those which stabilize the conformation of DNA. However, the irregular base composition of RNA precludes the incorporation of the entire molecule into a perfect double helix of this type, and the helical zones must coexist with random segments.

(4) Two alternative models for the structure of RNA may be postulated: (a) The nonhelical fraction of RNA may take the form of extended amorphous zones separating segments of nearly perfect helical organization. This model presupposes that extensive *complementary* sequences exist within the single strand. These tend to seek each other out to form perfect, or nearly perfect, helical regions. (b) Alternatively, nucleotides which do not participate in the helical organization may occur largely as numerous small *loops* external to the helically paired structures (Fig. 10-7). Implicit in this kind of model is the idea that RNA in solution automatically assumes the most stable configuration, which will in general be that which permits the maximum possible number of hydrogen bonds. For this to be possible RNA must possess some degree of lateral mobility, which permits a choice between various competitive configurations. The reversibility of the thermal denaturation of RNA provides strong evidence that this is the case.

There are no grounds for excluding entirely the structural contribution of all base pairs except those of the Watson-Crick type. The mutual compatibility of these pairings renders them the logical choice for the principal stabilizing factor. However, it is possible that occasional base interactions of a different kind may make a contribution to the stability of the molecular organization of RNA.

In any event, the helical content of viral and ribosomal RNA is insufficient to endow it with a molecular extension and asymmetry approaching those of DNA. In many ways the molecular configuration of RNA offers a parallel to that of the globular proteins. The deformability of RNA under electrostatic stress probably stems both from the expansion of preformed amorphous and flexible regions and from a partial rupture of the helical regions.

Thermal denaturation. It has already been mentioned that both the ultraviolet absorbancy and the optical rotation of RNA are temperature-dependent. Moreover, the degrees of change in these two properties are completely parallel and, when suitably normalized, fall upon the same curve (Fig. 10-6). The changes are reversible and rapid (13-16).

This process appears to be analogous to the thermal denaturation of DNA and undoubtedly reflects the same basic molecular event, namely the "melting" of the helical regions under thermal stress. As in the latter case, the process is an example of a helix-coil transition, whose mechanism is essentially equivalent to that discussed in section 5-5.

A point of divergence from the DNA case is the relative broadness of the thermal profile (Fig. 10-6). This is presumably a consequence of the short character of the helical zones of RNA, which renders its thermal transition of a relatively gradual nature.

The temperature corresponding to the midpoint of the thermal profile, at which 50% of the change is completed, is often (loosely) referred to as the *melting point*. Its value is definitely dependent upon the base composition of the RNA and generally increases with increasing guanine-cytosine content, as in the case of DNA (14).

Highly helical varieties of RNA. The preceding discussion refers primarily to the RNA of high molecular weight which has been isolated from ribosomes and from viruses. There is considerable evidence that *transfer* RNA has a much higher helical content and that its structure approximates a helical duplex of the DNA type.

Thus X-ray diffraction patterns of soluble RNA are consistent with a high degree of molecular organization and resemble those of the A form of DNA (16a). The most probable picture of the structure of transfer RNA is that of a single strand bent into a twisted hairpin shape and organized into a DNA-like conformation. Alternatively, it may be visualized as resembling a short segment of DNA with the two strands joined at one end by a loop.

At least one high molecular weight form of viral RNA is believed to have a highly helical conformation. The RNA of reovirus has a base composition which resembles that of DNA in that the adenine:uracil and guanine:cytosine ratios are close to unity (16b). This and the sharpness of its thermal melting profile suggest that its structure is of the DNA type.

Moreover, the X-ray diffraction patterns of this RNA are indicative of a highly organized molecular structure and resemble those of DNA and transfer RNA. Thus it is likely that the RNA of reovirus has a conformation of the DNA type (16b).

10-4 BIOSYNTHETIC POLYRIBONUCLEOTIDES

Synthesis by polyribonucleotide phosphorylase. A number of enzyme systems are known which can catalyze the polymerization of ribonucleotides. Perhaps the best characterized of these is polyribonucleotide phosphorylase, which has been isolated from numerous microorganisms, including *Azotobacter vinelandii* and *Micrococcus lysodeikticus* (17, 18).

This enzyme has a specificity quite different from that of the DNA-polymerase discussed in Chapter 9. It acts upon *ribo*nucleoside-5'-*di*-phosphates to produce, depending upon the composition of the reaction mixture, either polymers of a single nucleotide or copolymers of any composition. In contrast to DNA-polymerase, the base composition of the product is determined primarily by the composition of the reaction mixture and generally reflects that composition, at least roughly. The reaction is as follows:

(10-1) $\quad$ n XDP $\underset{\text{phosphorolysis}}{\overset{\text{polymerization}}{\rightleftarrows}}$ (XP)n + nP

nucleoside $\qquad\qquad\qquad\qquad$ polymer $\quad$ phosphate
diphosphate

Inorganic phosphate is split off during the forward reaction. This has provided a convenient means of monitoring the rate quantitatively.

For optimal activity the following components, in addition to enzyme, must be present in the reaction mixture:

(1) ribonucleoside diphosphate (ADP, UDP, CDP, GDP, or mixtures thereof)
(2) Mg^{++}
(3) a polyribonucleotide *primer*

In the complete absence of primer the velocity of the polymerization reaction is greatly reduced (20). Under these conditions, the product serves as its own primer, so that the reaction assumes an *autocatalytic* character, its rate increasing with the extent of the conversion.

Polyribonucleotides of both high and low molecular weight have been found to act as primers. In the former case, some degree of specificity occurs. Thus poly A of high molecular weight accelerates the polymerization of ADP, but not UDP.

Primer activity extends down to di- and trinucleotides, but not to mononucleotides. In this case specificity appears to be somewhat relaxed. Thus di- and triadenylic acids stimulate the formation of both polyriboadenylic acid (poly A) and polyribouridylic acid (poly U).

If the primer terminates in a *nonesterified* 3'-hydroxyl, it is chemically incorporated into the product at the terminus of the growing polyribonucleotide chain by a 3'-5' phosphodiester linkage. This is the case for adenylyl-5',3'-adenosine-5'-phosphate (pApA).

Di- and trinucleotides which terminate in a 3'-phosphate, and therefore lack a free 3'-hydroxyl, retain a capacity to stimulate polymerization, but are not incorporated into the product. Adenylyl-3'-5'-adenosine-3'-phosphate (ApAp) primes the formation of poly U, but does not itself combine with the polymer.

The mechanism of action of the primers remains incompletely understood. In contrast to the cases of DNA-polymerase (Chapter 9) or RNA-polymerase the primer has no directive influence upon the composition of the product. In practice, the generally impure enzyme preparations often contain sufficient polynucleotide material to satisfy any primer-requirement.

The polymerization reaction is reversible and approaches an equilibrium with time. In the presence of excess phosphate the enzyme also catalyzes the *phosphorolysis* of polyribonucleotides to the corresponding nucleoside diphosphates (2).

Nature of biosynthetic polyribonucleotides. Polymers containing only adenylic acid (poly A), uridylic acid (poly U), and cytidylic acid (poly C), as well as a mixed polymer containing all four nucleotides (poly AUGC), have been prepared in this way. Their primary structures have been examined by enzymatic structural probes in the same manner as has already been described in the case of natural RNA.

The biosynthetic polyribonucleotides obtained by the action of polyribonucleotide phosphorylase have been found to consist of linear chains of nucleotides united by 3'-5'-phosphodiester linkages. Their *primary* structure is thus entirely analogous to that of natural RNA.

The secondary structures of the biosynthetic polyribonucleotides depend upon their composition (20-25). Polyribouridylic acid (poly U) appears to be structureless in aqueous solution at room temperature, having a randomly coiled configuration. Polyriboadenylic acid (poly A) has at neutral and alkaline pH a partially helical conformation, which is somewhat analogous to that of natural RNA, although the nature of the base pairing which stabilizes the helical regions must of course be different. At acid pH, poly A undergoes a structural transition to a form which is much more highly ordered and helical (20, 21).

The copolymer of all four nucleotides in roughly equal proportions (poly AUGC) resembles natural RNA strongly in physical properties and secondary structure. Like the latter it possesses a fractional helical content, as reflected by optical rotation and ultraviolet hypochromism. It undergoes a thermal denaturation which parallels those of natural RNA's of similar composition (14). X-ray diffraction studies upon fibers of poly AUGC have indicated that the helical regions are of the DNA type and are presumably stabilized by the same hydrogen-bonded base pairs, with uracil replacing thymine.

An important difference exists between poly AUGC and natural RNA with regard to nucleotide sequence. The nucleotide sequence of poly AUGC is essentially *random*. In contrast, the linear order of bases for natural RNA's is highly specific. This is directly related to their biological function (section 9-7).

The similarity in configurational properties of poly AUGC and natural RNA indicates that the secondary structure of the latter is not dependent upon any regularity of base sequence. This suggests that the conformation of RNA may be immaterial to its biological function.

Other ribonucleotide copolymers of a wide range of compositions have been synthesized enzymatically. These have proved to be of great value in studies of the biological role of RNA (section 9-7).

Interaction of poly A and poly U. If solutions of poly A and poly U are mixed at neutral pH in the presence of electrolyte (ionic strength 0.1 or greater), a most interesting reaction occurs. A major decrease in absorbancy at 260 mμ occurs, whose magnitude depends upon the adenine:uracil ratio (23, 24, 25). If the ratio is one to one, the specific rotation at 589 mμ increases to a value close to that for the limiting case of a completely helical polyribonucleotide (300°) (14).

Parallel changes in the solution properties suggest that an interaction of poly A and poly U occurs to form a complex species of very different configuration. Thus the sedimentation coefficient, the molecular weight, and the intrinsic viscosity all show dramatic increases.

X-ray diffraction studies upon fibers of the equimolar complex have revealed that it has a highly helical conformation which is essentially equivalent to that of natural DNA (25). Thus the complex of poly A and poly U is not of a random character but involves the formation of a regular two-stranded helix (Fig. 10-8). This is stabilized by adenine-uracil hydrogen-bonded base pairs which are analogous to the adenine-thymine pairs of DNA.

The high degree of regularity shown by the A + U helical complex indicates that the bases of the two strands must be completely in register. Loops and other defects are eliminated by intramolecular rearrangement or "annealing" during the process of complex formation. The process is remarkably rapid, the change in ultraviolet absorbancy being complete in less than a second (23).

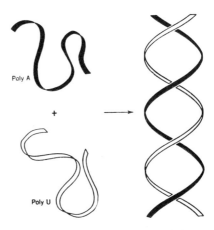

Fig. 10-8. Schematic representation of the interaction of poly A and poly U to form a doubly stranded helix.

The analogy to DNA extends to all details of the secondary structure. The two strands are anti-parallel. The adenine-uracil base pairs are packed into the core of the double helix with the phosphates on the periphery (Fig. 9-3). The bases are mutually parallel and roughly perpendicular to the helical axis. The separation of base pairs in the direction of the helical axis is 3.4 Å. The radius and pitch of the helix are similar to those of DNA (Fig. 9-3).

The (A + U) helical complex undergoes a thermal denaturation which resembles in all respects that of DNA (Chapter 9). It is reflected by a disruption of the helical organization and a separation into single strands.

The principal biological significance of the formation of the (A + U) helical duplex is its demonstration of the absence of any intrinsic kinetic, or other, barrier to the interconversion of single-stranded, randomly coiled polynucleotides and the highly ordered, multistranded helical forms. This has important implications for nucleic acid replication.

Examination of the molecular model of DNA (Figs. 9-3 and 9-5) reveals that a deep helical groove is present. The dimensions of this groove, which is of course also present in the (A + U) complex, are such as to accommodate a third polynucleotide strand.

While DNA itself is not known to add a third strand, such a process does occur in the case of the (A + U) helical duplex under certain conditions. If the ratio of uracil to adenine in the reaction mixture is increased beyond 1:1, then, depending on the ionic strength and temperature, varying amounts of a second helical complex are formed (24).

This has been shown by X-ray diffraction to be *triply stranded*, containing one strand of poly A and two of poly U. The second poly U strand is inserted in the helical groove of the (A + U) complex, which retains its original conformation and hydrogen bonding. Stabilization of the triple helix is attained by additional adenine-uracil hydrogen bonding of the type shown in Fig. 10-9. This is such as not to interfere with the hydrogen bonding of the (A + U) species.

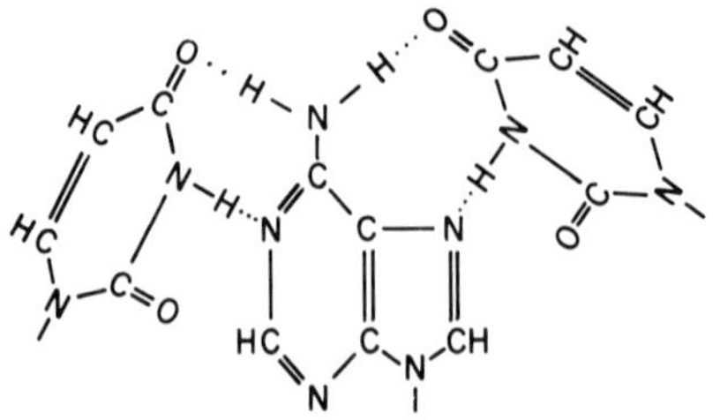

Fig. 10-9. Possible nature of the hydrogen bonding which stabilizes the (A 2U) triply stranded helix (24).

This does not complete the list of helix-coil transitions which have been observed for polyribonucleotides. All the polymers of a single nucleotide (poly A, poly U, poly C, poly G) can exist in helical forms which are stable under certain restricted conditions of pH, ionic strength, and temperature. The interested reader may find further details in the references cited at the end of this chapter (20-25).

Biological role of polyribonucleotide phosphorylase. Despite its widespread occurrence, the biological function of this enzyme remains uncertain. It does not appear likely that it is responsible for the synthesis of RNA which is active in protein synthesis, since there is no evidence that it can form polymers of other than random nucleotide sequence.

10-5 THE GUIDED SYNTHESIS OF RNA

RNA-polymerase. The most probable candidate at present for the role of synthesizing biologically active RNA is the enzyme RNA-polymerase, which has been detected in a number of mammalian and bacterial sources, including *Micrococcus lysodeikticus* and *Escherichia coli* (26, 27, 28).

The specificity of this enzyme is altogether different from that of polyribonucleotide phosphorylase and is somewhat analogous to that of the DNA-polymerase discussed in Chapter 9. Mixtures of all four *ribo-*

nucleoside *tri*phosphates, in the presence of $\overset{++}{Mg}$ or $\overset{++}{Mn}$ and a DNA *primer*, are converted to a polyribonucleotide according to the equation:

$$\left.\begin{array}{l} n\ pppA(ATP) \\ n\ pppG(GTP) \\ n\ pppC(CTP) \\ n\ pppU(UTP) \end{array}\right\} \leftrightarrows (Ap,\ Up,\ Cp,\ Gp)_n\ +\ \underset{\text{pyrophosphate}}{4n\ pp}$$

nucleoside triphosphates   RNA

As in the case of DNA-polymerase, pyrophosphate is split off. The optimal pH for the reaction is close to 7.

The biosynthetic polyribonucleotides formed have been shown to consist exclusively of linear polymers of ribonucleotides joined by 3'-5' phosphodiester linkages. Their primary structure is thus similar to that of natural RNA and the polymers produced by polyribonucleotide phosphorylase.

A still more interesting finding is that the primer DNA appears to have a definite directive influence upon the *nucleotide sequence* of the product. Ideally, it would be desirable to establish this point by a direct comparison of the sequences of the primer and the biosynthetic RNA. Since this is not possible at present, the less satisfactory approach has been adopted of comparing the relative frequency of occurrence of particular nearest neighbor nucleotide pairs in the primer and product (27).

In brief, the procedure is to prepare a biosynthetic RNA by the action of RNA-polymerase upon a mixture of nucleoside triphosphates, one of which is labeled with radioactive P^{32} phosphorus in the ester phosphate (ppp*X). Upon complete alkaline hydrolysis of the product, a mixture of nucleoside-2'- and 3'-phosphates is obtained. Alkaline hydrolysis splits each phosphodiester bond so as to leave the phosphate attached to the nucleoside esterified in the 2' or 3' position. In other words, the bond YpX is split so as to form Yp (2' and 3'). If nucleoside X was present as a labeled ppp*X triphosphate in the reaction mixture, then the Yp*X dinucleotide sequence will be hydrolyzed to yield *radioactive* Yp*.

Thus the relative occurrence of P^{32} in a particular nucleotide found in the hydrolyzate (as measured by its radioactivity) is a direct measure of the frequency with which the given nucleotide is followed by (linked by a 3'-5' phosphodiester bond to) the nucleotide which was originally labeled. In this manner, the relative frequency of occurrence of each of the possible nearest neighbor pairs can be determined.

For example, if radioactive cytidine triphosphate, ppp*C, is present in the reaction mixture, a possible nucleotide sequence in the polyribonucleotide product might be -pUpCpGpUp*CpUpA-. Complete alkaline hydrolysis of the RNA would convert this sequence to a mixture of Up, Gp, Cp, Ap, and (radioactive) Up*. The radioactivity of the Up (uridine-2'- and 3'-phosphate) present in the hydrolyzate is directly proportional to the frequency with which this residue is followed by cytidine in the biosynthetic polymer. By repeating this process for each of the ribonucleoside triphosphates in turn, the relative occurrence of the 16 possible nearest neighbor pairs may be determined.

By studies of this kind it has been found that the relative frequency of each of the possible dinucleotides in the biosynthetic RNA parallels fairly closely that of the corresponding dinucleotide in the DNA primer (with thymine replacing uracil). For example, the relative occurrence of UpA in the RNA synthesized is similar to that of TpA in the primer DNA, which has been determined in a similar manner (section 9-2). This relationship holds for all 16 possible dinucleotide combinations. The implication of this result is that the nucleotide sequences of primer and product are similar.

Mechanism of guided synthesis of RNA. The above finding is very significant with respect to both the mechanism of RNA synthesis and the general question of the transmission of genetic information from DNA to RNA.

To anticipate, the correspondence in nearest neighbor frequencies of the RNA and the primer DNA is exactly what would be predicted if the primer function of the latter consisted in serving as a *template* for the linear assembly of the ribonucleotides. This model presupposes that an *unwinding* of the DNA double helix precedes or accompanies the actual polymerization, so that the primer role is assumed by the individual polydeoxyribonucleotide strands. This picture is made more plausible by the observation that *denatured* DNA also serves as an effective primer.

According to this model, each DNA strand directs the synthesis of its *complementary* RNA strand. Each base in the DNA strand attracts and positions the corresponding member of its Watson-Crick pair (with uracil replacing thymine). Thus, as a hypothetical illustration, the sequence -TACGGAA- in a DNA strand would be translated into the sequence -AUGCCUU- in the RNA synthesized. If both DNA strands prime equally well, the relative frequency of nearest neighbor pairs will be preserved in the product. The postulated manner of attachment of the complementary RNA strand to its primer DNA strand is by the same interbase hydrogen bonding as is responsible for the stability of the DNA double helix itself. The mechanism of the unwinding of the native DNA primer, as well as that of the RNA-primer complex, remains conjectural. This model is very reminiscent of that postulated for DNA-polymerase.

The view that the nucleotide sequence of the biosynthetic RNA is complementary to that of the individual strands of the primer DNA has been further reinforced by the finding that a DNA-RNA *hybrid* can be formed by heating a mixture of primer and product above the denaturation temperature of the DNA and cooling *slowly* (28). The hybrid has been detected by CsCl density gradient ultracentrifugation (Chapter 9), utilizing the fact that its density is intermediate to those of the parent molecules.

The formation of hybrid is only comprehensible if separation of the DNA strands occurs upon denaturation and if some combination of the complementary DNA and RNA strands occurs upon cooling. The possibility that hybrid molecules might arise through some nonspecific interaction has been ruled out by the observation that hybridization does not occur for DNA's other than that used as a primer. This result provides strong support for the complementarity hypothesis and for the general mechanism of guided RNA synthesis summarized above.

It should be noted that the correspondence in base composition and sequence between primer and product is understandable only if:

(1) Both strands of DNA prime equally well (or the two strands are identical in base composition and sequence).

(2) The entire length of each strand is utilized.

The lack of correlation between the nucleotide contents of ribosomal RNA and DNA is not inconsistent with a similar mechanism for the biosynthesis of the former, since it is quite likely that only a part of the corresponding DNA strand is utilized in this case.

10-6 Protein Biosynthesis

General remarks. The guided synthesis of RNA by RNA-polymerase provides an obvious mechanism for the transcription of the information stored in DNA. While it would of course be premature to conclude that this is the sole mechanism operative in all biological systems, the direct demonstration of its existence is most encouraging with regard to the validity of the general scheme outlined in section 9-1 (29-35).

Most of the available evidence suggests that the nucleus is the primary site of RNA synthesis (5, 36). The complementarity of the base sequences of DNA and ribosomal RNA has already been discussed. The administration of radioactive RNA precursors to an intact animal has been shown to result in a sharp initial rise in the radioactivity of nuclear RNA, followed by a rapid decline as the cytoplasmic fractions become labeled (36). Moreover, RNA-polymerase appears to be localized to a large extent in the nucleus.

However, there is some evidence that RNA formation may not be confined absolutely to the nucleus and that some synthesis may occur in the cytoplasm (36). The significance of this and its relation to protein synthesis remain uncertain.

A considerable volume of information indicates that the ribosomes are the sites for the actual assembly of amino acids into proteins (29-37). Much work has been done with the *reticulocytes*, which synthesize hemoglobin. The administration of radioactive amino acids results in the rapid labeling of a protein which is readily extractable from the ribosomes of these cells (30). A high degree of labeling of this protein, believed to be hemoglobin, has been achieved under conditions where labeling of the structural proteins of the ribosomes is negligible. These and parallel experiments with other systems suggest that the rapidly synthesized proteins are not incorporated into the permanent ribosomal structure, but are manufactured for export to other parts of the cell. The metabolic turnover of the protein permanently associated with the ribosomes occurs at a much slower rate.

These observations, together with the invariable requirement for ribosomes in cell-free systems with synthetic capability, leave little doubt that a major part of protein synthesis *in vivo* occurs on the ribosomes. The freshly synthesized protein does not appear to be covalently attached to any permanent component of the ribosomes, in view of the ease with which it may be detached and released into the soluble fraction. Indeed, it is entirely possible that the nascent protein is present on the periphery of the ribosome (30).

The mechanism of the release of newly synthesized protein from the ribosome remains somewhat conjectural. The process does not appear to depend upon any physical disruption of the ribosomes. There is some evidence that the process may be under metabolic control.

At this point we are confronted with the problem of how the information stored in nuclear DNA is transmitted to the ribosomes to be translated into the structures of proteins. A number of grounds exist for excluding ribosomal RNA itself as the direct coding agent:

(1) Considerations related to the control of the rates of protein synthesis (section 10-7) require that the "messenger" from the genes be of relatively short lifetime. Ribosomal RNA appears to be entirely stable in growing cells and does not exchange nucleotides rapidly with other cellular fractions.

(2) The base composition of ribosomal RNA is remarkably constant in many species whose DNA compositions are very different.

(3) The available evidence indicates that ribosomal RNA can repre-

sent a transcript of only a limited portion of the corresponding DNA (section 10-1).

The existence of a metabolically unstable *messenger RNA*, which is formed in the nucleus and migrates to the ribosomes, has been postulated by Monod and co-workers (43) to account for the specificity and controlled nature of protein synthesis. Ribosomal RNA is not currently believed to direct the synthesis of protein, except perhaps for the structural proteins of the ribosomes themselves.

Direct evidence for the existence of such an RNA has been obtained by labeling growing *E. coli* cells for a very short time (10 to 30 seconds) with ^{14}C-uracil. Immediate ultracentrifugal examination of the RNA contents of such cells has shown the radioactivity to sediment at a velocity which does not correspond to that of soluble RNA or any of the components of ribosomal RNA. When short exposure to labeled uracil is followed by addition of excess unlabeled uracil, the new component disappears rapidly, its radioactivity becoming distributed throughout the ribosomal and soluble RNA fractions (43).

Still more compelling evidence for the role of messenger RNA is supplied by studies upon the infection of *E. coli* by T2 bacteriophage. These will be discussed in Chapter 11.

An outstanding gap in our present understanding of protein biosynthesis is the precise nature of the function of the ribosomes. They may serve to bind and orient messenger RNA so as to facilitate the action of the enzymes involved in the actual assembly of amino acids.

Some of the most dramatic recent advances in this area have stemmed from studies with cell-free systems (37-41). All such systems with synthetic capacity have included the following components: (1) ribosomes; (2) an *amino acid activating system,* including enzymes which catalyze both the formation of the acyl adenylates of the amino acids by reaction with ATP and their subsequent attachment to transfer RNA; (3) transfer RNA; and (4) messenger RNA plus a transfer enzyme system which catalyzes the transfer of amino acids from transfer RNA to the messenger RNA-ribosome system. ATP and GTP appear to be essential cofactors.

The amino acid-transfer RNA conjugates are in some manner bound specifically by messenger RNA. While direct proof is lacking, the binding may occur by the mutual recognition and attachment of complementary nucleotide patches in transfer and messenger RNA (Fig. 10-10). The exact mechanism of the process remains uncertain.

It is currently believed that several ribosomes may simultaneously utilize a single strand of messenger RNA, each ribosome traversing the en-

332   THE CHEMICAL FOUNDATIONS OF MOLECULAR BIOLOGY

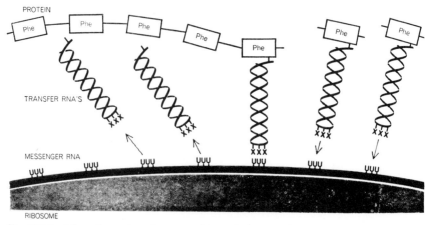

Fig. 10-10. Schematic version of one possible mechanism for the guided assembly of amino acids (42). The case illustrated is the mediation of polyphenylalanine synthesis by poly U. The UUU triplets shown are of course not contiguous. In actuality the bases of poly U are equidistant.

tire length of the strand so as to form a complete polypeptide. The term *polysome* has been given to such complexes containing more than one ribosome.

Activation of amino acids and the role of transfer RNA. As has already been mentioned, the incorporation of amino acids into protein requires their preliminary combination with molecules of transfer RNA, which is the term bestowed upon the active fraction of soluble RNA.

Most amino acids appear to combine with a single specific kind of transfer RNA, although there are instances of two different transfer RNA's which correspond to the same amino acid. Despite the specificity displayed by the different forms of transfer RNA, they all appear to possess the same terminal nucleotide sequence, which is the actual site of attachment of the amino acid (30, 34). This consists of two cytidylic residues followed by adenosine (-pCpCpA).

The addition of these three residues to the end of a polynucleotide chain of unknown base sequence is catalyzed by an enzyme (or enzyme system) present in the soluble cell fraction. The triphosphates of adenosine and cytidine are the precursors of the terminal grouping (34). The prevailing evidence indicates that the two CMP moieties of two molecules of CTP are first attached sequentially to the 3'-hydroxyl of the ribose of the terminal nucleotide, with a concomitant splitting out of pyrophosphate (Fig. 10-11). Then the terminal CMP is coupled with AMP by a further pyrophosphoryl cleavage of ATP (Fig. 10-11). The

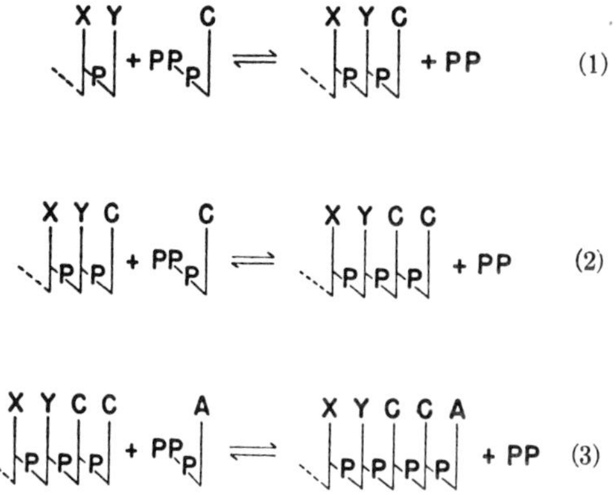

RIBONUCLEIC ACIDS AND THE BIOSYNTHESIS OF PROTEINS

$$\begin{array}{c} X\ Y \\ \Big|\ \Big| \\ \diagdown R\ R \end{array} + PP_R^C \rightleftharpoons \begin{array}{c} X\ Y\ C \\ \Big|\ \Big|\ \Big| \\ \diagdown R\ R\ R \end{array} + PP \qquad (1)$$

$$\begin{array}{c} X\ Y\ C \\ \Big|\ \Big|\ \Big| \\ \diagdown R\ R\ R \end{array} + PP_R^C \rightleftharpoons \begin{array}{c} X\ Y\ C\ C \\ \Big|\ \Big|\ \Big|\ \Big| \\ \diagdown R\ R\ R\ R \end{array} + PP \qquad (2)$$

$$\begin{array}{c} X\ Y\ C\ C \\ \Big|\ \Big|\ \Big|\ \Big| \\ \diagdown R\ R\ R\ R \end{array} + PP_R^A \rightleftharpoons \begin{array}{c} X\ Y\ C\ C\ A \\ \Big|\ \Big|\ \Big|\ \Big|\ \Big| \\ \diagdown R\ R\ R\ R\ R \end{array} + PP \qquad (3)$$

Fig. 10-11. Formation of the terminal trinucleotide sequence of transfer RNA.

process is now complete, and no further attachment occurs. The incorporation of the -pCpCpA terminal trinucleotide appears to be essential for the combination with amino acid (30, 34).

The conversion of each amino acid to a form suitable for incorporation into protein is believed to occur in two stages, both of which appear to be catalyzed by the same *activating enzyme* (30, 34). The activating enzymes do not appear to be localized in any particulate body, but occur in the soluble fraction. Their amino acid specificity, while high, may not be absolute (30). Thus there is evidence that the enzyme which catalyzes the activation of valine also activates threonine, although with a much lower efficiency. Enzymes which catalyze specifically the activation of methionine, tryptophan, tyrosine, alanine, threonine, serine, valine, isoleucine, leucine, and arginine have been isolated in purified form (30).

The initial stage of amino acid activation consists in the formation of an *acyl adenylate* with ATP. Pyrophosphate is split out by the reaction

(10-3) $\quad R-CH-CO\bar{O} + \text{adenosine}-O-\overset{\overset{O}{\|}}{\underset{O^-}{P}}-O-\overset{\overset{O}{\|}}{\underset{O^-}{P}}-O-\overset{\overset{O}{\|}}{\underset{O^-}{P}}-\bar{O}$
$\qquad\quad\ \ \underset{NH_3^+}{|}$

$\qquad\qquad\qquad\qquad\qquad\qquad\qquad\qquad\qquad\qquad\downarrow$

$\quad R-CH-\overset{\overset{O}{\|}}{C}-O-\overset{\overset{O}{\|}}{\underset{O^-}{P}}-O-\text{adenosine} + \bar{O}-\overset{\overset{O}{\|}}{\underset{O^-}{P}}-O-\overset{\overset{O}{\|}}{\underset{O^-}{P}}-\bar{O}$
$\quad\ \ \underset{NH_3^+}{|}$

The acyl adenylate intermediate remains attached to the enzyme, which catalyzes its subsequent attachment to transfer RNA (Fig. 10-12). The combination involves the formation of a covalent bond, probably

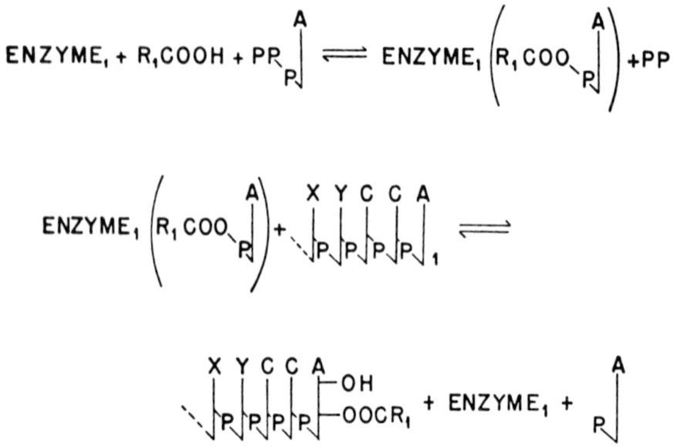

$$\text{ENZYME}_1 + \text{R}_1\text{COOH} + \text{PR}\!\!\begin{array}{c}\text{A}\\|\\\text{P}\end{array} \rightleftharpoons \text{ENZYME}_1 \left(\text{R}_1\text{COO}\!\!\begin{array}{c}\text{A}\\|\\\text{R}\end{array}\right) + \text{PP}$$

$$\text{ENZYME}_1 \left(\text{R}_1\text{COO}\!\!\begin{array}{c}\text{A}\\|\\\text{R}\end{array}\right) + \begin{array}{c}\text{X Y C C A}\\|\,|\,|\,|\,|\\\text{P P P P}\end{array}_1 \rightleftharpoons$$

$$\begin{array}{c}\text{X Y C C A}\\|\,|\,|\,|\,|\text{—OH}\\\text{P P P P —OOCR}_1\end{array} + \text{ENZYME}_1 + \begin{array}{c}\text{A}\\|\\\text{R}\end{array}$$

Fig. 10-12. Mechanism of attachment of activated amino acids to transfer RNA. It is not known whether attachment occurs to the 2′ or 3′ hydroxyl of the terminal adenosine in transfer RNA.

an ester linkage, between the α-carboxyl of the amino acid and the 2′(3′)-ribose hydroxyl of the terminal adenosine. Each amino acid appears to have a binding site in a specific transfer RNA. Since all active transfer RNA molecules are believed to have the same -pCpCpA terminal group, the enzyme specificity must be directed toward groups in the interior of the molecule.

Despite some conflicting data, the available evidence suggests that there is no complete species specificity for the activation of amino acids. Transfer RNA's from different species have in some cases been shown to be interchangeable. The general mechanism is probably quite universal.

Direction of protein synthesis by synthetic polynucleotides. While intact ribosomes are an essential component of all the cell-free systems thus far described which are capable of synthesizing protein, a question naturally arises as to what extent the process is dependent upon the actual configuration of the messenger RNA upon or within the ribosome. It is, for example, conceivable that the spatial conformation of the RNA is unimportant so long as the linear extension of the molecule is sufficient to render its nucleotide bases accessible. If this is the case, then it might be expected that an added polyribonucleotide which is not native to the system might also be able to assume the messenger RNA function by becoming superficially attached to the surface of the ribosome.

The experiments of Nirenberg and Matthaei and of Ochoa have settled this question and have revealed a new approach to the entire problem of deciphering the genetic code (37-42). Working with a cell-free system from *E. coli*, which contained whole ribosomes as well as the other essential components for protein synthesis cited in the preceding section, they demonstrated that the incorporation of labeled amino acids into protein was definitely stimulated by the addition of RNA either from *E. coli* itself or from an altogether different source, such as yeast or TMV.

The exciting observation was then made that stimulation of incorporation also resulted from the addition of synthetic polynucleotides and that, moreover, a definite specificity existed as to the particular amino acids whose incorporation was accelerated by a polynucleotide of known composition. Thus, polyribouridylic acid (poly U) enhanced the incorporation of only a single amino acid, phenylalanine. Incubation of phenylalanine and poly U with the cell-free system described above resulted in the production of a new polypeptide, which analysis revealed to be polyphenylalanine.

This result provided a gratifying confirmation of the template function of messenger RNA. Further experiments showed that the addition of polyribonucleotide copolymers of variable composition could stimulate the incorporation of a variety of amino acids. The specificity of the added polyribonucleotide was conferred by its composition and hence presumably by the frequency of occurrence of particular nucleotide sequences.

It may be recalled that the polyribonucleotides produced by the action of polyribonucleotide phosphorylase are of essentially random sequence (section 10-5). Thus the frequency of occurrence of a particular nucleotide sequence in a given copolymer is governed entirely by statistical factors and can be computed a priori without direct sequence analysis. For example, if one considers the distribution of trinucleotide sequences in a 1:1 UC copolymer, the fraction of such sequences which are of the UUU type is equal to $\frac{1}{2} \times \frac{1}{2} \times \frac{1}{2} = \frac{1}{8}$. The efficiency with which particular amino acids were incorporated into protein was found to be correlated with the frequency of occurrence of one or more such specific sequences.

Thus for the first time it has been possible to make a direct correlation between the composition of a messenger RNA and that of the polypeptide whose synthesis was guided by it. This has led to efforts to use this system to establish the nature of the genetic code itself. A discussion of this aspect should however be preceded by an account of its theoretical background.

The genetic code. As has been mentioned earlier, many of the difficulties associated with the deciphering of the genetic code arise from the fact that there are 20 amino acids and only 4 nucleotides. Thus specificity of sequence is feasible only if each amino acid is coded for by a sequence of several nucleotides. If specificity is to be absolute, each amino acid must correspond to a different nucleotide sequence. This would be impossible if each sequence were only two nucleotides long. There would then be a total of only 4^2 or 16 possible different binary combinations, which is less than the total number of amino acids.

If each sequence is three nucleotides long, a total of 4^3 or 64 different combinations are possible. This is more than enough and thereby presents new problems. If the idea is rejected that a given amino acid might be coded for by more than one nucleotide triplet, then it follows that a large number of the possible combinations must be "nonsense" triplets and must be recognizable as such by the protein-synthesizing system. If this is indeed the case, then the further questions arise as to whether the triplets are overlapping and if, as seems likely, they are not, then at what point occurs the beginning of the message.

The earlier theories advanced as to the nature of the genetic code generally included the assumption that the code was *nondegenerate;* that is, that no two different nucleotide triplets could code for the same amino acid. There are actually no real grounds for this assumption, which has turned out to be incorrect.

Current thinking has tentatively favored the idea that the messenger RNA template is simply divided into groups of meaningful, nonoverlapping triplets beginning at a fixed point, presumably the chain terminus. Other, more complicated coding schemes have been proposed, but have found no substantial experimental support.

Nirenberg and Matthaei, as well as Ochoa and co-workers, have approached the problem of the genetic code by observations upon the selective stimulation of the incorporation into polypeptide of various amino acids by the polyribonucleotide copolymers mentioned earlier. It is assumed that the code is of a triplet type. Upon this basis the frequency of occurrence of each of the nucleotide triplets occurring in a particular copolymer has been quantitatively correlated with the relative efficiency of incorporation of each amino acid (Table 10-3).

In this way a tentative genetic code has been proposed (Table 10-3). Several features of this provisional code deserve comment.

The code is degenerate; that is, two or more coding units can direct the incorporation of a single amino acid. For example, the utilization of leucine is stimulated by both UUC and UUG.

TABLE 10-3. SUMMARY OF CODING UNITS

Amino Acid	RNA Code Words*			
Alanine	CCG	UCG'	ACG'	
Arginine	CGC	AGA	UGC'	CGA'
Asparagine	ACA	AUA	ACU'	
Aspartic acid	GUA	GCA'	GAA'	
Cysteine	UUG			
Glutamic acid	GAA	GAU'	GAC'	
Glutamine	AAC	AGA	AGU'	
Glycine	UGG	AGG	CGG	
Histidine	ACC	ACU'		
Isoleucine	UAU	UAA		
Leucine	UUG	UUC	UCC	UUA
Lysine	AAA	AAU		
Methionine	UGA			
Phenylalanine	UUU	CUU		
Proline	CCC	CCU	CCA	CCG'
Serine	UGU	UCC	UCG'	ACG
Threonine	CAC	CAA		
Tryptophan	GGU			
Tyrosine	AUU			
Valine	UGU	UGA'		

* Arbitrary nucleotide sequence.
' Probable.

The degeneracy of the code may well be related to the occasional occurrence of two different transfer RNA's which correspond to the same amino acid.

Present status of the problem. The tentative nature of the genetic code cited in Table 10-3 should be emphasized. The explosive evolution of the subject at the time of writing shows no signs of abating, and it is possible that extensive revisions in the formulation of the coding units may be necessary.

However, the fact is inescapable that a significant breakthrough in the problem of protein biosynthesis has at last occurred. A high order of probability can now be assigned to the central hypothesis cited in section 9-1.

If it is granted that the basic mechanism for the guided formation of the primary structures of proteins is now beginning to be understood, there remains the problem of how the specific folding into the organized secondary and tertiary structures is accomplished. The hypothesis has been advanced that the molecular architecture of proteins is somehow fixed by their amino acid sequence and that an explicit mechanism for the specific folding of each protein is unnecessary. This "predestination" hypothesis finds some support from the demonstrated reformation of the enzymatically active form of ribonuclease from a state in which the organized structure is almost, or entirely, lost (Chapter 7). However, the question must be regarded as being not entirely settled.

10-7 ENZYME INDUCTION

General remarks. One very important aspect of protein synthesis has been omitted entirely from the preceding discussion. This is the question of how the synthesis of particular proteins is controlled and made to conform to the needs of the cell (43, 44, 45). That some kind of regulatory system must be present is indicated by the specialization of cells with respect to protein synthesis. For example, the genes responsible for the production of hemoglobin must be present in every cell of the body. However, the synthesis of hemoglobin is confined to certain specialized cells, while the cells of most tissues produce no hemoglobin whatsoever.

Perhaps the most compelling evidence for a regulatory mechanism is provided by the phenomenon of *enzyme induction*. When $E.$ $coli$ is grown in the absence of a β-galactoside, only traces of the enzyme *β-galactosidase* are formed by this organism. The addition of a galactoside to the culture medium results in a stimulus of the production of this enzyme by a factor of 10^4. Only galactosides produce this effect. Nevertheless, the synthesis of β-galactosidase has been traced to a specific gene, which has been definitely located on chromosomal maps (Chapter 1). The existence of this gene becomes apparent only in the presence of a galactoside.

Evidence of another kind has been derived from enzyme *repression*. The synthesis of *tryptophan synthetase* by $E.$ $coli$ is likewise dependent upon an explicit gene, which has been clearly identified and located on the chromosome map. The enzyme is formed only when the bacteria are grown on a tryptophan-free medium. The addition of tryptophan completely suppresses its formation.

These findings are not easy to reconcile with the traditional picture of a one-to-one relationship between genes and enzymes, with each gene acting entirely independently.

Structural and regulatory genes. Monod and co-workers have proposed an ingenious mechanism capable of accounting for these findings. While it cannot of course be regarded as established with finality, it is certainly provocative enough to justify description in detail (43, 44, 45). The postulates are as follows:

(1) The molecular structure of each protein is completely specified by *structural* genes, one of which accounts for each type of polypeptide chain if more than one is present. These act by guiding the assembly of ribonucleotides to form a specific transcript of their deoxyribonucleotide sequence. The transcript, messenger RNA, is subsequently detached from the DNA strand and migrates to the ribosomes, where it directs the synthesis of protein.

(2) The formation of messenger RNA is an oriented and sequential process which can be initiated only at certain points on the DNA strand. These points, or segments, of initiation are called *operators*. In many instances the transcription of several adjacent genes may depend upon a single operator. The genes whose action is controlled by the same operator are collectively termed the corresponding *operon* (Fig. 10-13).

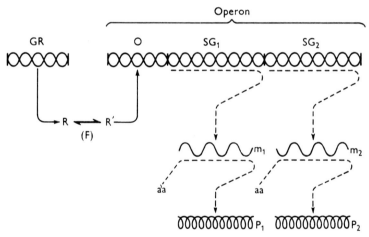

Fig. 10-13. Monod's model for the regulation of enzyme synthesis (43). GR = regulator gene; O = operator; SG_1, SG_2 = structural genes; m_1, m_2 = messenger RNA's made by SG_1 and SG_2; P_1, P_2 = proteins made by m_1 and m_2; R = repressor converted to R' in presence of effector (F).

(3) In addition to structural genes, the chromosome is assumed to contain genes of different function called *regulator* genes. The nucleotide sequence of a regulator gene is identical with all, or part, of that of an operator, according to the original version of the theory. The regulator acts by forming a *repressor* molecule which is somehow able to combine

reversibly with the operator. This association blocks the initiation of messenger RNA synthesis on the operon and thereby inhibits the formation of the proteins controlled by the structural genes of the operon. The nature of the repressor and the mechanism of its combination with the operator remain conjectural. In the original version of this theory, it was assumed to be an RNA transcript of the regulator which combined with the operator by virtue of its complementary nucleotide sequence. However, the actual mechanism may be much more complex.

(4) The repressor (R) is further postulated to react specifically by an unknown mechanism with certain small molecules, designated as effectors (F).

(10-4) $$R + F \leftrightarrows R' + F'$$

(5) For inducible enzyme systems only the intact repressor (R) is supposed to be capable of interaction with the operator. The reaction with an effector, such as a galactoside in the example cited earlier, blocks the action of the repressor and permits enzyme synthesis to occur. For repressible enzyme systems only the modified repressor (R') is active, and the presence of effector (for example, tryptophan in the case of tryptophan synthetase) inhibits transcription.

Much of the evidence for a mechanism of the above type is genetic and depends upon the predictable effects of mutations in structural and regulator genes. Mutations in a structural gene will effect exclusively the corresponding enzyme molecule. Mutations of a regulator gene may effect the syntheses of several distinct proteins which are controlled by different structural genes. Mutations which eliminate the regulator function should generally result in rapid and uncontrolled synthesis of the corresponding protein. In addition, structural and regulatory mutations should be genetically resolvable and should be located at different points of the genetic map.

Studies with $E. \ coli$ are entirely consistent with these predictions. Numerous regulatory mutations have been identified which affect the synthesis of several enzymes, including β-galactosidase, to the same degree. They all are located outside of the structural gene areas of the enzymes on genetic maps.

Genetic evidence for the operator has also been obtained from the same $E. \ coli$ system. Mutations have been identified which block the synthesis of several enzymes and which are located outside the regulator gene area on genetic maps (43).

Perhaps the least satisfactory aspect of the Monod scheme is the nature of the interaction between effector and repressor. In particular, the origin of the high degree of specificity manifested by the effector remains highly speculative. Indeed, there is no *direct* evidence for the originally postulated polynucleotide character of the repressor, and it may well be of a protein nature.

Finally, it should be recognized that the postulated metabolic instability and transient existence of messenger RNA are integral parts of the general scheme. If messenger RNA were stable, it could of course be utilized repeatedly, and the response of the systems to the addition of effectors would be much less rapid.

10-8 THE LINEAR ASSEMBLY OF THE POLYPEPTIDE CHAINS OF HEMOGLOBIN

In the preceding sections the question of precisely how the amino acids are assembled on the messenger RNA template has been left open. A number of mechanisms may easily be postulated, between which it is difficult to make a choice *a priori*. The formation of peptide bonds might begin at one end of the chain and proceed to the other in an orderly, stepwise manner. Alternatively, peptide bond formation might occur simultaneously between all neighboring amino acids on a loaded template. It is also conceivable that bond formation might be initiated at several different points of the chain and that the resultant peptide segments subsequently coalesce to form the complete chain.

Dintzis has attempted to resolve this problem by examining the radioactive labeling of freshly synthesized hemoglobin. Rabbit reticulocytes were selected as the synthesizing system. These account for most of the red blood cells of rabbits made anemic by repeated injections of phenylhydrazine. The protein synthesis occurring in these immature cells is confined almost entirely to the production of hemoglobin. The synthetic activity of the reticulocytes survives their isolation from blood, and they continue to form hemoglobin for many hours after being placed in an incubation medium, provided that the essential amino acids are made available.

It will be recalled (Chapter 6) that hemoglobin contains two kinds of polypeptide chain, each half-molecule possessing one of each type. These are assembled on the ribosomes of the reticulocytes. At any moment hemoglobin chains in all stages of formation will be present on the ribosomes. The completed chains are incorporated into hemoglobin molecules, which are released by the ribosomes into the soluble fraction.

The approach of Dintzis was to add radioactive leucine to the suspension of reticulocytes and observe the distribution of radioactivity in different parts of the polypeptide chains of freshly synthesized hemoglobin (46, 47). Hemoglobin formed at varying intervals after the introduction of labeled leucine was split into its constituent polypeptide chains, which were separated chromatographically. The isolated chains were digested with trypsin to a series of peptides, which were isolated by paper electrophoresis and paper chromatography. Nine leucine-containing peptides from the digest of each polypeptide chain were selected for study. These were identified by their position on the chromatogram, designated arbitrarily by numbers, and assayed for radioactivity.

Each of the possible models for polypeptide synthesis permits a definite prediction as to the distribution of radioactivity in the product. In particular, the *sequential* model, which postulates a stepwise assembly of amino acids starting from one end, predicts a very uneven labeling of polypeptide chains formed at short times after the introduction of radioactive leucine.

Each of the polypeptide chains of hemoglobin may be represented by:

$$A_1A_2A_3 \ldots A_l/B_1B_2B_3 \ldots B_m/, \ldots, /G_1G_2G_3 \ldots G_n/, \ldots$$

where the diagonals indicate the points of attack by trypsin and $A_1A_2A_3 \ldots A_l$ are the amino acids of the tryptic fragment originating from the NH_2-terminal end of the chain, and so forth. Or, in condensed form, the chain may be represented as:

$$ABCDEFG \ldots$$

where the letters stand for tryptic fragments.

If it is assumed that linear assembly begins at A_1, then according to the sequential model the peptides are synthesized in the order A, B, C, D, E, F, G.

At zero time, immediately prior to the addition of labeled leucine, the unfinished chains present on the ribosomes will include all stages of synthesis:

$$\begin{array}{c} A \\ AB \\ ABC \\ \cdot \\ \cdot \\ \cdot \\ ABCDEFG \\ \cdot \\ \cdot \end{array}$$

All peptide units formed after the introduction of radioactive leucine will be labeled. The completed chains corresponding to the above cases may be represented by:

$$AB*C*D*E*F*G* \cdots$$
$$ABC*D*E*F*G* \cdots$$
$$ABCD*E*F*G* \cdots$$
.
.
.
$$ABCDEFG \cdots$$
.
.
.

where the asterisks indicate the presence of radioactivity. Since the probability that a particular fragment is already incorporated at zero time into an incomplete chain decreases from A to G, it would be expected on the sequential model that, in hemoglobin formed *at short-time intervals after addition of labeled leucine,* the degree of labeling of the fragments would increase from A to G.

With increasing time, the fraction of the completed hemoglobin which originates from chains which were partially synthesized on the ribosomes at zero time decreases progressively. Since chains whose assembly is initiated at times subsequent to the addition of label will have a uniform distribution of radioactivity, it may be predicted that the gradient of radioactivity from peptide A to peptide G will decrease with time until uniform labeling is ultimately approached.

The above predicted behavior corresponds completely with the actual findings of Dintzis (46). For short incubation times, the degree of labeling of the 9 peptides of each chain followed a definite order, which was preserved at subsequent times. With increasing time, the gradient of radioactivity progressively declined until, at very long times, all peptides were labeled to the same extent.

Dintzis also examined the labeling of *incomplete* hemoglobin present on the ribosomes as a function of time. In this case, the sequential model predicts that, at very short times, the labeling will be uniform, since each incomplete chain will have added only a short radioactive segment and the probability that the section added will be A, B, and so on is constant. At long times, when the ribosomes have become saturated with labeled leucine, a gradient of radioactivity should reappear. However this is opposite in direction to that of completed hemoglobin, since in a given population of ribosomes there will always be more A segments than B, more B than C, and so forth.

Again the experimental observations were in accord with the predictions of the sequential mechanism. The labeling of unfinished polypeptide was initially uniform. A gradient developed with time, whose direction was opposite to that found for completed hemoglobin chains.

These results are consistent with, and provide strong support for, the conclusion that the individual polypeptide chains of hemoglobin are assembled in a stepwise manner beginning at one terminus. Pending the availability of further evidence upon other systems, it is reasonable to postulate that this mechanism is of general occurrence. Dintzis has recently completed the proof for the case of hemoglobin by correlating the tryptic fragments with the known amino acid sequences of the polypeptide chains (47). The assembly of the chains was found to commence at the NH_2-terminal end.

GENERAL REFERENCES

1. *The Nucleic Acids*, E. Chargaff and J. Davidson, Academic Press, New York (1955).
2. *Polynucleotides*, R. Steiner and R. Beers, Elsevier, Amsterdam (1961).
3. *Proteins and Nucleic Acids*, M. Perutz, Elsevier, Amsterdam (1962).
4. *Ribonucleoproteins and Ribonucleic Acids*, F. Allen, Elsevier, Amsterdam (1962).
5. *The Biological Role of Ribonucleic Acids*, J. Brachet, Elsevier, Amsterdam (1960).
6. *The Chemistry of Nucleic Acids*, D. Jordan, Butterworth, Washington (1960).

SPECIFIC REFERENCES

Ribosomes

7. *Progress in Biophysics*, K. McQuillen, vol. 12, p. 67, Pergamon Press, London (1962).

Primary structure of RNA

8. D. Brown and A. Todd in reference 1, vol. 1, chap. 12.
9. E. Volkin and W. Cohn, *Fed. Proc.*, **11**, 303 (1952).
10. L. Heppel, R. Markham, and R. Hilmoe, *Nature*, **171**, 1152 (1952).
11. W. Cohn and E. Volkin, *Arch. Biochem. Biophys.*, **35**, 465 (1952).
12. W. Cohn, *Proc. Third Int. Congr. Biochem.*, p. 152, Brussels (1955).

Secondary structure of RNA

13. P. Doty in *Structure and Biosynthesis of Macromolecules*, Biochemical Society Symposia, no. 21, p. 8, Cambridge University Press, London (1962).
14. P. Doty, H. Boedtker, J. Fresco, H. Haselkorn, and M. Litt, *Proc. Natl. Acad. Sci. U. S.*, **45**, 482 (1959).
15. B. Hall and P. Doty, "Microsomal Particles and Protein Synthesis," *Wash. Acad. Sci.* (1958).
16. J. Fresco, B. Alberts, and P. Doty, *Nature*, **188**, 98 (1960).
16a. M. Spencer, W. Fuller, M. Wilkins, and G. Brown, *Nature*, **194**, 1014 (1962).
16b. R. Langridge and P. Gomatos, *Science*, **141**, 694 (1963).

Polyribonucleotide phosphorylase

17. M. Grunberg-Manago, P. Ortiz, and S. Ochoa, *Biochim. et Biophys. Acta*, **20**, 269 (1956).
18. R. Beers, *Biochem. J.*, **66**, 686 (1957).
19. Reference 2, chaps. 4 and 5.

Biosynthetic polyribonucleotides

20. R. Beers and R. Steiner, *Nature*, **179**, 1076 (1957).
21. J. Fresco, *Trans. N. Y. Acad. Sci.*, **21**, 653 (1959).
22. A. Rich in *The Chemical Basis of Heredity*, W. McElroy and B. Glass (eds.), p. 557, Johns Hopkins Press, Baltimore (1957).
23. G. Felsenfeld and A. Rich, *Biochim. et Biophys. Acta*, **26**, 457 (1957).
24. G. Felsenfeld, D. Davies, and A. Rich, *J. Am. Chem. Soc.*, **79**, 2023 (1957).
25. A. Rich and D. Davies, *J. Am. Chem. Soc.*, **78**, 3548 (1956).

RNA-polymerase

26. S. Weiss, *Proc. Natl. Acad. Sci. U. S.*, **46**, 1020 (1960).
27. S. Weiss and T. Nakamoto, *Proc. Natl. Acad. Sci. U. S.*, **47**, 694 (1961).
28. E. Geiduschek, T. Nakamoto, and S. Weiss, *Proc. Natl. Acad. Sci. U. S.*, **47**, 1405 (1961).

Protein synthesis

29. A. Meister, *Rev. Mod. Phys.*, **31**, 210 (1959).
30. M. Simpson, *Ann. Rev. Biochem*, **31**, 333 (1962).
31. M. Nirenberg and J. Matthaei, *Proc. Natl. Acad. Sci. U. S.* **47**, 1588 (1961).
32. F. Crick, *Sci. Amer.*, **207**, 66 (1962).
33. J. Hurwitz and J. Furth, *Sci. Amer.*, **206**, 41 (1962).
34. M. Hoagland in *The Nucleic Acids*, E. Chargaff and J. Davidson (eds.), vol. III, p. 349, Academic Press, New York (1960).
35. S. Spiegelman, *Fed. Proc.*, **22**, 36 (1963).
36. J. Davidson in *Structure and Biosynthesis of Macromolecules*, Biochemical Society Symposia, no. 21, p. 29, Cambridge University Press, London (1962).

37. S. Ochoa, *Fed. Proc.*, **22**, 62 (1962).
38. J. Matthaei and M. Nirenberg, *Proc. Natl. Acad. Sci. U. S.*, **47**, 1580 (1961).
39. M. Nirenberg, J. Matthaei, and O. Jones, *Proc. Natl. Acad. Sci. U. S.*, **48**, 104 (1962).
40. A. Tsugita, H. Fraenkel-Conrat, M. Nirenberg, and J. Matthaei, *Proc. Natl. Acad. Sci. U. S.*, **48**, 846 (1962).
41. J. Matthaei, O. Jones, R. Martin, and M. Nirenberg, *Proc. Natl. Acad. Sci. U. S.*, **48**, 666 (1962).
42. M. Nirenberg, *Sci. Amer.*, **208**, 80 (1963).

Enzyme induction

43. J. Monod, F. Jacob, and F. Gros in *Structure and Biosynthesis of Macromolecules*, Biochemical Society Symposia, no. 21, p. 104, Cambridge University Press, London (1962).
44. M. Cohn and J. Monod in *Adaptation in Microorganisms*, p. 132, Cambridge University Press, London (1953).
45. J. Monod and M. Cohn, *VIth International Congress of Microbiology*, Symposium on Microbial Metabolism, p. 42, Rome (1953).

Hemoglobin synthesis

46. H. Dintzis, *Proc. Natl. Acad. Sci. U. S.*, **47**, 247 (1961).
47. M. Naughton and H. Dintzis, *Proc. Natl. Acad. Sci. U. S.*, **48**, 1822 (1962).

11

The Nucleic Acids of Viruses as Carriers of Biological Information

11-1 GENERAL REMARKS

The nature of viruses. As the smallest biological units containing all the information necessary for their own replication, the viruses have been the center of intense interest for many years (1, 2, 3). Despite the heterogeneous character of the agents grouped under the designation of viruses, they share a number of properties. There is first their size. Although highly variable, the diameter is in most cases between 10 and 200 mμ (Fig. 11-1). This exceeds the range of visibility of ordinary microscopes so that electron microscopy is required for the direct observation of most viruses.

A second general characteristic of viruses is their totally parasitic nature. All known viruses have an extremely close relationship with the living cell, outside of which they cannot reproduce. There are no instances of successful cultivation of viruses in a cell-free medium.

The chemical nature of viruses is rather simple in comparison with bacteria or the cells of higher organisms. The very small viruses generally contain only nucleic acid and protein, which consists of only a few, or even a single, molecular species. The nucleic acid may be either DNA or RNA, but not both (Table 11-1).

As is well known, viruses are important as pathogenic agents. As such, they are responsible for a host of diseases affecting human beings, including probably some forms of cancer. In addition to the animal viruses, there exist numerous varieties which attack plants or bacteria. The latter are known as *bacteriophages*.

In general, viruses consist of a core of nucleic acid and an external

TABLE 11-1. SIZES AND NUCLEIC ACID CONTENTS OF SOME IMPORTANT VIRUSES

Virus	Dimension (angstroms)	Nucleic Acid
Adenovirus	700	?
Herpes	1000	DNA
Polyoma	450	DNA
$\phi \times 174$	230	DNA
Tobacco mosaic	3000 × 170	RNA
Influenza	90	RNA
T-even bacteriophage	1000 × 800 (head)	DNA
Orf	2600 × 1600	?
Vaccinia	3000 × 2400	DNA

shell of protein. In some cases, particularly for the larger viruses, the entire particle is enclosed in a membrane, or envelope. The protein case serves as a protective coating and in some instances appears to assist in the penetration of the cell walls of the host (2). The viral nucleic acid enters the cell and diverts its synthetic apparatus to the production of new complete virus. The study of such processes has provided much of the most suggestive information as to the nature of the genetic code and the mechanism of protein synthesis.

Structural classification of viruses. A striking feature of many of the smaller viruses when examined by the electron microscope is their regular and

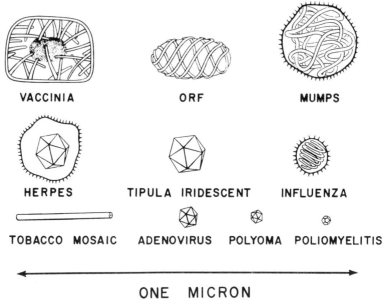

Fig. 11-1. Sizes and shapes of some well-known viruses (4).

symmetrical appearance. When observed at low magnification they often seem to possess spherical symmetry. Examination at higher magnifications shows the presence of distinct structural subunits arranged in a regular pattern (4-8).

In general, the protein component of the smaller viruses consists of a limited number of molecular species. These, singly or in clusters, form the morphologically distinct subunits, or *capsomeres*, of the shell (*capsid*).

A recurrent structural feature of the smaller viruses is the arrangement of their capsomeres in an icosahedral pattern. The icosahedron, which has 20 faces, belongs to the class of polyhedrons which possess cubic symmetry. Other members of this class include the tetrahedron (four faces) and the dodecahedron (12 faces). Each of the 20 faces of the icosahedron is an equilateral triangle (Fig. 11-2).

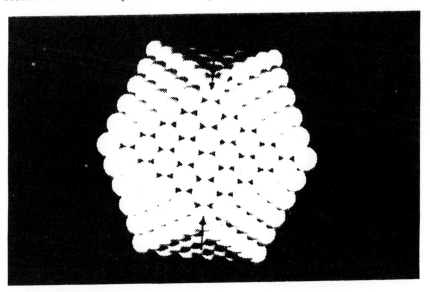

Fig. 11-2. Arrangement of spherical subelements to form an icosahedron (2).

The examination of the architecture of the smaller viruses has required the use of special electron microscopic techniques. In earlier studies using the shadow technique (section 4-2), it was often possible to observe the bumpy character of the virus surface and infer the presence of discrete subunits. The use of negative staining (section 4-2) has permitted the actual counting of the subunits.

For example, the bacterial virus $\phi \times 174$, already mentioned in Chapter 9, appears to consist of 12 capsomeres (4) arranged in icosahedral symmetry (Fig. 11-3). At the highest magnifications each capsomere is seen to consist of smaller subunits arranged in a ringlike structure.

The *herpes* virus, the infective agent of "cold sores," contains 162 capsomeres arranged in an icosahedral pattern (Fig. 11-4). These consist of smaller subunits arranged to form elongated hollow prisms of which 12 are pentagonal and 150 hexagonal (5). The former are located at the corners and the latter on the faces and edges of the particle (Fig. 11-4). The combination of pentagonal and hexagonal units is geometrically consistent with the enclosure of space, which cannot be accomplished by hexagons alone.

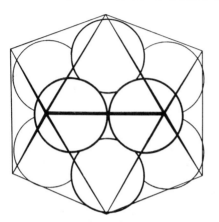

Fig. 11-3. Schematic model of ϕ X 174 virus.

The *adenovirus*, which is an agent of respiratory diseases in man, likewise has an icosahedral structure (6). However, the number and shape of the capsomeres are quite different from the case of the herpes virus (Fig. 11-5). There are a total of 252 capsomeres, of which 12 are located at corners and the balance on edges or faces. The former have five nearest neighbors and the latter six.

The *polyoma* virus, which produces cancer in rodents, probably contains 42 capsomeres arranged in icosahedral symmetry (2). A possible arrangement of hexagonal and pentagonal units is shown in Fig. 11-6.

Not all viruses possess icosahedral symmetry. The architecture of a second major category is based upon a helix. Tobacco mosaic virus (section 11-2) is the best characterized example of this class.

Certain more complicated animal viruses also possess helical symmetry. Electron microscopic observation suggests that *influenza* virus contains an internal component consisting of subunits arranged in a compact spiral (Fig. 11-6), which is

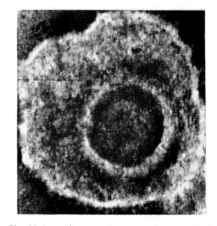

Fig. 11-4a. Electron microscope photograph of herpes virus, showing envelope which it sometimes possesses (4).

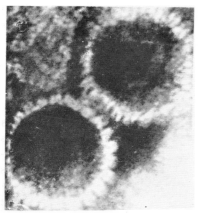
Fig. 11-4b. Herpes virus without envelope, showing capsomeres (4).

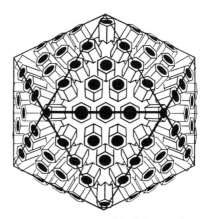
Fig. 11-4c. Schematic model of herpes virus (4).

surrounded by an external membrane (8). Numerous surface projections extend from the latter (Fig. 11-7). These are believed to figure in the attachment of the virus to the surface of attacked cells.

There exists a third broad group of large viruses which do not exhibit a simple symmetry. This class includes the "T-even" bacteriophages (section 11-4). The viruses of vaccinia, orf, and cowpox likewise are of this type.

The vaccinia virus contains a dense interior region surrounded by a set of distinguishable membranes. The over-all structure is quite complex

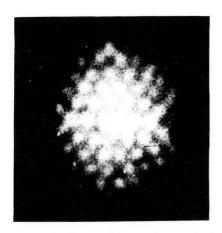
Fig. 11-5a. Electron microscope photograph of adenovirus (2).

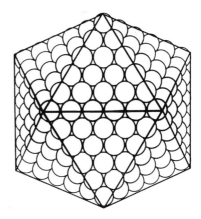
Fig. 11-5b. Schematic model of adenovirus.

(4). The orf virus (Fig. 11-8) appears to contain tubular components arranged in a criss-cross pattern (4).

11-2 THE INFECTIOUS RNA OF TOBACCO MOSAIC VIRUS

Characteristics of the virus. Tobacco mosaic virus (TMV) is the infectious agent of a well-known disease of tobacco plants (9, 10). The symptoms of this disease include a characteristic light and dark pattern, as well as a deformation of the leaves (Fig. 11-9). Because the disease may be transmitted to plants of different families, it has a rather low host specificity.

Fig. 11-6. Schematic model of polyoma virus.

The cells of the tobacco leaf are protected by a cellulose cell wall, which forms a barrier to penetration by foreign objects. In practice, the infection process is facilitated by mechanical injury to the leaves, as may be produced by rubbing with the finger or a spatula, to permit entry of the virus. The number of necrotic spots, or lesions, increases with the concentration of virus and provides the basis of a method for assaying the virus content.

The purified virus is easily isolated from the sap of infected plants by high speed centrifugation. Up to 2 grams of virus can be prepared in this way from a liter of sap. The virus may be obtained in the form of paracrystalline needles.

Structure of the virus. Electron microscopic examination of purified TMV shows the individual particles to be rods or cylinders about 3000 Å long and 75 Å in radius (Fig. 4-2). The molecular weight, as obtained by light-scattering, is close to 40×10^6 (9, 10, 11).

The chemical composition is remarkably simple, even for a virus. The only nucleic acid present is RNA which accounts for about 5% of the mass of the virus. If all the RNA were present as a single molecule, it would have a molecular weight of about 2×10^6. Physical measurements, including light-scattering, have shown that carefully prepared RNA from this source has a molecular weight not far from 2×10^6. Thus it is likely

NUCLEIC ACIDS OF VIRUSES AS CARRIERS OF INFORMATION 353

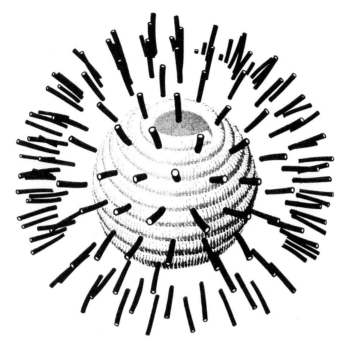

Fig. 11-7. Schematic model of influenza virus (4).

Fig. 11-8. The orf virus (4).

Fig. 11-9. Appearance of a tobacco leaf infected with tobacco mosaic virus (2).

that the RNA of this virus does indeed occur as a single molecule (10).

The balance of the particle consists of 2130 identical globular protein units, of molecular weight 17,000. The viral RNA is present as a central core, entirely surrounded by protein (10). Under certain conditions it is possible to strip away part of the protein coat so as to expose part of the RNA. This may be accomplished through limited treatment with detergents. Electron microscopic examination of the products of such partial deproteinization reveals the RNA as a filament protruding from the remainder of the virus cylinder, rather like the fuse of a firecracker. Treatment with ribonuclease removes the exposed RNA.

X-ray diffraction has permitted a closer look at the detailed structure of the virus particle (12, 13, 14). The spatial organization is based upon the versatile and ubiquitous helix. The RNA is coiled into a single chain helix of radius 40 Å (Fig. 11-10), which extends the entire length of

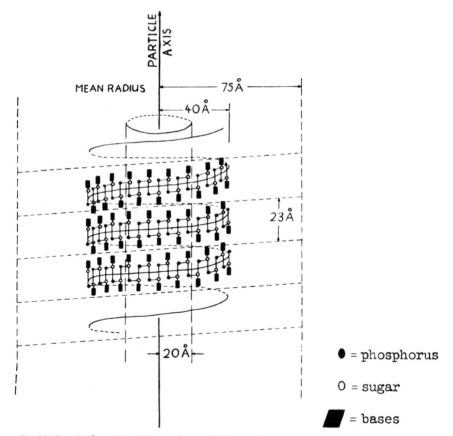

Fig. 11-10. Configuration of the single-stranded RNA of TMV within the intact virus particle (10).

the particle. The spacing of consecutive turns is such as to make stabilization of the helix by base pairing of the DNA type unlikely. Apparently, stabilization of the RNA conformation within the intact virus is achieved primarily by interactions with the protein subunits.

The structurally equivalent subunits of TMV are arranged in a helical array on the periphery of the particle (Fig. 11-11). The maximum radius is close to 85 Å. The pitch of the helix is 23 Å. The RNA helix is fitted into the protein array in a very compact manner. A hole of radius about 20 Å runs down the center of the particle.

Isolation of viral RNA. The separation of the viral RNA from the protein in such a way as to preserve its biological activity requires a procedure mild enough to avoid rupture of phosphodiester bonds, thereby maintaining the integrity of its primary structure (10, 15, 16). The most successful technique thus far developed involves emulsifying the virus with

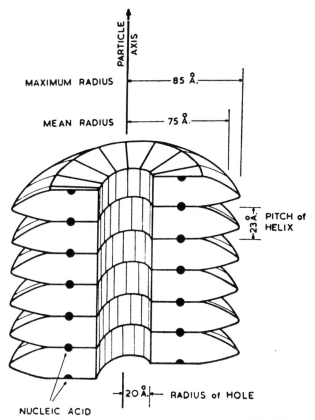

Fig. 11-11a. Arrangement of the protein subunits of TMV (10).

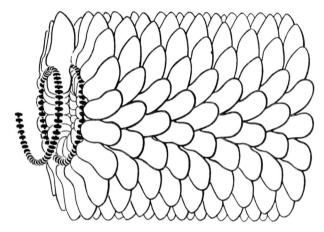

Fig. 11-11b. Appearance of a TMV particle with part of the protein coat stripped away, exposing the RNA strand (4).

a phenol-water mixture. This results in a suspension of small phenol droplets throughout the mixture, thereby producing a very large water-phenol interfacial area. Denaturation of the viral protein occurs at this interface. The denatured protein is insoluble.

The emulsion is then broken by centrifugation into a denser phenol phase, and a less dense aqueous phase. The denatured protein collects at the interface, while the viral RNA remains in solution in the aqueous phase. By several repetitions of this process a preparation of RNA which is essentially free of protein may be obtained.

Infectivity of protein-free RNA. Viral RNA prepared as described above has an infectivity, when rubbed onto tobacco leaves, ranging up to about 0.5% of that of the same weight of RNA incorporated into the intact virus. The disease induced by injection of RNA has symptoms entirely similar to those produced by infection with intact virus. Moreover, intact virus may be isolated from the sap of infected plants which has all the characteristics of normal TMV.

It has been demonstrated conclusively that the residual infectivity arises from the RNA itself and is not due to contamination with intact virus. Thus the action of ribonuclease, which does not attack intact TMV, results in the loss of all activity.

The reason for the low activity of viral RNA, as compared with intact TMV, remains uncertain. It may arise from unavoidable alteration of the molecule during the isolation process, or from an enhanced susceptibility to hydrolytic enzymes present in the tobacco plant.

The property of infectivity appears to require the preservation of all, or almost all, of the *primary* structure of the native RNA. One or two ribonuclease-induced breaks per molecule appear to be sufficient for the loss of all activity. In contrast, there is no evidence for the dependence of infectivity upon the maintenance of the secondary structure.

The RNA-free viral protein has been examined for infectivity and found to be completely inert.

Reconstitution of virus. If a solution of protein-free RNA is mixed with viral protein *in vitro*, a spontaneous combination occurs to produce a particle which is indistinguishable from natural TMV when examined by electron microscopy (10). The RNA content is about 5%, as in the case of natural RNA.

The reconstituted virus has an infectivity which is considerably greater than that of free RNA. This is probably a consequence of its greater resistance to ribonuclease degradation. Indeed, the currently prevalent opinion is that the function of viral protein may be essentially passive in nature, being primarily concerned with the protection of the RNA from enzymatic destruction.

If RNA from a particular strain of TMV is combined with protein obtained from a different strain, a reconstituted particle having the usual characteristics of TMV is obtained. However, the infectivity characteristics and the nature of the disease symptoms are determined entirely by the RNA. Moreover, the fresh virus synthesized is exclusively of the strain from which the RNA was obtained.

The interaction of virus protein with RNA in the reconstitution is not highly specific. Synthetic polyribonucleotides of molecular weight 10^4 to 10^5 can combine with TMV protein to form rods of the same apparent width as TMV, but of variable length. Such rods are of course biologically inert.

Mechanism of the infection process. The synthesis of intact virus does not begin until about 20 to 30 hours after the initiation of infection (17). The period is somewhat shorter if inoculation is made with free RNA rather than with intact TMV, suggesting that a part of the latent period in the latter case is occupied with the stripping off of the protein coat.

A large amount of free viral RNA is synthesized prior to the appearance of intact virus. The concentration of free RNA actually passes through a maximum, decreasing as the incorporation into virus begins.

The synthesis of viral protein does not begin until several hours after RNA synthesis begins. Once combination of RNA and protein commences, the production of complete virus proceeds very rapidly. After about 100 hours, all the RNA is packed into protein. The concentration of virus levels off when a concentration of about 10^{-3} grams virus per gram of tobacco leaves is attained.

Inactivation of infectious RNA. The infectious RNA of TMV may be inactivated by irradiation with X-rays in the dry state. The inactivation process obeys first-order kinetics and appears to correspond to an actual scission of the RNA strand to form inactive polynucleotides of lower molecular weight (10).

Inactivation may also be accomplished by the action of a number of chemical agents. Thus treatment with formaldehyde (HCHO), which reacts with the primary amino groups of adenine, cytosine, and guanine, results in inactivation. In general, most reagents which are known to alter the bases of RNA will abolish the property of infectivity.

It is also possible to inhibit the replication of infectious RNA by incubation of infected tobacco leaves with various chemical analogs of natural bases. These are incorporated into freshly synthesized RNA to yield a product with diminished capacity for multiplication. Among the "unnatural" bases capable of subverting the replication of infectious RNA are 2-thiouracil, 5-fluorouracil, and 5-bromouracil.

The production of mutants. As has been discussed in Chapter 8, the action of nitrous acid (HNO_2) upon nucleotides results in the deamination of cytosine to form uracil and of adenine and guanine to form hypoxanthine and xanthine, respectively. This deamination also occurs upon treatment of RNA with HNO_2.

It has been demonstrated that the limited exposure of the infectious RNA of TMV, or TMV itself, to HNO_2 can result in the production of "mutants" whose infectious properties are different from those of the parent strain (18). This has been shown with a strain of TMV normally incapable of producing necrotic lesions upon the Java variety of tobacco. Treatment of its RNA with HNO_2 results in the appearance of necrotic lesions after inoculation of the Java plant. The concentration of active mutants is dependent upon the time and conditions of exposure to HNO_2.

Thus in this case a definite correlation exists between alteration of the nucleotides of RNA and the appearance of a new biological property.

The production of viable mutants is accompanied by a parallel inactivation of the virus. Whether a particular chemical mutation is lethal will presumably depend upon the number and location of the bases which are altered.

Effect of mutation upon composition of protein subunit. Further insight into the function of the RNA of TMV as a determinant of protein structure has been gained by a comparison of the amino acid composition of the protein component of normal and mutant TMV.

Fraenkel-Conrat and Tsugita have utilized the action of nitrous acid to produce mutants distinguishable by their altered infectivity. The mutant selected, designated by the number 171, produced only local lesions in Java tobacco (19).

A comparison of the over-all amino acid compositions of the protein subunit of mutant 171 with the corresponding figures for the common variety of TMV disclosed significant differences. The number of residues per molecule of aspartic acid, threonine, and proline decreased by one, while the number of alanines, serines, and leucines showed a corresponding increase of one residue each.

It was moreover possible to pinpoint the location of one of the altered residues. The COOH-terminal sequence of the protein of common TMV is -Gly-Pro-Ala-Thr. Hydrolysis by carboxypeptidase is thus confined to the terminal Ala-Thr bond, since further hydrolysis is blocked by the presence of a proline residue (section 3-4). In the case of mutant 171, stepwise hydrolysis proceeds up the chain; threonine, alanine, *leucine*, and glycine are split off in that order.

Hence it appeared that the COOH-terminal sequence had been altered from -Gly-Pro-Ala-Thr to -Gly-Leu-Ala-Thr.

The results with mutant TMV are in basic accord with the hypothesis that the primary structure of the protein subunit of TMV is dictated by the nucleotide sequence of its RNA and provide strong evidence for the general mechanism of protein synthesis outlined earlier.

The observed replacements of particular amino acids resulting from nitrous acid treatment of the infectious RNA of TMV provides an independent means of checking the tentative genetic code discussed in Chapter 11. In the example cited above, proline has been replaced by leucine. According to the provisional code, the coding units for proline and leucine are UCC and UUC, respectively. Since nitrous acid treatment converts cytosine to uracil, it is entirely feasible that UCC could be converted to UUC in this way. Hence this result is consistent with the proposed genetic code.

In subsequent work a number of amino acid replacements have been identified and correlated with the expected changes in the corresponding coding triplets. The agreement, while imperfect, is good enough to be encouraging with respect to the validity of the proposed genetic code.

Significance of results with the infectious RNA of TMV. The demonstration that a single RNA *molecule* of well-defined composition could mediate its own reproduction, as well as that of its protein partner, was a fundamental turning point in the evolution of theories of the replication of biological systems. In brief, the following conclusions could be drawn:

(1) The infectious RNA of TMV contains within itself all the biological information required to instruct the host organism in the details of its own synthesis and that of the viral protein.

(2) The only feasible way in which this information can be stored in the RNA molecule is by virtue of its specific nucleotide sequence.

(3) Therefore, the RNA molecule must be able to serve as a kind of *template* for the assembly both of nucleotides to form a new RNA molecule and of amino acids to form the protein subunits. It is not as yet possible to specify whether this is accomplished directly by the enzyme systems already present in the host or whether the directed synthesis of some new enzyme or enzymes occurs first.

In any event, the specific nucleotide sequence of the infectious RNA is translated first into the specific nucleotide sequence of new RNA and subsequently into the specific amino acid sequence of viral protein. The replication of viral RNA probably proceeds by a mechanism very similar to that involved in the semiconservative replication of single strands of separated DNA.

11-3 INFECTIOUS RNA'S OF OTHER VIRUSES

Tobacco ringspot virus. This is a plant virus, which produces a characteristic disease of tobacco plants. Unlike TMV, the intact virus particle has a spherical shape. The RNA of this virus has been isolated by methods similar to those employed in the case of TMV and has been shown by analogous techniques to have a residual infectivity. Treatment with ribonuclease destroys all activity.

The animal viruses. A number of animal viruses are of the RNA type. While investigations of their structure are much less advanced than in the case of TMV, there is definite evidence that, in the examples thus far examined, the RNA is in the interior of the particle, being encased in protein.

Infectious RNA has been isolated from a number of animal viruses. These include *equine encephalomyelitis, poliomyelitis,* and *foot and mouth disease* virus. The proof that the infectivity was due to RNA itself has in each case been accomplished by methods similar to those utilized in the case of TMV.

11-4 THE T-EVEN BACTERIAL VIRUSES

General remarks. The bacterial viruses, or *bacteriophages*, are parisitic microorganisms whose hosts are bacterial cells. Like other viruses, they are completely parasitic and are unable to maintain metabolism or to replicate outside of the host. They have the property of diverting the metabolism of the host to their own purposes, so as to harness it almost exclusively for their own replication (21, 22).

The bacteriophages include numerous species of different size, composition, and properties. Much of the available information as to the process of bacteriophage infection has been derived from studies with a series of seven phages which attack a common host, strain B of *E. coli*. These seven, which are known as the *T phages*, were isolated independently but were later shown to fall into related classes. One of these, which fortuitously includes T2, T4, and T6 (the *T-even* phages), has characteristics which render its members particularly favorable subjects for study. Thus by far the most detailed information is available for these three viruses, and the bulk of the discussion to follow will be centered about them.

Electron microscopic examination of the T-even phages shows them to be shaped somewhat like tadpoles, consisting of a bipyramidal hexagonal head and an elongated tail (3, 4, 21, 22) (Figs. 11-12 and 11-13). The tail appears to consist of a helical contractile sheath surrounding a central hollow core. At the end of the core there is a plate to which about six tail fibers are attached.

The "molecular weight" of the T-even phages is from 200 to 500×10^6. About 25% to 50% of the total mass consists of DNA. This is confined to the head, where it occurs in the interior and is encased by protein.

If the T-even phage is transferred to a medium of low ionic strength, a swelling of the nucleic acid core occurs which results in the rupture of the protein case and the release of the DNA into solution.

In addition to DNA several proteins are known to be present. These include the head protein, which encloses the DNA, and the protein of the tail fibers, as well as several other tail proteins. There also appears to be an enzyme present in the tail which can attack the bacterial cell wall. The virus particle also contains the polyamines putrescine and spermidine. Tightly bound ATP and dATP are present in the tail.

The mechanics of infection. In the absence of host cells the T-even phages are completely inert and show no sign of reproduction or metabolism. Growth occurs only within the host cells of *Escherichia coli* (*E. coli*).

362   THE CHEMICAL FOUNDATIONS OF MOLECULAR BIOLOGY

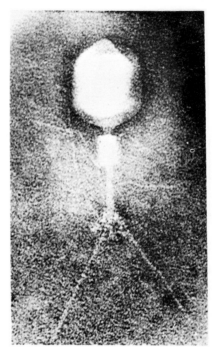

Fig. 11-12.   Electron microscope photograph of a T2 bacteriophage (2).

The mode of infection is best described as like that of a microsyringe (20, 21, 22, 23). When introduced into a culture of *E. coli*, the T2 phage particles become attached to the bacterial cells by their tails (Fig. 11-13). Initial contact with the bacterial wall is probably made by the plate structure and tail fibers. After contact has been made the helical sheath contracts (Fig. 11-13), becoming much shorter and thicker. The mechanism of contraction is still rather poorly understood.

The attachment of the phage particles is followed by an enzymatic attack upon the bacterial cell wall. The next step is the injection of the DNA contents of the phage head into the bacterium. This appears to be a rapid and "all-or-none" phenomenon. Precisely how the bulky DNA molecule is squeezed through the relatively narrow core of the tail in a short time remains a matter for conjecture.

Almost all the viral protein remains outside the bacterial cell, and there is no evidence for the participation of the protein in the subsequent events.

Phases of the infection process. During the initial phase of the period following injection of the DNA, no infectious material can be located either

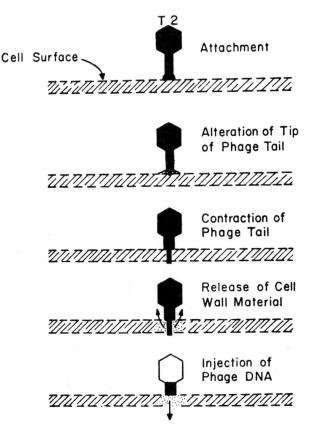

Fig. 11-13a. Schematic version of infection process.

within the bacterium or attached to its surface, and the phage is said to be in a *vegetative* state. This initial period is referred to as the "*eclipse period*," which lasts for about 10 minutes (20, 21, 22).

After introduction of viral DNA into the bacterial cell, the synthesis of *bacterial* DNA and RNA ceases. However, synthesis occurs of certain proteins essential to the phage replication. These are not found in mature phage and almost certainly include enzymes needed to make specific viral components. Synthesis of RNA is also observed although RNA is not found in mature phage.

After about 6 minutes, the synthesis of new viral DNA begins. The rate of synthesis augments rapidly until a limiting rate of about 10 phage DNA units per minute is attained after about 10 minutes.

After about 8 minutes, detectable synthesis of phage protein begins. At about 10 minutes after the initiation of infection, the first mature

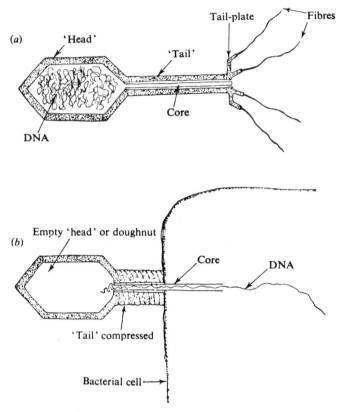

Fig. 11-13b. Contraction of sheath in initial stages of infection (2).

phage appears within the bacterial cell. At this time a reservoir of about 50 units of phage DNA and 10 to 20 units of phage protein is present. As the assembly of intact phage progresses, this reservoir is continually drawn upon. However, the rates of DNA and protein synthesis appear to be sufficient to maintain it at a roughly constant level.

The formation of mature phage appears to be irreversible. Once phage protein and DNA are incorporated into complete virus, they are effectively inaccessible to the dynamic equilibrium existing in the cell.

The accumulation of intact phage continues until rupture of the cell wall occurs after about 21 minutes. This is accompanied by the release of phage into solution. Up to 100 phage particles may be released at this time.

Characteristics of the phage DNA. The base composition of the DNA of T-even phage is unusual in that cytosine is replaced entirely by 5-hydroxymethyl cytosine. The Watson-Crick base ratios are preserved, the contents of guanine and 5-hydroxymethyl cytosine being equivalent. The physical properties of phage DNA are entirely similar to those of DNA from other sources, and there is no reason to doubt that the secondary structures are identical, with 5-hydroxymethyl cytosine forming the same hydrogen bonds with guanine as does cytosine. The 5-hydroxymethyl cytosines are often substituted in the hydroxymethyl group by one or two glucose units.

If all the DNA of a T2 bacteriophage occurred as a single molecule, it would have a molecular weight close to 130×10^6. Because molecules of this order of size are very difficult to study with the conventional techniques, an *autoradiographic* approach has been developed (24).

If virus is grown on bacterial cells which have multipled in a medium containing a high specific activity of P^{32}, bacteriophage particles may be prepared with as many as 100 to 200 P^{32} atoms per particle. This occurs almost entirely in the phage DNA. When such labeled particles are embedded in a photographic emulsion, each β particle arising from the radioactive decay of a P^{32} atom will produce a visible track in the emulsion, originating at the phage particle. The cluster of such tracks originating from a single particle after prolonged standing is called a "star."

The average number of tracks per star is a measure of the number of P^{32} atoms in the corresponding particle and hence, if labeling is uniform, of the size of its DNA component. By comparative studies of the star sizes of intact phage and of its DNA which has been liberated by disrupting the virus, it has been possible to obtain a measure of the molecular size of phage DNA.

The best current estimates, based on autoradiographic and other evidence, indicate that the DNA of T2 bacteriophage is a single molecule whose molecular weight is of the order of 130×10^6.

It is worthy of mention that the length of completely extended phage DNA is greater by orders of magnitude than the dimensions of the phage head into which it is packed. Clearly it cannot exist within the phage as a perfect Watson-Crick double helix. The means whereby the molecule is endowed with sufficient flexibility to permit its folding into a compact shape is uncertain. It is possible that occasional interruptions in the complimentary base pairing provide amorphous "hinge points," which allow folding to occur.

Fate of injected DNA. By allowing phage to multiply within a host which has been grown in a culture medium containing derivatives of radioactive phosphorus, it is possible to obtain phage preparations whose DNA is labeled with P^{32} containing nucleotides. If this labeled phage is allowed to infect bacteria grown in a normal medium, it is possible to study the fate of injected DNA by tracer experiments.

Three main conclusions have emerged from such studies (20, 25):

(1) The bulk of the DNA ultimately appears in the new phage. A recovery of at least 60% of the initial radioactivity has been reported. Because of inevitable losses, this is actually a lower limit and it is probable that the true transfer may approach 100%.

(2) Prior to incorporation into mature phage, the injected DNA appears to persist as large independent units within the bacterial cell. There is no indication of any enduring attachment of original DNA to bacterial structures or for any major dispersal or fragmentation of the molecules.

(3) The original, or *parental*, DNA is not distributed evenly among the progeny phage, but is passed on, at least partially, as large "chunks" comprising a significant fraction of the mass of the original molecule. It is not possible at present to specify the size of the transmitted fragments more exactly, as there is considerable conflict in the existing evidence. It is of course probable that the transmitted fragment represents a definite portion of the polynucleotide structure of the parental DNA.

Only a minor part of the DNA of newly produced phage can arise from that of the infecting particle. The balance is derived from the breakdown and utilization of host DNA, as well as from material present in the culture medium at the time of infection.

RNA synthesis following infection. The *net* synthesis of RNA by the host bacterial cell ceases soon after the initiation of the infection process. However an important degree of metabolic turnover of RNA persists. Moreover, the new RNA synthesized has been shown to have a base composition analogous to that of the DNA of T2 bacteriophage (26).

The newly synthesized RNA appears to be associated to a large extent with the *E. coli* ribosomes and migrates with them upon ultracentrifugation. However, the attachment to the ribosomes is much weaker than that of ribosomal RNA and the binding is largely abolished under conditions of low Mg^{++} concentration. The distribution of sedimentation coefficients is quite different from that of ribosomal RNA. A rather broad distribution is present, ranging between 5 and 30 S.

The similarity of the base ratios of T2 RNA and DNA suggests that the analogy may extend further to a detailed correspondence of nucleotide sequence. This has indeed proved to be the case (26). Thermal denaturation of a mixture of T2 RNA and DNA, followed by slow cooling, has been shown to result in the formation of molecular hybrids which can be detected by CsCl banding. The process is quite specific, and no hybridization occurs with DNA from other sources.

If the capacity for hybridization is accepted as positive evidence for complementarity of base sequence, then the newly synthesized RNA may be regarded as a collection of molecules representing complementary transcripts of limited regions of T2 DNA and possessing an analogous over-all base ratio.

The RNA which is newly synthesized in the course of infection by T2 bacteriophage has many of the properties expected for messenger RNA, including metabolic instability, polydispersity of molecular size, and complementarity of base sequence to that of phage DNA. It is thus entirely reasonable to suggest that it may be responsible for the guided synthesis of phage protein, as well as any essential enzymes not present in normal *E. coli*.

Enzymes produced during infection. Several new enzymes are known to be produced during the early period of infection with the T-even phages. An enzyme which, in the presence of formaldehyde and tetrahydrofolic acid, converts deoxycytidylic acid to its 5-hydroxymethyl derivative (dHMP) has been demonstrated. This enzyme has not been detected in extracts of normal bacterial cells, or in bacteriophage itself (20).

A *5-hydroxymethyldeoxycytidylic kinase*, which converts dHMP to the triphosphate (dHTP), has also been identified. Since the nucleoside triphosphates are believed to be the precursors of phage DNA, the synthesis of this enzyme, as well as that which forms dHMP from dCMP, is probably an essential preliminary to the extensive production of new DNA.

Other enzymes formed during infection are responsible for labeling 5-hydroxymethyl cytosine with glucose.

Radioactive suicide of phage. This account of phage replication would be incomplete without mentioning one set of experiments which cannot be readily fitted into the currently prevalent picture of DNA replication, which favors the semiconservative mechanism (Chapter 9). Stent and colleagues have examined the lethal effects of labeling phage with P^{32} (27). The P^{32} isotope decays to form a sulfur atom. Because the latter cannot substitute for phosphorus in the phosphodiester linkage, each such radioactive decay process must produce a break in the phosphodiester backbone of the labeled DNA.

This process frequently results in inactivation of the phage and is often referred to as radioactive "suicide" of the virus. In isolated phage, inactivation occurs with an efficiency of about 1/10 for each radioactive disintegration. This figure is surprisingly low if phage replication occurs exclusively by a semiconservative mechanism.

Stent and co-workers infected P^{32}-labeled cells with P^{32}-labeled phage in a P^{32}-containing medium. In this manner a high radioisotope content of the progeny phage was assured.

Samples of infected cells were frozen at $-196°$ at various times after infection and were stored in this state of suspended animation, in which all metabolic processes were effectively halted. After storage, the cells were thawed and assayed for viable phage. The puzzling result was obtained that samples frozen about 10 minutes after infection, in the late eclipse period, showed no change in their capacity to produce active progeny during the period of storage, although radioactive decay proceeded at an undiminished rate.

These experiments indicate a remarkable insensitivity to radioactive destruction during the later stages of the eclipse period, suggesting that the genetic information has somehow been transcribed in a form which can survive breakage of the sugar-phosphate backbone. The interpretation of these results is still controversial. They emphasize the complexity of the problem of phage replication and the dangers of premature generalization.

Genetic aspects of phage replication. In the case of T4 bacteriophage, an important start has been made in refining genetic analysis to the point where a correlation of explicit properties of the mature organism with the molecular topography of its DNA is now in sight (28, 29, 30).

The bacteriophages have particular advantages as subjects for this kind of study. It is possible to start with a single phase particle. Moreover, each phage particle contains only a single copy of its genetic information, in contrast to higher organisms for which duplicate copies are always present. In the language of classical genetics phage is said to be *haploid*. This greatly simplifies the problem of genetic analysis.

A further advantage is derived from the rapidity of phage replication. Thus a new generation of phage arises within minutes, as compared with days in the case of insects and years in the case of mammals. Thus a large number of generations can be examined within a short time.

The general experimental technique is as follows. An agar plate culture of *E. coli* bacteria is inoculated with a dilute suspension of T4 bacteriophage. This is followed by the development with time of clear spots, or *plaques*, in the turbid culture. These represent the progressive lysis of *E. coli* by successive generations of phage. If inoculation is made with a sufficiently dilute solution, the plaques are well separated and each plaque may be identified with the progeny of a *single* original infecting phage. As successive generations of the initial phage undergo replication, the plaque expands radially. Moreover, the size and shape of the plaques are hereditary characteristics of the phage which can be easily differentiated and recorded.

In the case of T4 phage, spontaneous changes, or *mutations*, often arise in the genetic apparatus which result in altered hereditary characteristics. A particular class of mutants called *rII mutants* can be identified readily from the appearance of their plaque (Fig. 11-14).

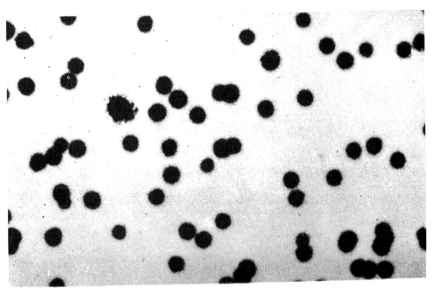

Fig. 11-14. Plaques of T4 phage on a culture of strain B E. coli (30).

T4 phage of the standard, or *wild*, type can produce plaques on either of two *E. coli* host strains, B or K. However, the rII mutants can reproduce on B cultures, on which they produce a characteristic large plaque, but cannot replicate on cultures of strain K.

Nevertheless, an rII mutant can replicate normally on bacterial strain K provided that the cell is simultaneously infected with a phage particle of the standard type. This may be accomplished by carrying out the infection in liquid cultures. If the ratio of phage to bacteria is high ($> 3:1$), there is an excellent chance that many bacteria will undergo simultaneous infection by both types.

It is evident that the DNA of the standard phage particle can perform some necessary function within the bacterial cells of the K type which cannot be accomplished by the mutants. It has been postulated that this function resides in a small portion of the DNA, which has been designated as the rII region.

The detailed topography of the rII region may be examined by utilizing the process known as *genetic recombination* (Figs. 11-15, 11-16, and

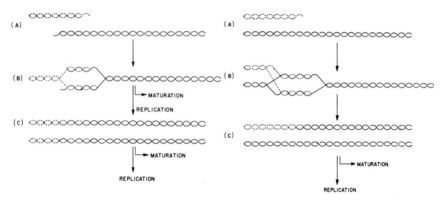

Fig. 11-15. Two alternative mechanisms for recombination involving breakage. The model on the left represents recombination by breakage and rejoining. That on the right illustrates recombination by breakage and copying (31).

11-17). In practice, two independently arising rII mutants are isolated by removal from mutant plaques visible on B and *crossed* with each other by simultaneous infection of a liquid culture of B cells.

In this case the vast majority of the progeny arising within the bacterial cell will be produced by ordinary replication processes and will resemble one or other parent. Such normal progeny will lack the capacity to grow on strain K. However, if the two mutants arise from alterations in different parts of the DNA molecule there exists a finite probability that a few normal individuals may be regenerated by *recombination* of normal sections of the molecule. Such viable recombinants may be readily detected by plaque formation on a culture of the K strain. In this manner a single recombinant may be detected among a billion progeny and very precise values obtained for the probability of recombination.

Mechanism of recombination. The mechanism of recombination on a molecular level is still incompletely understood. Two general models have been proposed. The process may consist of the rupture of two defective DNA molecules, followed by the rejoining of their nondefective portions

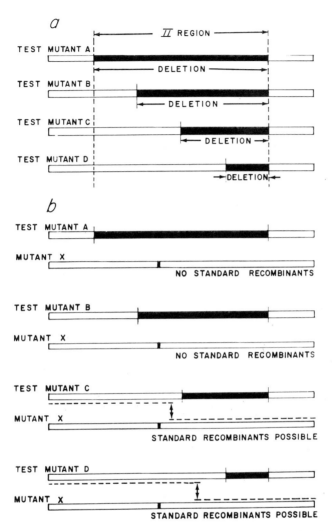

Fig. 11-16. Schematic version of recombination. The dark areas represent altered or deleted portions of the DNA molecule. Recombination is possible if these do not overlap (30).

to form a normal molecule, which is then replicated. Alternatively, the replication process may somehow copy only the good portions of the two molecules. This hypothetical mechanism has been called "copy choice."

Meselson and Weigle have made an experiment which has an important bearing upon the general mechanism of recombination (31). The system examined was not one of the T-even series, but belongs to the

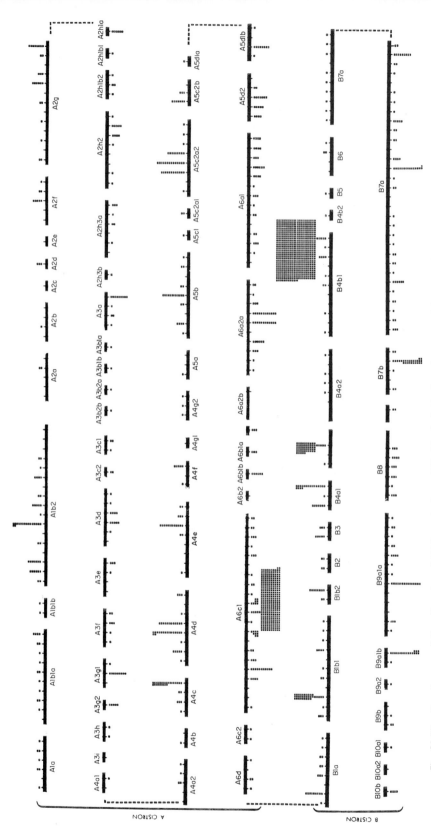

Fig. 11-17. Genetic map of the rII region of T4 bacteriophage (30). The individual combinations of letters and numbers represent distinguishable mutations. The inversion of alternate rows is made to emphasize that the region is a continuous molecular strand. Each spontaneous mutation at a site is represented by a black square. Sites without squares represent artificial chemically produced mutations.

class of *temperate* phages (section 11-5). Crosses were made between λ phage and a different strain of λ phage heavily labeled with the isotopes C^{13} and N^{15}. CsCl density gradient centrifugation was used to determine the distribution of labeled parental DNA among the parental genotypes and recombinants of the progeny.

The results indicated that discrete amounts of original parental DNA appeared in recombinant phage. This suggests that recombination occurs by breakage of parental DNA, followed by the reconstruction of complete genetic units from the fragments. The finding of a particular recombinant containing substantially *more* than 50% of original parental DNA indicated that recombination by breakage may occur without separation of the two strands. If replication is semiconservative, then the phage chromosome need not replicate in order to undergo recombination by breakage.

These experiments do not distinguish between two alternative mechanisms involving breakage. The chromosomal fragments produced could be incorporated into complete chromosomes either by direct joining or by completion of a fragment by copying the missing portion from the homologous region of a chromosome of different parental origin (Fig. 11-15). The latter alternative has been called *break and copy*. The possibility of some copy choice recombination is not completely ruled out.

Genetic mapping of the rII region. The feasibility of regenerating a viable phage by genetic recombination depends upon the extent of the defective zones in the DNA of the two mutants and their respective positions within their DNA. If the defective regions *overlap,* so that there exists a particular zone which is altered in *both* mutants, then it is obvious that no recombinatory process can produce a normal DNA unit. In this case no phage of the standard type will result from crossing in a liquid culture of strain B *E. coli*.

If, on the other hand, the defective zones do not overlap, then it is possible to assemble a normal DNA molecule by a linear union of the native portions of both mutants, and there is a finite probability that this will occur by genetic recombination upon crossing. This has provided the basis for the genetic mapping of the rII region.

By *mapping* in this sense there is meant the identification of the maximum possible number of distinguishable mutations and their mutual location. Because of the linear character of the genetic carrier, which is a fragment of a DNA molecule, it is to be expected *a priori* that the map of the identifiable mutations will have the form of a linear sequence. This expectation has been amply confirmed in practice, and no evidence for any nonlinear regions, such as loops, has been obtained.

Each mutation is designated by a combination of letters and numbers indicating the origin and order of discovery. Each mutation is identified with a particular site, or collection of sites, which is named after it. The completed genetic map has the form of a linear sequence of such sites each of which gives rise to a recognizable mutation. It is not of course possible as yet to make the final step of identifying the sites in terms of explicit nucleotide sequences.

The first stage of genetic mapping requires the use of mutations containing relatively major alterations, which probably correspond to *deletions,* of the rII genetic carrier. These are obtained from the class of mutants which do not have the property of spontaneous reversion. This automatically includes those in which large alterations or deletions have occurred (Fig. 11-16).

Suppose for example that four different nonreverting mutants, labeled A, B, C, and D, have been isolated. By crossing experiments it is found that A and B, A and C, D and C show recombination, but that A and D, D and B, B and C do not. It follows that the latter three pairs must overlap and hence are located next to each other. Thus the linear sequence must be ADBC.

By the use of gross mutations of this type, a rather rough outline of the genetic map may be obtained. To reach a higher order of resolution requires the use of the class of mutants which do possess the property of spontaneous reversion. These represent relatively minor alterations possibly reflecting changes in only a single nucleotide pair. Mutants of this class are often referred to as *point mutants.*

The point mutant is first located within one of the major mutation zones by the recombination test. By making use of overlapping mutations its position can be further pinpointed. Thus if point mutation a fails to recombine with either of the major overlapping mutations A and D, then it must lie in the region of overlap. Finally, point mutants lying within the same region are tested for recombination with each other. A pair which shows recombination is concluded to consist of point mutants at different sites. Each site is named after the mutant identifying it. The last step in the mapping is the ordering of the sites within each segment. This requires quantitative measurements of recombination frequency. The greater the separation of point mutants, the greater the probability of recombination. If breaks at all points are equally likely, the probability is inversely proportional to the separation. Thus, if point mutant a recombines with c twice as frequently as with b, then the order is abc.

In some ways the process of genetic mapping is analogous to that of the primary structure determination of a protein. However, the result of such analysis is a linear sequence of numbers standing for sites, at each of which a distinct point mutation has been identified. Each site presumably consists of one or more nucleotide pairs (Fig. 11-17).

There appear to be many sites at which spontaneous mutations do not occur. To fill in the genetic map more completely it is necessary to resort to chemical mutagenic agents which can induce mutations at the more refractory sites. Among the more effective of these are the reagents nitrous acid and hydroxylamine, the dye proflavine, and the unnatural bases 5-bromouracil and 2-amino purine. Many of the sites shown in Fig. 11-17 were identified by these means.

To what exactly does a site correspond? In all probability the sites identified with point mutations represent very short nucleotide sequences and possibly single nucleotide pairs.

If the number of times a spontaneous mutation has occurred at each site is indicated on such a map, a chart is obtained showing the relative frequency of mutations as a function of linear position within the sequence. Such a chart (Fig. 11-17) shows that the occurrence of spontaneous mutations is far from uniform. A disproportionate share of the mutations occur at several "hot spots," such as the one at B4b1.

A different aspect of the problem is to determine the functional capacity of the different parts of the rII region. For example, two mutants with alterations in different zones may both be able to grow upon simultaneous infection of a bacterium of strain K provided that the altered zones are *complementary* and, acting *in combination,* can fulfill the requirements for growth.

This kind of experiment should be differentiated from those involving recombination. It is not the production of a standard form which is being tested for in this case, but the cooperative replication of two mutants in their *original* forms.

This approach rests upon the fact that simultaneous growth of two rII mutants will be possible if mutations are not present in the same functional region in both cases. The upshot of these experiments is that the entire rII region may be divided into two linear functional units, which have been called *cistrons*. Simultaneous growth of two mutants can occur *only if their respective mutations occur in different cistrons.* The two phage rII mutants must possess, between them, an intact A cistron and an intact B cistron. If the mutations of both occur in the A cistron, or in the B cistron, growth does not occur.

It is natural to raise the question of just what area on the genetic map of Fig. 11-17 represents a classical *gene*. Is it a single site, a cistron, or the entire rII region? The question is really meaningless since the genes of classical genetics represent only gross properties and cannot be defined in molecular terms. Since the property in question is the ability to reproduce in a strain K culture, the most nearly correct answer is that each cistron is a single gene which is operationally analogous to the "genes" governing eye color, and so on, in higher organisms.

Significance of studies with T-even phage. The results discussed above provide one of the strongest arguments in favor of the central hypothesis cited earlier as to the mechanism of the transmission of genetic information. The reproduction of phage has been shown to be a consequence of the injection of phage DNA. The fact that the synthesis of new RNA precedes that of phage protein is consistent with the idea that this RNA is an essential intermediary for protein synthesis. Finally, in the case of T4 phage, it has been possible to refine genetic mapping to the point where it approaches a primary structure determination, although in a nonchemical language.

Nonsense mutations. As already mentioned in Chapter 10, the currently favored picture of the gene represents it as marked off into *nonoverlapping* meaningful nucleotide triplets, which are read from a fixed point. If this model is correct then, in order for a complete polypeptide chain to be synthesized, each nucleotide triplet, as read from the reference point, must correspond to an amino acid. Thus the sequence

$$/a^*_1 a_2 a_3 / a_4 a_5 a_6 / a_7 a_8 a_9 / a_{10} a_{11} a_{12} /\text{---}$$

where the reference point is indicated by an asterisk is divided into four triplets. If a mutation converts one of these into a triplet which does not code for an amino acid (a *nonsense* triplet), then the continuity is broken and a complete polypeptide cannot be produced. Since there are 64 possible nucleotide triplets and only 20 amino acids, the possibilities for nonsense triplets are numerous.

By utilizing a mutant with special properties it has been possible to identify nonsense mutations in the A cistron of the rII region of T4 bacteriophage.

As was discussed in the preceding section, the A and B cistrons normally act independently, and a defect in one cistron does not influence the functioning of the other. Each cistron may be regarded as producing a specific messenger RNA, which is in turn translated into a polypeptide chain (Fig. 11-18).

An unusual mutant form (r 1589 of T4 phage) has been identified in which the two cistrons are effectively joined together (32, 33). In the r 1589 mutant a segment appears to be deleted which includes the boundary region between the two cistrons and a portion of each. The deletion is sufficient to abolish the A function but not the B function. Hence the r 1589 mutant will complement with any strain which has an intact A cistron. Apparently, the deleted tip of the B cistron is not essential (32, 33).

However, because of the absence of the cistron-dividing element, the B fragment no longer functions independently of the A. Thus the B function may be abolished by crossing certain deletions into the A fragment by genetic recombination. The same result was produced by certain mutations induced by the dye proflavine, which is believed to produce the addition or deletion of a *single* nucleotide (32, 33).

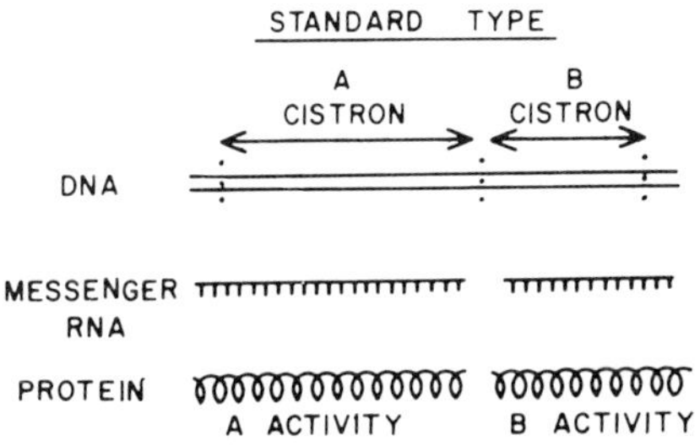

Fig. 11-18. Schematic version of the rII region of a standard type phage, showing the independent functioning of the A and B cistrons (35).

This result is understandable if the nucleotide sequence is read in successive coding units starting from a fixed point (32). If the A and B cistrons are joined so as to produce a single molecule of messenger RNA, then any deletion in the A cistron whose length is not equal to that of a single coding unit, or an integral multiple thereof, would cause a shift in the reading frame and completely disrupt the translation of the B cistron (32).

If the mutation in the A cistron is of the substitution type, in which a particular nucleotide is replaced by another *without a net change in the total number of nucleotides,* the B function may also be eliminated. If such a mutation converts a meaningful triplet into a nonsense triplet, then synthesis of the polypeptide would not continue beyond that point. If the mutation converts a meaningful triplet to a second meaningful triplet which codes for a different amino acid, then the synthesis of the polypeptide would not be interrupted and the function of the B cistron would be preserved. Mutations of this type which alter rather than block protein synthesis have been called *missense* mutations (Fig. 11-19).

Substitution mutations of the *transition* type, for which one base pair is exchanged for another without change in the orientation of purine and pyrimidine with respect to the two DNA chains, can be identified by their reversion induced by certain base analogs, such as 2-amino purine (32, 33).

If a mutation in the A cistron has been identified as belonging to the above class, it may be classified as a nonsense or missense mutation by insertion in series with r 1589 by genetic recombination. The double mutant is tested for B cistron activity. If this is present, then the muta-

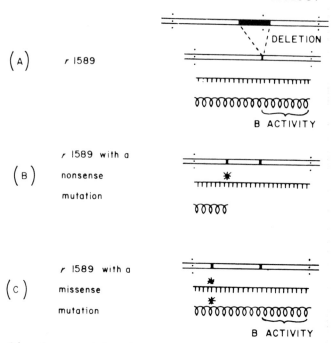

Fig. 11-19. Schematic representation of an r 1589 mutant. A deletion has refused the two cistrons, so that they no longer act independently (35).

tion is missense, and the coding unit corresponds to a different amino acid from the original. If the mutation is nonsense, then B activity will be blocked.

Certain rII mutants of T4 phage are said to be *ambivalent*. These are inactive in one *E. coli* strain but active in another. The bacterial host may be modified through mutation to result in activation of an entire group of rII phage mutants. Such a modified bacterium is said to possess a *suppressor* mutation, which eliminates an ambivalent set of phage mutations (33).

Benzer and Champe have studied a group of such ambivalent mutations which were inserted in the A fragment in series with r 1589 by recombination (33). These were tested for B cistron activity in two *E. coli* strains, KB and KB-3. Activity was absent when KB was the host but present for the suppressor strain KB-3.

Thus a set of mutations has been identified which behave as nonsense mutations for one host strain (KB) and as missense mutations for a second strain (KB-3). These results have been interpreted in terms of the presence of a new or modified transfer RNA in KB-3 which can recognize the altered nucleotide triplet in the mutant phage and translate it into an amino acid. This modified transfer RNA is apparently missing from KB.

This explanation is consistent with the degenerate character of the genetic code. The suppressor mutation thus may be regarded as a hereditary alteration of the code.

11-5 THE TEMPERATE BACTERIOPHAGES

The lysogenic state. The T series of bacteriophages includes only "virulent" agents which kill the host cells which support their replication. Another important class of bacteriophage, the *temperate* viruses, has an entirely different type of parasitic relationship with the host (20).

The infection of a susceptible bacterial strain by a stock of temperate phage results in the invasion and destruction of a fraction of the infected cells. However, an important portion does not undergo immediate lysis but instead attains a state of coexistence with the infecting viruses, in which a genetic element containing the hereditary characteristics of the parental phage particle is added to the genetic complement of the host or its progeny. This new genetic element, which is referred to as *prophage,* undergoes reproduction simultaneously with the bacterial cell and is thereby passed on to the descendants of the original host cells. A bacterial culture which carries a prophage in this manner is said to be *lysogenic.*

The lysogenic culture has some properties which are not present in the original. In particular, the spontaneous conversion of the prophage from a latent to an active form, which initiates unrestricted virus synthesis and produces lysis of the cell, occurs at a low but finite rate. The presence of prophage also appears to endow the lysogenic cell with a high degree of resistance to infection by other phages of the same species and sometimes to invasion by unrelated phages.

Some temperate phages have the capacity of undergoing a transition to an active form in the presence of appropriate agents, resulting in a mass lysis of the host cells. The most effective agents are generally mutagenic, including X-ray and ultraviolet irradiation and nitrogen mustards. Temperate phages of this type are said to be *inducible.*

The most extensive investigations upon the properties of temperate phage have been made with four species which attack strains of *E. coli* and *Shigella.* These are called lambda (λ), P1, P2, and P22. Like the members of the T series they appear to possess a head and tail, whose approximate dimensions are cited in Table 11-2.

The process of lysogenization. The relative occurrence of lysis and lyso-

TABLE 11-2. DIMENSIONS OF SOME BACTERIOPHAGES

Phage	Head Dimension	Tail Dimension
T1	50 mμ	150 × 10 mμ
T2, T4, T6	95 × 65 mμ	100 × 25 mμ
λ	55 mμ	140 × 10 mμ
P1	86 mμ	210 × 20 mμ
P2	66 mμ	125 × 20 mμ

genization depends upon the culture conditions, including the composition of the medium and the temperature. In addition, the presence of inhibitors of protein synthesis tends to favor establishment of the lysogenic state. Such inhibitors, including 5-hydroxyuridine and various amino acid analogs, appear to prevent some essential step leading to lysis. Their effectiveness is enhanced if they are added some minutes after infection.

It is generally assumed that the infective process is analogous to that of T2 bacteriophage and involves the injection of phage DNA into the host cell. Experiments with P^{32}-labeled phage DNA have shown that most of the phosphorus incorporated initially into the infected cells is transmitted to successive generations (34).

Many temperate phages can undergo mutation to forms with greatly diminished ability to enter the prophage state, so that the invaded bacterial cells are almost always lysed. These are referred to as *clear* mutants, or λ_c. A second class of mutants exists which invariably lyse the infected cell and in addition can overcome the immunity conferred by the presence of prophage. These are known as *virulent* mutants, or λ_{vir}.

While the evidence is incompletely conclusive, there is reason to believe that the formation of relatively stable lysogenic cells does not occur at once but is usually preceded by a prolonged period in which the phage and bacterial genetic units exist in a state of loose association. This may persist for several cell generations during which replication of the phage DNA must occur (34).

Under certain conditions a cooperative action of temperate phages can occur. Thus a simultaneous infection by two different clear mutants may result in a frequency of lysogenization which is greater than that of either mutant acting singly. This occurs in many cases where the prophage is not a recombinant but represents one of the parental types.

This cooperative action is reminiscent of that exhibited by the two cistrons of the rII region of T2 bacteriophage. This result suggests strongly that the phage particle must carry out certain functions prior to lysogenization.

Induction. The process of induction of a lysogenic culture by ultraviolet radiation has some features which are not present in the case of infection by T-even phage. The synthesis of RNA and proteins continues until lysis, although at decreasing rates. However, cell division is inhibited.

The doses of ultraviolet radiation required for induction are sufficient to block DNA synthesis completely for a time. Following this initial period, whose duration depends upon the dose and upon the conditions of irradiation, DNA synthesis resumes at an accelerated rate.

In the case of induced lysogenic cells of *B. megaterium*, exposure to ribonuclease specifically inhibits the synthesis of phage protein relative to total protein synthesis. If ribonuclease is added immediately after induction, the synthesis of phage antigens may be blocked entirely.

Transduction. Among the most interesting attributes of the temperate bacteriophages is their capacity to transmit genetic information from one bacterial cell to another. When spontaneous rupture of lysogenic bacteria occurs, the liberated phage particles can attack new host cells of a different strain and, in the process, transfer, or *transduce,* some of the genetic characteristics of the original strain. Only a single phenotypic characteristic is generally transmitted at a time, although instances are known of simultaneous transduction of several markers.

Transduction may be visualized as reflecting an exchange of elements of the viral and bacterial chromosomes, followed by the transport of the acquired genetic elements to the new host. The process is very reminiscent of *transformation.* Transduction has been employed by several investigators for genetic mapping.

GENERAL REFERENCES

1. *The Viruses,* F. Burnet and W. Stanley, Academic Press, New York (1959).
2. *Viruses,* K. Smith, Cambridge University Press, London (1962).
3. *Plant Viruses,* K. Smith, Methuen, London (1960).

SPECIFIC REFERENCES

Structure of viruses

4. R. Horne, *Sci. Amer.,* **208,** 48 (1962).
5. P. Wildy, W. Russell, and R. Horne, *Virology,* **12,** 204 (1960).
6. R. Horne, S. Brenner, A. Waterson, and P. Wildy, *J. Molec. Biol.,* **1,** 84 (1959).

7. P. Wildy, M. Stoker, I. Macpherson, and R. Horne, *Virology*, **11**, 444 (1960).
8. R. Horne, A. Waterson, P. Wildy, and A. Farnham, *Virology*, **11**, 79 (1960).

Tobacco mosaic virus

9. Reference 1, vol. 2, p. 1.
10. H. Schuster in *The Nucleic Acids*, E. Chargaff and J. Davidson (eds.), vol. III, p. 245, Academic Press, New York (1960).
11. C. Hall, *J. Am. Chem. Soc.*, **80**, 2556 (1958).
12. J. Watson, *Biochim. et Biophys. Acta*, **13**, 10 (1954).
13. R. Franklin and A. Klug, *Biochim. et Biophys. Acta*, **19**, 403 (1956).
14. R. Franklin, *Nature*, **177**, 929 (1956).
15. H. Fraenkel-Conrat, B. Singer, and R. Williams, *Biochim. et Biophys. Acta*, **25**, 87 (1957).
16. H. Schuster, G. Schramm, and W. Zillig, *Z. Naturforsch*, **11b**, 339 (1956).
17. G. Schramm and R. Engler, *Nature*, **181**, 916 (1958).
18. A. Gierer and K. Mundry, *Nature*, **182**, 1457 (1958).
19. A. Tsugita and H. Fraenkel-Conrat, *Proc. Natl. Acad. Sci. U. S.*, **46**, 636 (1960).

The bacteriophages

20. R. Sinsheimer in *The Nucleic Acids*, E. Chargaff and J. Davidson (eds.), vol. III, p. 187, Academic Press, New York (1960).
21. G. Stent in reference 1, vol. 2, p. 237.
22. R. Williams, *Rev. Mod. Phys.*, **31**, 233 (1959).
23. C. Levinthal, *Rev. Mod. Phys.*, **31**, 227 (1959).
24. C. Levinthal and C. Thomas, *Biochim. et Biophys. Acta*, **23**, 453 (1957).
25. A. Hershey and E. Burgi, *Cold Spring Harbor Symposia Quant. Biol.*, **21**, 91 (1956).
26. S. Spiegelman, *Fed. Proc.*, **22**, 36 (1963).
27. *Advances in Virus Research*, G. Stent, vol. 4, p. 95, Academic Press, New York (1958).
28. S. Benzer, *Proc. Natl. Acad. Sci. U. S.*, **41**, 344 (1955).
29. S. Benzer in *The Chemical Basis of Heredity*, W. McElroy and B. Glass (eds.), p. 70, Johns Hopkins Press, Baltimore (1957).
30. S. Benzer, *Sci. Amer.*, **207**, 70 (1959).
31. M. Meselson and J. Weigle, *Proc. Natl. Acad. Sci. U. S.*, **47**, 857 (1961).
32. F. Crick, L. Barnett, S. Benzer, and R. Watts-Tobin, *Nature*, **192**, 1227 (1961).
33. S. Benzer and S. Champe, *Proc. Natl. Acad. Sci. U. S.*, **48**, 1114 (1962).
34. S. Goodgal, *Biochim. et Biophys. Acta*, **19**, 333 (1956).
35. S. Luria, D. Fraser, J. Adams, and J. Buras, *Cold Spring Harbor Symposia Quant. Biol.*, **23**, 71 (1958).

12

The Carbohydrates and Their Biosynthesis

12-1 THE MONOSACCHARIDES

General remarks. As a group the carbohydrates are by far the most abundant organic substances present in nature. They account for most of the organic structure of plants and occur to some extent in all animals.

In animals, the carbohydrates function primarily as a source of energy for the maintenance of essential processes. The free energy stored in carbohydrates is converted to a usable form by the reactions of oxidative metabolism (Chapter 13). In addition, many important substances are partly carbohydrate in nature, including the nucleotides and numerous coenzymes. In plants, the function of maintaining the structural organization is largely assumed by polymeric carbohydrates, especially cellulose.

The basic carbohydrate molecular units are the *monosaccharides*, or *simple sugars*, of which the more complex carbohydrates are polymers and to which they may be converted by exhaustive hydrolysis.

The monosaccharides may be regarded as polyhydric alcohols which contain a reducing carbonyl group of either an aldehyde or keto nature. The two series are referred to as *aldoses* and *ketoses*, respectively. The simplest member of the aldose series is *glyceric aldehyde*, or *glyceraldehyde* (Fig. 12-1), which contains a single asymmetric carbon and exists as D- and L-mirror images or enantiomorphs. The parent ketose is *dihydroxyacetone* (1,3-dihydroxy-2-propanone) (Fig. 12-2), which contains no center of asymmetry and exists as a single stereoisomer.

The two series may be visualized as being built up systematically by the addition of —CHOH groups to the three-carbon prototype (Figs.

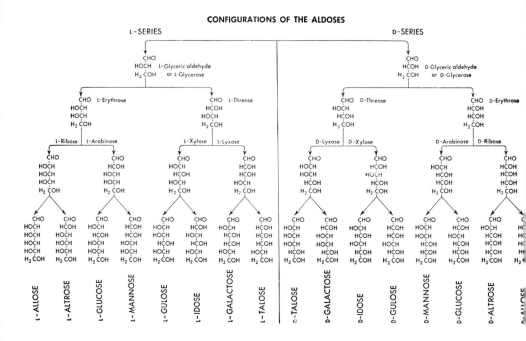

Fig. 12-1. The aldose monosaccharides.

12-1 and 12-2). Sugars containing 3, 4, 5, 6, or 7 carbon atoms are designated as *trioses, tetroses, pentoses, hexoses,* and *heptoses,* respectively.

With the addition of each —CHOH group, the number of asymmetric centers increases by one and the number of possible stereoisomers by a factor of two. The total number of stereoisomers for an n-carbon sugar is equal to 2^{n-2} and 2^{n-3} for the aldose and ketose series, respectively, and represent 2^{n-3} and 2^{n-4} pairs of D- and L-enantiomorphs. Non-enantiomorphic stereoisomers have different physical properties and are regarded as distinct and different sugars.

The configuration of the monosaccharides is assigned by their relationship to the simplest member of the series containing an asymmetric center (glyceraldehyde or erythrulose). If the configuration of the CHOH *next to the terminal —CH_2OH group* is like that of D-glyceraldehyde or D-erythrulose the sugar is said to be in the D-configuration. Similarly,

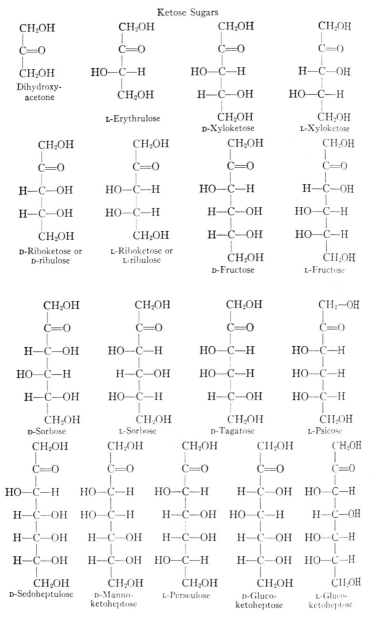

Fig. 12-2. The more important ketose monosaccharides.

if the configuration of the penultimate CHOH group is like that of the L-form of the prototype, the sugar is assigned an L-configuration. In terms of the linear structural formulas of Figs. 12-1 and 12-2, the D- and L-isomers are indicated by placing the penultimate —OH on the right or left, respectively. As in the case of the amino acids, the members of each pair of enantiomorphs have optical rotations which are opposite in sign.

Tautomeric forms of the sugars. While many of the chemical properties of the monosaccharides are consistent with the presence of free carbonyl groups, as occur in the linear structures shown in Figs. 12-1 and 12-2, it is well established that the pentoses and higher sugars exist in solution predominantly as tautomeric ring forms in dynamic equilibrium with a small amount of the linear form. The ring structures correspond to six-membered (*pyranose*) or five-membered (*furanose*) cyclic acetals, in which the carbonyl group of the linear form is replaced by a —CHOH group (Fig. 12-3).

An additional center of asymmetry is present in the ring forms, which may exist in two configurations designated as α and β (Fig. 12-3). These generally have specific rotations which are quite different. Because of the dynamic equilibrium existing between the various tautomeric forms, the higher sugars exist in solution as mixtures of the α- and β-isomers with a small amount of the linear form.

$$\text{HO—CH} \leftrightarrows \text{HC=O} \leftrightarrows \text{HC—OH}$$

β-form $\qquad\qquad\qquad$ α-form

The additional hydroxyl group present in the ring forms has properties

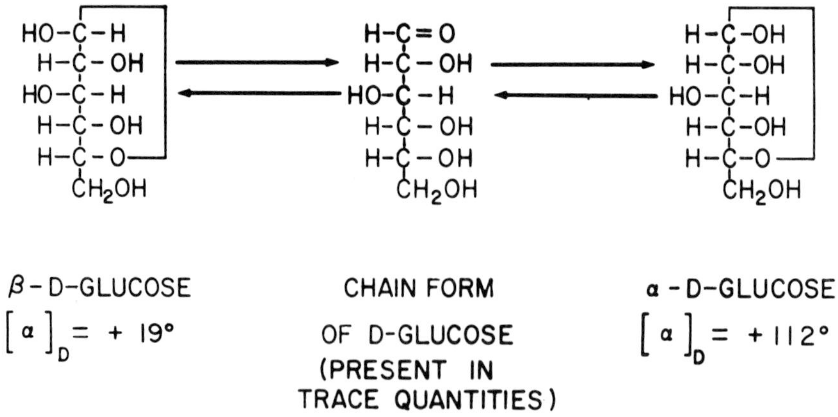

```
HO-C-H              H-C=O              H-C-OH
H-C-OH              H-C-OH             H-C-OH
HO-C-H     ─→      HO-C-H      ─→     HO-C-H
H-C-OH      ←─      H-C-OH      ←─     H-C-OH
H-C-O─┐             H-C-OH             H-C-O─┐
CH₂OH               CH₂OH              CH₂OH
```

β-D-GLUCOSE $\qquad\qquad$ CHAIN FORM $\qquad\qquad$ α-D-GLUCOSE

$[\alpha]_D = +19°$ $\qquad\qquad$ OF D-GLUCOSE $\qquad\qquad$ $[\alpha]_D = +112°$

$\qquad\qquad\qquad\qquad$ (PRESENT IN
$\qquad\qquad\qquad\qquad$ TRACE QUANTITIES)

Fig. 12-3a. Equilibrium between the ring and linear forms of D-glucose.

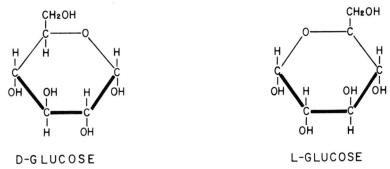

D-GLUCOSE **L-GLUCOSE**

Fig. 12-3b. Stereochemical formulas for ring forms of D- and L-glucose.

```
     H   OCH₃                CH₃O   H
      \ /                      \ /
       C                        C
       |                        |
    H—C—OH                   H—C—OH
       |                        |
    HO—C—H    O              HO—C—H    O
       |                        |
    H—C—OH                   H—C—OH
       |                        |
    H—C                      H—C
       |                        |
      CH₂OH                    CH₂OH

    (α)_D = +159°            (α)_D = −34°
```

Fig. 12-3c. The methyl glucosides: (left) methyl α-D-glucopyranoside; (right) methyl β-D-glucopyranoside.

somewhat different from the other secondary hydroxyls. In particular it is readily converted to a methyl derivative called a *glucoside* by treatment with acid methanol (Fig. 12-3). Conversion to the glucoside abolishes the dynamic equilibrium and freezes the α- and β-forms in their respective configurations.

Sugar derivatives. In the *deoxy* sugars the oxygen of a hydroxyl group is removed, leaving the hydrogen. One very important example, 2-deoxy-D-ribose (2-deoxy-D-erythro-pentose), has already been described as a constituent of deoxyribonucleotides (Chapter 8).

Amino groups may be substituted for the hydroxyl groups of monosaccharides to form the *amino sugars*. Two examples are 2-amino-2-deoxy-glucose (glucosamine) and 2-amino-2-deoxy-galactose (galactosamine) (Fig. 12-4). These exist in ring forms analogous to those of their parent monosaccharides.

$$
\begin{array}{c}
\text{H}-\text{C}=\text{O} \\
\text{H}-\text{C}-\text{NH}_2 \\
\text{HO}-\text{C}-\text{H} \\
\text{H}-\text{C}-\text{OH} \\
\text{H}-\text{C}-\text{OH} \\
\text{CH}_2\text{OH}
\end{array}
\qquad
\begin{array}{c}
\text{H}-\text{C}=\text{O} \\
\text{H}-\text{C}-\text{NH}_2 \\
\text{HO}-\text{C}-\text{H} \\
\text{HO}-\text{C}-\text{H} \\
\text{H}-\text{C}-\text{OH} \\
\text{CH}_2\text{OH}
\end{array}
$$

2-AMINO-2-DEOXY-D-GLUCOSE (GLUCOSAMINE)   2-AMINO-2-DEOXY-D-GALACTOSE (GALACTOSAMINE)

Fig. 12-4. Two amino sugars.

The aldehyde groups of members of the aldose series are easily oxidized to the corresponding carboxyl derivatives, which are called *aldonic* acids. Oxidation of the terminal —CH_2OH group produces the corresponding *uronic* acid (Fig. 12-5).

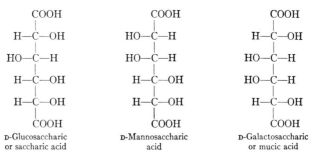

Fig. 12-5. Aldose derivatives in which both the aldehyde and terminal hydroxyl groups have been oxidized to carboxyls.

The phosphate esters of the monosaccharides are very important as intermediates in their synthesis and interconversion in biological systems. Indeed, the biosynthesis of sugars and the initial stages of their oxidative metabolism (section 13-3) proceed almost entirely by way of phosphorylated intermediates. The 1- and 6-phosphates of glucose (Fig. 12-6) and fructose are of particular importance (section 13-3).

The oligosaccharides. The oligosaccharides are short chains of up to four monosaccharide units. The only oligosaccharides of general biological importance belong to the class of *disaccharides*. These may be further grouped into *reducing* disaccharides, which contain aldehyde or keto groups and *nonreducing* disaccharides whose carbonyl groups have been eliminated through involvement in the intersugar linkages.

$$\text{D-GLUCOSE 6-PHOSPHATE} \qquad \text{D-GLUCOSE 1-PHOSPHATE}$$

Fig. 12-6. The 1- and 6-phosphates of glucose.

Among the more important of the reducing disaccharides are *maltose*, *lactose*, and *cellobiose* (Fig. 12-7). Maltose and cellobiose are both dimers of glucose joined by a 1,4 glucosidic bond (Fig. 12-7). The only difference is in the configuration about the C_1 atom involved in the glucosidic bond. This is in the α configuration for maltose and in the β configuration for cellobiose. Lactose is a β-glucoside containing glucose and galactose.

The most familiar example of a nonreducing disaccharide is of course *sucrose* (Fig. 12-7). This consists of one molecule each of glucose and fructose, joined by a glucosidic bond between the C_1 hydroxyl of the former and the C_2 hydroxyl of the latter. The configuration about the C atom is α in the case of glucose and β in the case of fructose.

12-2 THE POLYSACCHARIDES

The starches. The starches (4, 7, 9) have a widespread occurrence as reserve carbohydrates in plants. They are found in particularly heavy concentration in grains and seeds, tubers, and many fruits. The starches are usually stored in the form of granules, whose shape and internal structure are characteristic of the particular source.

The primary biological function of starch appears to be as a nutritional reservoir for plants. It is of course an important foodstuff for animals, being converted to glucose prior to utilization.

The contents of starch granules may be grouped into two main fractions which may be distinguished by their solubility properties. If the granules are leached with water of progressively higher temperature, about 10 to 20% of the starch is extracted fairly easily. This more soluble

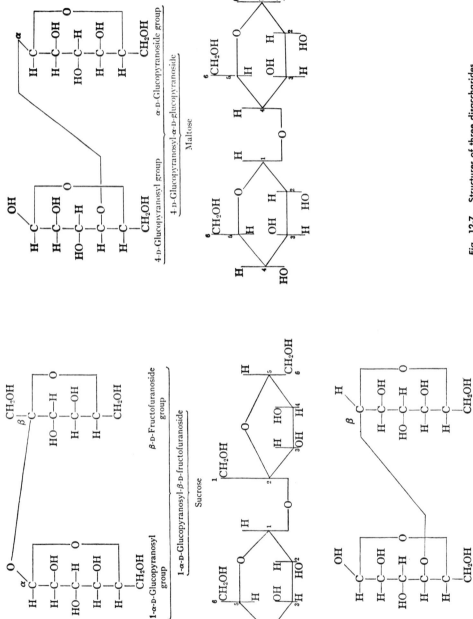

Fig. 12-7. Structures of three disaccharides.

fraction is called *amylose*. The residual, less soluble fraction is called *amylopectin*. However, the isolation of highly purified samples of either component generally requires chemical fractionation.

Both molecules are polymers of the pyranose form of D-glucose, to which they may be converted by exhaustive acid hydrolysis. The mode of linkage and the degree of branching have been determined by methylation studies (Fig. 12-8).

By treatment with a suitable methylating agent, such as dimethyl sulfate, all the available hydroxyls (those not involved in interglucose linkages) may be converted to their methyl (—OCH_3) derivatives. These are stable enough to survive complete acid hydrolysis to monomer units (except for that at the C_1 position of the terminal glucose at the reducing [aldehyde] end, which is hydrolyzed to the corresponding hydroxyl).

Exhaustive methylation of amylose, followed by acid hydrolysis, yields only 2,3,6-trimethyl glucose with a small amount of 2,3,4,6-tetramethyl glucose from the (nonreducing) terminal units (Fig. 12-8), indicating that the 2,3, and 6 hydroxyls of the vast majority of glucose units are not involved in bonding. Since there are only a total of five hydroxyls per glucose unit, the availability of three of these indicates that each unit participates in only two interglucose linkages and hence that branching is absent. The existence of branches would be reflected by the presence of dimethyl glucose in the hydrolysate.

Further evidence for the linear nature of amylose is derived from the correspondence of the molecular weights estimated from the mole fraction of terminal groups with those obtained directly by physical measurements. This indicates that only one terminal unit is present per molecule.

Since only one reducing group is present per amylose molecule, it is clear that the interglucose linkages must involve the C_1 hydroxyls. From the methylation studies it is known that the nonreducing terminal unit, which is converted to 2,3,4,6-tetramethyl glucose in the hydrolysate, must be a *glucopyranose* since its C_5 hydroxyl is blocked and its C_4 hydroxyl is free. (The terminal glucose from the reducing end is converted to 2,3,6-trimethyl glucose.)

Since both the C_4 and C_5 hydroxyls are blocked for the remaining units, it follows that the glucosidic linkages must be 1, 4 and that the glucose units must be in the pyranose form. Optical rotatory studies have further shown that the glucose residues are in the α-configuration.

In summary, the primary structure of amylose corresponds to linear polymers of α-D-glucopyranose, joined by 1,4 glucosidic linkages.

Amylopectin is likewise a polymer of α-D-glucopyranose. However, hydrolysis of a completely methylated derivative yields, in addition to 2,3,4,6-tetramethyl glucose and 2,3,6-trimethyl glucose, a considerable

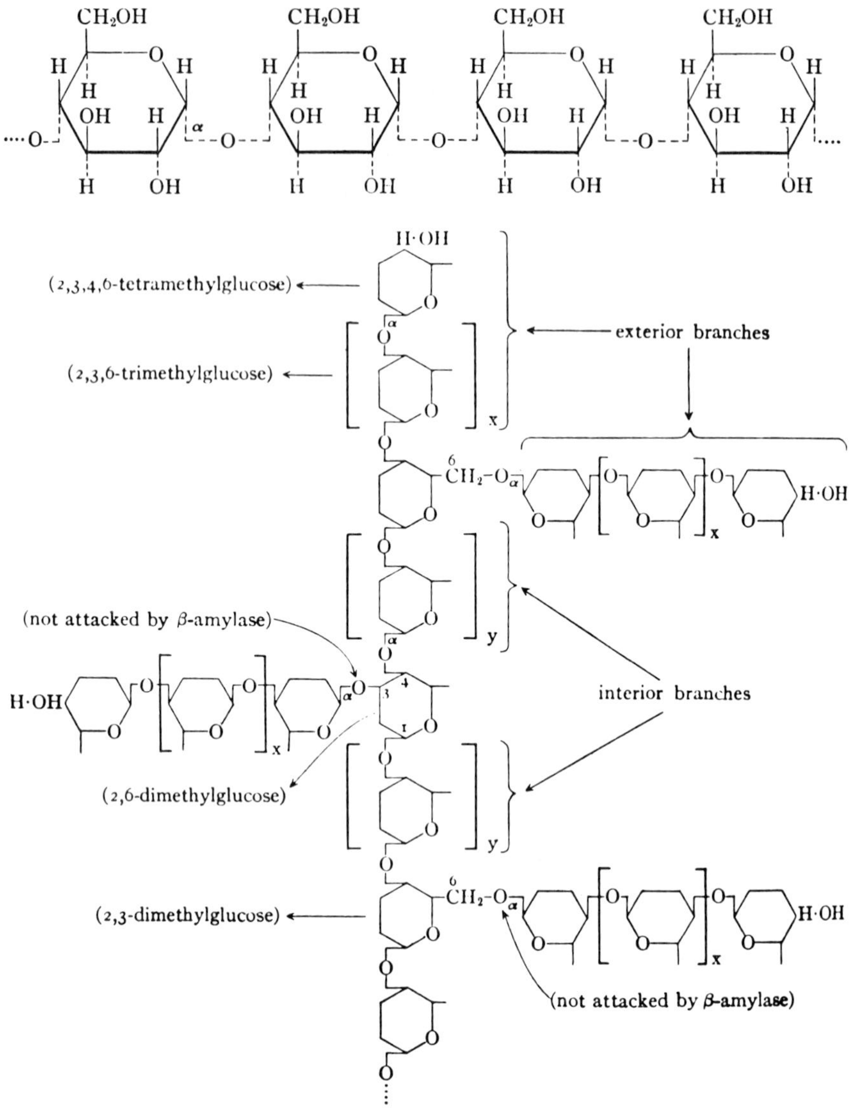

Fig. 12-8. Structures of amylose and amylopectin, as determined by methylation studies.

amount of 2,3-dimethyl glucose. Moreover, the mole fraction of end-groups ($\sim 4\%$) is much larger than in the case of amylose, and the apparent molecular weight computed from end-group analysis is lower than the true value by orders of magnitude.

The evidence cited above is fairly conclusive in establishing a highly branched structure for amylopectin. The branches are attached by 1,6 glucosidic linkages (Fig. 12-8). The sections of linear polysaccharide between branch points are quite short, usually about 25 glucose units long.

Physical studies upon solutions of either starch component are doubly handicapped by the danger of hydrolytic degradation during extraction and by a pronounced tendency to aggregate in solution. Extreme precautions are necessary to obtain information on the true molecular units rather than artifacts produced by association or degradation. Many investigators have preferred to convert amylose or amylopectin to a derivative with less tendency to aggregate, such as the acetate.

Both amylose and amylopectin are highly polydisperse. Amylose fractions of molecular weights ranging from 10^4 to 10^6 have been isolated. The molecular weight of amylopectin appears to be of a higher order of magnitude. Values of the order of 10^7 to 10^8 have been reported for unfractionated preparations of amylopectin and its derivatives.

While there has been some degree of controversy as to the molecular configuration of amylose in solution, the bulk of the evidence indicates that its viscosity properties in aqueous solution are those of a system of flexible randomly coiled polymers. The same may be said of amylopectin, which appears to have a more compact configuration than amylose, as would be anticipated from its highly branched character.

Amylose combines with iodine in aqueous solution to form a blue complex of interesting properties. The absorption spectrum is very different from that of free iodine. X-ray diffraction and other evidence has led to the proposal of a *helical* conformation for the complex (Fig. 12-9). This

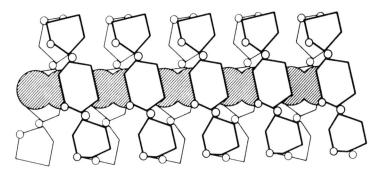

Fig. 12-9. Proposed structure for the iodine complex of amylose.

has six glucose units per turn. The I_2 molecules are in the core of the helix (Fig. 12-9) in a ratio of one I_2 per six glucose residues.

Glycogen. Glycogen is biochemically of the first importance as the principal reserve carbohydrate of animal tissues. Glycogen is a biological precursor of glucose derivatives and as such is a source of chemical energy for cellular processes (section 13-3).

Like amylopectin, glycogen is a branched polymer of glucopyranose units joined by 1,4 α-glucosidic linkages (1,6). However, methylation studies have revealed that the degree of branching is much higher than for amylopectin. Thus a representative glycogen has about one terminal glucose for each 11 glucose units, as compared with a ratio of about 1 to 25 in the latter case. In both cases, branches arise by 1,6 glucosidic linkages.

In summary, glycogen is a branched structure whose external branches are six to seven glucose residues long, while the separation of branch points along the chain averages about three glucose units. The molecular weight appears to be very high, about 10^7.

Cellulose. Cellulose is the principal structural polymer of plants and as such is a basic element of their cellular architecture. Like starch and glycogen it is a polymer of glucose, but with strikingly different physical properties (4, 5, 7, 8, 9).

Methylation studies have shown that cellulose consists of linear polymers of glucose units joined by 1,4 β-glucosidic linkages (Fig. 12-10). In

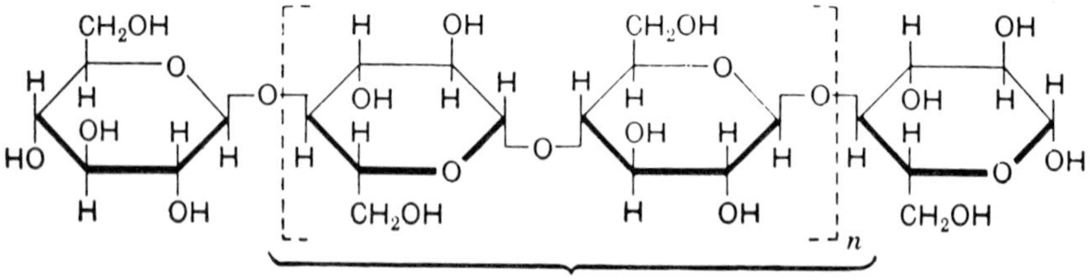

Repeating cellobiose unit of cellulose
Fig. 12-10. Primary structure of cellulose.

contrast to starch and glycogen, there is no evidence for any important degree of branching.

The organization of cellulose molecules within the plant cell wall corresponds to a very complex pattern, of which X-ray and electron microscopic observations permit only an incomplete description. In brief, bundles of cellulose molecules are organized into fibrils, in which amorphous and crystalline zones alternate (Fig. 12-11).

Hyaluronic acid. This mixed polysaccharide is produced by both bacteria and animals and has been isolated from a wide variety of biological sources, including the vitreous humor of the eye, connective tissue, umbilical cord, synovial fluid, and encapulated strains of streptococci.

THE CARBOHYDRATES AND THEIR BIOSYNTHESIS 395

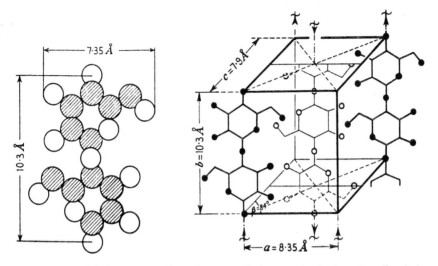

Fig. 12-11a. Model showing mutual arrangement of glucose units in the unit cells of the crystalline regions of cellulose, as deduced from X-ray diffraction.

Fig. 12-11b. Electron microscope picture of cellulose fibrils from cotton hairs.

The primary biological role of hyaluronic acid appears to stem from the high and non-Newtonian viscosity of its aqueous solutions. It appears to be responsible for the lubricant and shock absorbing properties of synovial fluid. The presence of hyaluronic acid in interfibrillar regions has also led to conjecture that it serves as a barrier to bacterial infection.

Hyaluronic acid consists of equimolar amounts of D-glucuronic acid and the N-acetyl derivative of D-glucosamine. The best picture of its primary structure is that of *linear* polymers of these two units in alternating sequence.

The polyelectrolyte character of hyaluronic acid endows it with a much greater relative molecular extension in solution than the neutral polysaccharides. As a consequence, its solutions have much higher viscosities, which is probably related to its biological function.

12-3 THE BIOSYNTHESIS OF CARBOHYDRATES

General remarks. Living cells may be grouped into two broad categories with respect to their mechanisms for extracting energy from their surroundings, so as to carry out the various energy-consuming processes essential for the maintenance of life (section 13-1). Cells of the first kind, called *heterotrophic,* require a supply of chemical energy from an external source, in the form of organic molecules of considerable complexity, such as proteins, fats, and carbohydrates. The chemical energy stored in these foodstuffs is converted to usable form by the processes of oxidative metabolism (Chapter 13). Heterotrophic cells include all mammalian cells and those of higher animals in general.

Cells of the second kind, called *autotrophic,* satisfy their energy requirements by direct utilization of the radiant energy of sunlight. The primary examples are the cells occurring in the leaves of green plants, together with those of certain bacteria. The complex set of reactions whereby radiant energy is trapped and harnessed for the synthesis of carbohydrates are referred to collectively as *photosynthesis* (10-16).

The over-all result of photosynthesis may be loosely regarded as the opposite of the respiration of heterotrophic organisms. The latter consume molecular oxygen and use it to oxidize carbohydrates to CO_2 and H_2O. Photosynthetic organisms incorporate H_2O and atmospheric CO_2 into carbohydrates and release excess oxygen as O_2. Since this process is thermodynamically unfavored (section 13-1) and must surmount a considerable energetic barrier, an external source of energy is required. This is supplied by radiant energy, which is trapped by the chlorophyll-containing structures of green plants and applied to the synthesis of cofactors essential for CO_2 incorporation.

The importance of photosynthesis extends beyond its function in supplying the needs of plants. Since animals satisfy their nutritional requirements from plants, directly or indirectly, virtually all forms of life are ultimately dependent upon photosynthesis for the continuation of biological functions.

In the higher photosynthetic organisms, including green plants, the absorption of radiant energy occurs in specialized subcellular particles called *chloroplasts* (Fig. 12-12). These are self-contained units capable

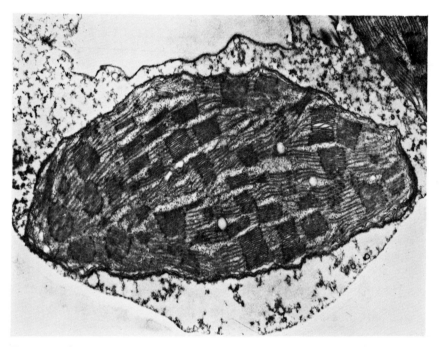

Fig. 12-12. Electron microscope picture of a chloroplast showing grana (reference 2, Chapter 13).

of carrying out the entire photosynthetic operation. The absorbing pigment is chlorophyll (Fig. 12-13), which, like heme, is a porphyrin derivative. The chlorophyll of green algae and the higher plants consists of two related components, chlorophyll a and chlorophyll b, both of which are esters of dibasic acids. The two esterifying alcohols are methanol and phytol ($C_{20}H_{39}OH$). Both forms of chlorophyll occur as $\overset{++}{Mg}$ complexes (Fig. 12-13).

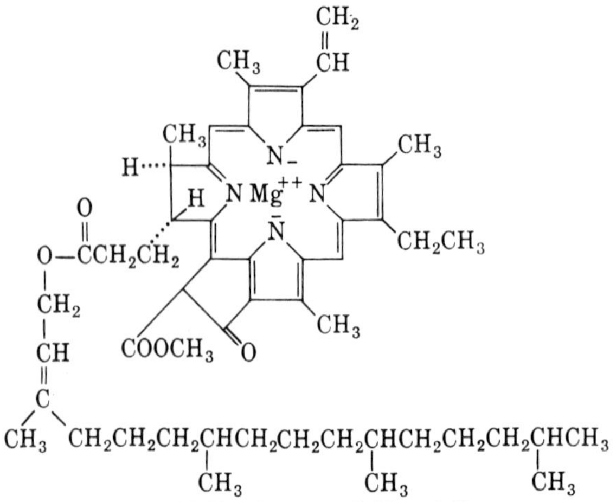

Fig. 12-13. Structure of chlorophyll a.

Chlorophyll may be extracted from chloroplasts by organic solvents. A strong and characteristic visible absorption spectrum is observed, with peaks at about 670 mμ, 430 mμ, and 410 mμ (for chlorophyll a) depending upon the solvent. Solutions of chlorophyll are strongly fluorescent.

The initial event of the photosynthetic process is the absorption of quanta of radiant energy by chlorophyll. The subsequent events may be divided into two sets of reactions, only the first of which is light-dependent.

In the primary phase of photosynthesis the radiant energy trapped in chlorophyll serves to synthesize two compounds essential for the actual incorporation of CO_2 into carbohydrate. The first of these, ATP, is produced by a *photophosphorylation* of ADP. The second cofactor is the reduced form of triphosphopyridine nucleotide (TPNH), which arises by a photochemical reduction of the oxidized (TPN) form. (The structure of TPN is described in section 13-2 and Fig. 13-5.)

Both reactions, the phosphorylation of ADP and the reduction of TPN, consume energy and in the language of thermodynamics (Appendix B) are said to be strongly *endergonic* (section 13-1). They are only enabled to occur naturally if this energy loss is compensated for by some form of energy supply, which in this case is photochemical.

The second phase of photosynthesis, in which CO_2 is incorporated into carbohydrate, is only indirectly dependent upon the primary photochemical event and proceeds in the absence of illumination. Indeed, the assimilation of CO_2 has been found to proceed in cells devoid of chloroplasts, provided that ATP and TPNH are made available.

The role of ATP in the secondary phase of photosynthesis is as a carrier of chemical potential and a powerful phosphorylating agent. In the former capacity its dephosphorylation may be coupled with an energetically unfavored reaction so that the over-all process becomes thermodynamically feasible (section 13-1).

TPNH (Fig. 13-5) acts as a very powerful reducing agent (section 13-2) and as such figures in the initial fixation of CO_2 and the reduction of carboxylate to aldehyde.

The chloroplasts. The photosynthetic apparatus of higher plants is localized within definite structural elements which are confined by membranes. These are called *chloroplasts*.

Electron microscopic examination of the chloroplasts reveals a highly developed internal organization. In particular, a large number of closely spaced layers or *lamellae* are set in a finely granular matrix (Fig. 12-14). In addition, there are in most cases well-defined regions (the *grana*) in which the lamellae are more densely and closely packed (Fig. 12-14). The

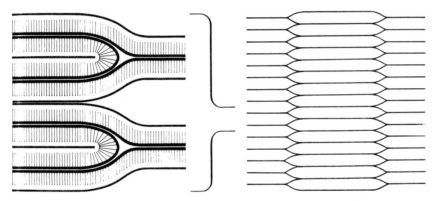

Fig. 12-14. Schematic version of chloroplast substructure (reference 2, Chapter 13).

concentration of lamellar surface is about twice as high within the grana as elsewhere. This arises from a pairing of the lamellae in these regions (Fig. 12-14).

The individual lamellae, of thickness about 130 Å, have a double layer structure, within which the flat chlorophyll molecules are believed to be stacked in piles (Fig. 12-14). The nature of the primary phase of photosynthesis is such as to make it highly probable that the structural organization of the grana is intimately related to its efficiency.

The primary phase of photosynthesis. The initial stage of photosynthesis, which can be accomplished in its entirety by isolated chloroplasts, is concerned with the formation of two compounds of central importance for subsequent events. These are ATP and TPNH. Neither process requires the reduction or incorporation of CO_2. These two substances represent the first stable, chemically well-defined products of the energy conversion occurring in chloroplasts.

The currently prevalent view of the generation of ATP by green plants represents it as arising from the phosphorylation of ADP, a reaction which is made possible by the utilization of the radiant energy absorbed by chlorophyll (14, 15, 16).

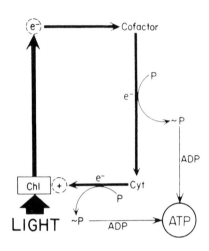

Fig. 12-15. Arnon's model for cyclic photophosphorylation (3). It should be recognized that this picture is probably oversimplified.

(12-1) $\quad P + ADP \xrightarrow{light} ATP$

In contrast to the processes of oxidative phosphorylation (Chapter 13), the photophosphorylation of ADP in no way depends upon the conversion of a chemical substrate; the only "substrate" consumed is light.

The mechanism for the photophosphorylation of ADP which will be discussed here is that developed by Arnon (Fig. 12-15), in which the chloroplast is regarded as a closed catalytic system. Several distinct photochemical pathways have been postulated to occur in chloroplasts. The first of these, referred to as *cyclic photophosphorylation*, has ATP as its sole product.

Arnon's model for cyclic photophosphorylation, while noncommital in detail, contains the following general steps:

(1) Chlorophyll is raised from its normal, *ground* state to a higher energy level by the absorption of a quantum of radiant energy.

(12-2) $\quad\quad\quad\quad\quad Chl + h\gamma \rightarrow Chl^*$

(2) The "excited" chlorophyll molecule loses an electron to become oxidized, thereby forming a "hole" or "odd ion," which will avidly accept another electron so as to return to its ground state.

(12-3) $\quad\quad\quad\quad\quad Chl^* \rightarrow Chl^+ + \epsilon^-$

It should be recognized that this version is probably quite oversimplified. Other pigments besides chlorophyll are known to participate in photosynthesis, and it is possible that additional steps involving energy transfer may occur between 12-1 and 12-3.

(3) The expelled electron is captured by, and reduces, a cofactor of uncertain identity. It is possible that several cofactors may operate in series or that alternative pathways may exist.

(12-4) $$X + e^- \rightarrow X^-$$

(4) The cofactor is in turn reoxidized by passing on its electron to the *cytochrome* system of chloroplasts. The cytochromes (section 13-2) are a family of related heme-containing proteins whose prosthetic (heme) groups can exist in oxidized or reduced states denoted by Cyt-Fe^{+++} and Cyt-Fe^{++}, respectively.

(12-5) $$X^- + \text{Cyt-Fe}^{+++} \rightarrow \text{Cyt-Fe}^{++} + X$$

The number of cytochromes operative and the order of their reaction remain uncertain.

(5) Finally, the "terminal" cytochrome is reoxidized by transfer of its electron to Chl^+, thereby completing the cycle.

(12-6) $$\text{Cyt-Fe}^{++} + \text{Chl}^+ \rightarrow \text{Cyt-Fe}^{+++} + \text{Chl}$$

The most important feature of this model is the *indirect* mechanism for the recapture of its electron by Chl^+. The return to the ground state results in an important release of energy. In solutions or suspensions of purified chlorophyll, the return to the ground state occurs abruptly by the emission of fluorescent radiation and collisional inactivation. In chloroplasts, the process is indirect and stepwise, so that energy is released as manageable "packets."

The actual phosphorylation of ADP occurs in coupled reactions which are "driven" (whose energy requirements are supplied by) the processes occurring in stages (4) and (5). The nature and mode of action of the phosphorylating enzymes are unknown, as is the detailed mechanism of coupling.

The nature of the cofactor, or cofactors, is likewise uncertain. Since phosphorylation is strongly stimulated by the addition of vitamin K (Fig. 12-16) or flavin mononucleotide (section 13-2), both of which are present in chloroplasts and both of which have redox potentials (Appendix D) in a suitable range, these compounds are likely candidates for this function. However, the process is also stimulated by nonphysiological compounds, suggesting that alternate pathways may exist.*

---

* An iron-containing protein, *ferredoxin,* has recently been identified in chloroplasts and may be the initial electron acceptor (K. Tagawa, H. Tsujimoto, and D. Arnon, *Proc. Natl. Acad. Sci. U. S.,* **49,** 567 (1963).

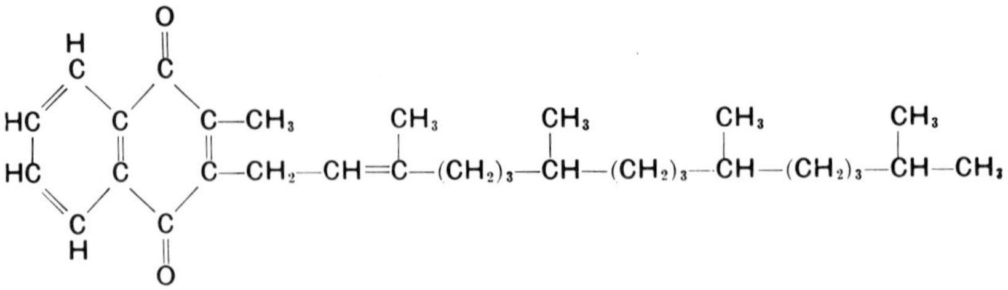

Fig. 12-16. Structure of vitamin K.

The over-all reaction may be written:

(12-7)
$$2\text{Chl}^+ + 2\overset{++}{\text{Fe}}\text{-Cyt} + \text{ADP} + \text{P} \rightleftarrows \text{ATP} + 2\text{Chl} + 2\overset{+++}{\text{Fe}}\text{-Cyt}$$

The terminal cytochrome is thus left in the oxidized state, to be reduced and then reoxidized in another turn of the cycle.

The production of reduced TPN (TPNH) by chloroplasts does not occur by cyclic photophosphorylation, but by an alternative mechanism in which ATP and TPNH are both produced. This process, which is referred to as *noncyclic photophosphorylation,* differs from the cyclic process in that water is consumed.

The chemical events occurring in noncyclic photophosphorylation may be summarized by:

(12-8)
$$2\text{TPN} + 2\text{ADP} + 2\text{P} + 2\text{H}_2\text{O} \rightleftarrows 2\text{TPNH} + \text{O}_2 + 2\text{ATP}$$

Under suitable experimental conditions the evolution of one mole of oxygen is accompanied by the reduction of two moles of TPN and the esterification of two moles of phosphate (Fig. 12-17). This process represents the sole known instance where ATP synthesis is coupled with an *evolution* of oxygen and a *reduction* of TPN. This is the precise opposite of the events of oxidative phosphorylation (section 13-6).

The initial steps are believed to be the same as for cyclic photophosphorylation (equations 12-1 and 12-2). The reduction of TPN is catalyzed by *TPN reductase* (Fig. 12-17). The electron expelled from chlorophyll is captured by TPN, together with a proton liberated by the splitting of water, to form TPNH.

In the tentative mechanism proposed by Arnon, the electron returned to Chl$^+$ is derived from a hydroxyl ion produced by the splitting of H$_2$O. As in the case of cyclic photophosphorylation, the cytochrome system serves as electron carrier. The transit of the electron through the

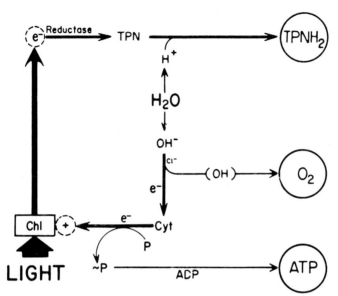

Fig. 12-17. Arnon's model for noncyclic photophosphorylation (3).

cytochromes is coupled to an enzymatic process which phosphorylates ADP (Fig. 12-17).

The feasibility of this model depends upon whether the cytochrome system has a sufficient redox potential to drive the reaction:

(12-9) $$2H_2O \rightarrow O_2 + 4H^+ + 4e^-$$

Unfortunately, our knowledge of the redox potentials of cytochromes occurring in chloroplasts is still incomplete.

The secondary phase of photosynthesis. The strictly photochemical aspects of the synthesis of carbohydrates conclude with the processes discussed in the preceding section. The incorporation of CO_2 and the actual assembly of carbohydrates depend only indirectly upon the primary photochemical reaction and can, in fact, proceed in cells devoid of chlorophyll, provided that the essential cofactors are made available.

The most important factor in the interconversion of organic molecules occurring in the later stages of photosynthesis is the *carbon reduction cycle,* in which most of the CO_2 is assimilated (Fig. 12-18). The identification of the intricate set of reactions which make up this process was largely accomplished by the work of Calvin and his associates (10-13). Much of the experimental work was done with the green algae *Chlorella pyrenoidosa.* However, the basic mechanism is probably quite general.

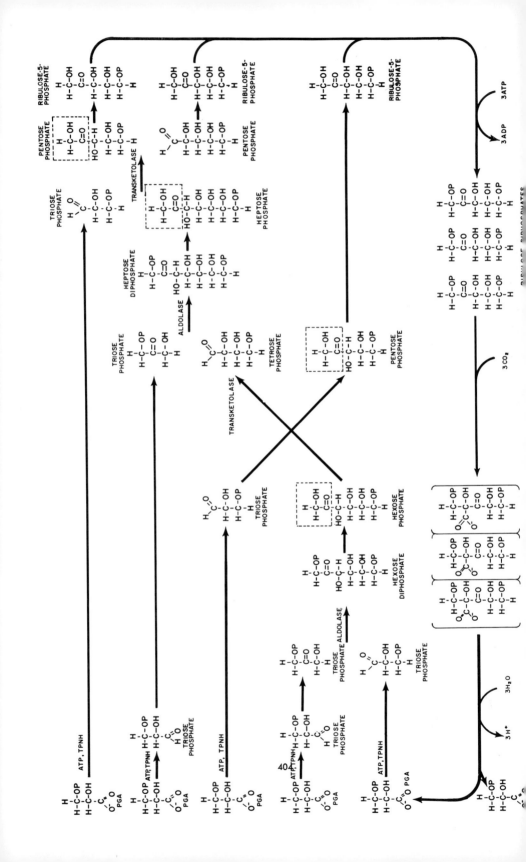

The experimental approach adopted depended upon the labeling of intermediates with the radioactive C^{14} isotope. Growing suspensions of algae were supplied with labeled CO_2. This was done by adding labeled $\overline{HCO}_3$ to the medium. The algae, after varying periods of exposure, were killed, and their metabolism stopped, by addition to a high concentration of methanol. In this manner, the period of CO_2 fixation could be limited and sharply defined.

An extract of the killed algae was then prepared and analyzed by two-dimensional paper chromatography. The distribution of radioactivity on the chromatogram was determined by placing it in contact with an X-ray film. The presence of radioactive spots was indicated by the appearance of darkened areas on the film (Fig. 12-19). A quantitative determina-

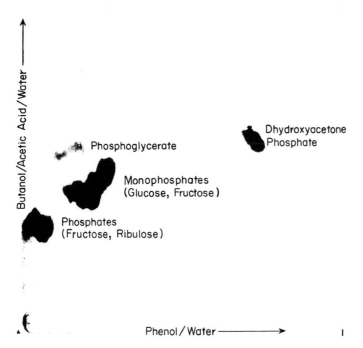

Fig. 12-19. Two-dimensional paper chromatogram showing earlier products of photosynthesis (10).

tion could be made by eluting each spot and measuring its total radioactivity with a Geiger-Müller counter. Identification of the spots was made by comparison with known compounds.

If the time of exposure to $C^{14}O_2$ was kept very short, most of the C^{14} which was incorporated appeared in the earliest intermediates. With increasing time a progressive labeling of subsequent intermediates occurred.

The earliest compound to be labeled was 3-phosphoglyceric acid (PGA). Quantitative measurements indicated that at very short times C^{14} was incorporated into PGA at a rate much more rapid than that for any other intermediate. It was therefore concluded that PGA was the earliest *stable* product of CO_2 fixation. Since the C^{14} was found to be present in the carboxyl group, there was a definite implication that the assimilation of $C^{14}O_2$ proceeded via a carboxylation.

A variant of this approach permitted identification of steps dependent upon the products of the light reaction. If the illumination was shut off *simultaneously* with the introduction of $C^{14}O_2$, it was found that PGA formation continued for some time with little conversion to subsequent sugar intermediates. If illumination was continued, labeled PGA did not accumulate and radioactivity subsequently appeared in compounds identified as the triose phosphates 3-phosphoglyceraldehyde and dihydroxyacetone phosphate. This suggested that the latter arose by reduction of PGA and isomerization. If reduction were TPNH-dependent, the reaction would proceed in the dark only until the limited supply of this cofactor was exhausted, since the production of TPNH would cease in the absence of illumination.

If photosynthesis was allowed to proceed until uniform labeling of all intermediates was attained and the illumination then turned off, the subsequent change with time in their concentrations showed a striking variance. Labeled PGA continued to accumulate, while the concentration of the labeled pentose derivative ribulose-1,5-diphosphate decayed rapidly. The levels of the other intermediates remained relatively stable. This indicated that the synthesis of ribulose-1,5-diphosphate, but not its conversion, depended upon a cofactor, presumably ATP, which was produced by the light reaction. This would be expected if this compound were formed by a phosphorylation which utilized ATP.

Moreover, the simultaneous drop in ribulose-1,5-diphosphate and accumulation of PGA, together with the location of PGA at the earliest stage of CO_2 incorporation, suggested that *the fixation of CO_2 occurred by a carboxylation of the ribulose derivative to form an unstable intermediate which decomposed to form PGA* (Fig. 12-18).

This view was subsequently confirmed by the complementary experiment of suddenly withdrawing the CO_2 while leaving the light on. In this case, the PGA concentration declined rapidly while the ribulose-1,5-diphosphate accumulated.

A quantitative comparison of the rates of carboxylation and total carbon uptake indicated that at least 70% of C^{14} incorporation occurred at this stage. The balance may be assimilated by alternative pathways as yet unidentified. The enzyme catalyzing the addition of CO_2 to ribulose-1,5-diphosphate is believed to be *carboxydismutase*.

PGA is the point of departure for the subsequent conversions which ultimately regenerate ribulose-1,5-diphosphate and complete the cycle. The complete pathway was traced out by identification of the intermediates and determination of the order of their appearance by procedures similar to those already described. Only the result will be cited here.

The PGA formed by carboxylation is transformed in two ways. A fraction is diverted to other pathways while the remainder is reduced to 3-phosphoglyceraldehyde in a coupled reaction requiring ATP and TPNH (Fig. 12-18). Part of the 3-phosphoglyceraldehyde is isomerized to dihydroxyacetone phosphate (DHAP).

In the subsequent reactions of the cycle, five molecules of triose phosphate are condensed to three molecules of pentose phosphate by a complicated series of interconversions (Fig. 12-18), which may be summarized as follows:

(1a) Part of the 3-phosphoglyceraldehyde condenses with one molecule of DHAP to form fructose-1,6-diphosphate. This is the reverse of a reaction important in the glycolysis phase of the oxidative metabolism of carbohydrates (section 13-3) and is likewise catalyzed by *aldolase*.

(1b) Fructose-1,6-diphosphate, after a preliminary dephosphorylation to fructose-6-phosphate, is converted to the four-carbon compound erythrose-4-phosphate, by the removal of two carbon atoms by *transketolase*.

(1c) The terminal CH_2OH-CO-grouping of fructose is transferred to a 3-phosphoglyceraldehyde molecule by transketolase to form the pentose derivative xylulose-5-phosphate.

(2) The erythrose-4-phosphate formed in 1b condenses with a molecule of DHAP to form the seven-carbon sedoheptulose-1,7-diphosphate in a reaction catalyzed by aldolase.

(3) After a preliminary dephosphorylation the two terminal carbons of sedoheptulose are transferred to a 3-phosphoglyceraldehyde by transketolase to form one molecule each of xylulose-5-phosphate and ribose-5-phosphate.

(4) The two molecules of xylulose-5-phosphate and one of ribose-5-phosphate, which are formed in (1) to (3), are isomerized to ribulose-5-phosphate, the common terminal product.

(5) Ribulose-5-phosphate is phosphorylated by ATP to form ribulose-1,5-diphosphate, which may in turn combine with CO_2 in another revolution of the cycle.

The carbon reduction cycle includes one phosphorylation reaction (ribulose-5-phosphate → ribulose-1,5-diphosphate), which consumes a molecule of ATP; a reduction (PGA → 3-phosphoglyceraldehyde), which consumes one molecule each of ATP and TPNH, and the carboxylation already described. The remaining reactions are either isomerizations or group transfers which do not require these cofactors. The *net* result of each complete turn of the cycle is the incorporation of three molecules of CO_2 and the production of one three-carbon compound.

Formation of other sugars. Eleven different phosphorylated sugar derivatives have been identified as intermediates of the carbon reduction cycle. In addition, glucose-1-phosphate and glucose-6-phosphate, which do not lie on the pathways of the cycle, are rapidly labeled soon after the introduction of $C^{14}O_2$. In general, the monosaccharides occur only as phosphorylated derivatives in photosynthetic organisms.

While unequivocal proof of the origin of the glucose derivatives has yet to be obtained, they are believed to arise from isomerization of the fructose-6-phosphate which is an intermediate of the carbon cycle. Enzymes are known which catalyze the interconversion of the 6-phosphates of glucose and fructose (*phosphohexose isomerase*) and the 1- and 6-phosphates of glucose (*phosphoglucomutase*). If these are operative in chloroplasts, the formation of the glucose derivatives is easily accounted for. Similar isomerzations could also explain the origin of the mannose and galactose phosphates which have also been detected.

Sucrose is the first *free* sugar to be labeled to any extent. It probably arises by some form of condensation reaction involving the corresponding phosphorylated monosaccharides, which is followed or accompanied by the splitting off of phosphate.

12-5 BIOSYNTHESIS OF THE POLYSACCHARIDES

Uridine diphosphate D-glucose (UDPG). The biosynthesis of complex polysaccharides proceeds by a process called *transglycosylation*, in which the carbohydrate moiety of a glycosyl donor is transferred to an acceptor, forming a new glycoside. The glycosyl donor may be a sugar phosphate, as in the formation of glycogen from glucose-1-phosphate by phosphorylase, or a sugar nucleotide, as UDPG, which has been identified as a substrate for the synthesis for many polysaccharides (17, 18, 19, 20).

The structure of UDPG is well established. It contains one residue each of UDP and α-D-glucose, joined by esterification of the terminal phosphate and the C_1 hydroxyl of the glucose (Fig. 12-20).

UDPG first came to be recognized as an agent for glucose transfer

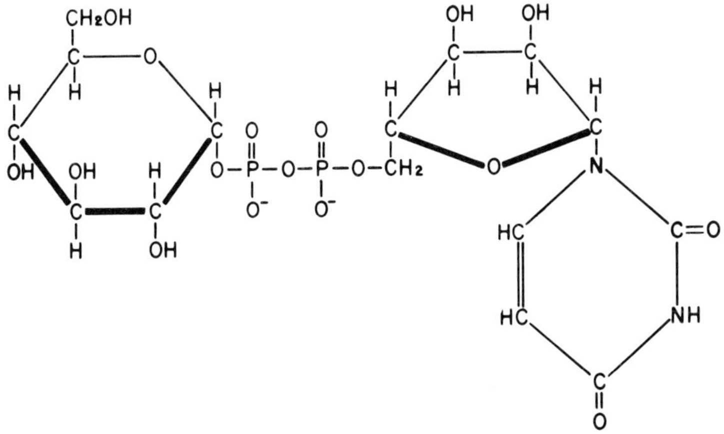

Fig. 12-20. Structure of UDPG.

in connection with studies of galactose utilization in yeast, where it is an intermediate in the reversible interconversion of the 1-phosphates of glucose and galactose.

(12-10) galactose-1-phosphate + UDPG

glucose-1-phosphate + UDP − galactose

UDP − galactose ⇆ UDP − glucose

UDPG itself appears to be formed by a transfer of a uridyl group from UTP to a glucose-1-phosphate, with a splitting out of pyrophosphate.

(12-11) UTP + glucose-1-phosphate

UDPG + PP

Pyrophosphorylases capable of carrying out this reaction are widely distributed in nature.

The general nature of the role of UDPG in the biosynthesis of polysaccharides has been established by recent investigations. The bacterium *Acetobacter xylinum* produces a polysaccharide identical with plant cellulose. Cell-free extracts from this microorganism have shown to incorporate C^{14}-labeled glucose from UDPG into cellulose, indicating that UDPG is a probable precursor.

Similarly, an enzyme preparation from liver has been obtained which catalyzes the transfer of glucose from UDPG to a glycogen primer, according to the equation:

(12-12) UDPG + (glucose)$_n$ ⇆ UDP + (glucose)$_{n+1}$

It is likely that this reaction represents an alternative to the phosphorylase pathway for the synthesis of glycogen. Similar findings have been made in the case of starch.

Analogs of UDPG appear to be utilized in the biosynthesis of acid and mucopolysaccharides. Thus a number of higher plants have been shown to contain enzymes which convert UDPG to UDPG-galacturonic acid, which is incorporated into pectin. Similarly, UDP-N-acetyl-D-glucosamine and UDP-galacturonic acid have been shown to be precursors of hyaluronic acid.

Biosynthesis of glycogen. The most important route for the biosynthesis of glycogen appears to involve a mechanism somewhat different from that summarized above. The enzyme *phosphorylase* catalyzes the transfer of glucose units from glucose-1-phosphate to glycogen, with the splitting out of inorganic phosphate.

$$(12\text{-}13) \quad \begin{array}{c} \text{glucose-1-phosphate} \\ + \\ (\text{glucose})_n \end{array} \rightarrow \begin{array}{c} (\text{glucose})_{n+1} \\ + \\ P \end{array}$$

The reaction *in vitro* is primer-dependent and does not proceed unless some glycogen is present initially. Phosphorylase exists in two forms, an active dimer (phosphorylase a) and an inactive monomer (phosphorylase b), which becomes active in the presence of 5'-AMP.

The mutual relationship of the two alternative pathways for glycogen synthesis remains obscure. It is possible that this is related to the physiological regulation of glycogen synthesis and utilization.

GENERAL REFERENCES

1. *Principles of Biochemistry*, A. White, P. Handler, E. Smith, and D. Stetten, McGraw-Hill, New York (1954).
2. *Molecular Biochemistry*, E. Kosower, McGraw-Hill, New York (1962).
3. *Light and Life*, W. McElroy and B. Glass, Johns Hopkins Press, Baltimore (1961).

SPECIFIC REFERENCES

Polysaccharides

4. E. Hirst in *Structure and Synthesis of Macromolecules*, Biochemistry Society Symposia, no. 21, p. 45, Cambridge University Press, London (1962).
5. *Cellulose and Cellulose Derivatives*, E. Ott, Interscience, New York (1943).

6. "The Chemistry of Glycogen," K. Meyer in *Advances in Enzymology*, vol. 3, p. 109, Academic Press, New York (1943).
7. *Chemistry of the Carbohydrates*, W. Pigman, and R. Goepp, Academic Press, New York (1948).
8. "Formation and Patterns of Cellulose Fibrils," A. Frey-Wyssling in Symposium on Biocolloids, *J. of Cell. and Comp. Phys.*, **49**, suppl. 1, p. 63 (1957).
9. *Colloid Science*, A. Alexander and P. Johnson, vol. 2, p. 766, Oxford University Press, New York (1949).

Photosynthesis

10. *The Photosynthesis of Carbon Compounds*, M. Calvin and J. Bassham, W. A. Benjamin, New York (1962).
11. M. Calvin and P. Massini, *Experimentia*, **8**, 445 (1952).
12. J. Bassham, A. Benson, L. Kay, A. Harris, A. Wilson, and M. Calvin. *J. Am. Chem. Soc.*, **76**, 1760 (1954).
13. J. Bassham and M. Kirk, *Biochim. et Biophys. Acta*, **43**, 447 (1960).
14. D. Arnon, *Nature*, **184**, 10 (1959).
15. D. Arnon, *Sci. Amer.*, **203**, 105 (1960).
16. D. Arnon in reference 3, p. 489.

Biosynthesis of polysaccharides

17. W. Hassid in reference 4, p. 63.
18. L. Leloir, C. Cardini, and E. Cabib in *Comparative Biochemistry*, M. Florkin and H. Mason (eds.), vol. 2, p. 97, Academic Press, New York (1960).
19. L. Leloir and C. Cardini, *J. Am. Chem. Soc.*, **79**, 6340 (1957).
20. M. Rongine de Fekete, M. Leloir, and C. Cardini, *Nature*, **187**, 918 (1960).

13

Energy Transformations by Biological Systems

13-1 ENERGY STORAGE AND UTILIZATION

General remarks. The maintenance of life in biological systems is dependent upon the continuation of a number of functions, all of which require energy. Some of the most important of these are the biosynthesis of proteins and nucleic acids, the performance of mechanical work by muscle contraction, the active transport of materials against diffusion gradients, and the phenomena associated with nerve conduction (1-3).

The introduction of the thermodynamic concept of *free energy* (F) is essential to an intelligible discussion of energy transformation in living systems (Appendix B). This has the dimensions of energy and is usually expressed as calories. For present purposes the free energy may be regarded as that fraction of the total energy which is potentially available for external work, or some other specific application, rather than dissipation as thermal motion of the surroundings. The free energy change (ΔF) accompanying a particular chemical reaction represents the maximum energy thereby made available for useful work. *All spontaneously occurring chemical processes involve a net decrease in the free energy of the system* ($\Delta F < 0$).

Every chemical compound may be regarded as possessing a well-defined amount of free energy for a particular set of conditions. When reactants are converted into products, the quantity ΔF, which is equal to the free energy of the products minus the free energy of the reactants, is positive if energy must be put into the system and negative if energy is released to the environment. Only in the latter case will the reaction proceed by itself. It may be said that reactions can only go "down hill" of their own accord, from compounds of higher to those of lower free energy. *Processes of positive ΔF must be supplied with free energy greater than ΔF from another source if they are to proceed.*

Formally, the free energy is defined in terms of heat content, or enthalpy (H), and entropy (S) by (Appendix B):

(13-1) $$F = H - TS$$

For a reversible chemical reaction of the type

(13-2) $$aA + bB + \cdots \leftrightarrows cC + dD + \cdots$$

the free energy change accompanying the conversion of one mole of each of the reactants into products is given by:

(13-3) $$\Delta F = \Delta F^\circ + RT \ln \frac{C^c D^d \cdots}{A^a B^b \cdots}$$

Here ΔF°, *the standard free energy change,* is defined as the value of ΔF for *standard* conditions (concentrations of all species equal to one mole per liter; $T = 25°C$; pressure = one atmosphere). ΔF° is related to the equilibrium constant (K) by

(13-4) $$\Delta F^\circ = -RT \ln K$$

When the concentrations of all species correspond to chemical equilibrium, ΔF is equal to zero and no net conversion occurs.

For a net chemical reaction to occur, ΔF must be negative. The standard free energies of many biologically important compounds have been determined and tabulated. The use of equation (13-3) permits prediction of whether a particular reaction will occur for known concentrations of reactants and products.

Readers unfamiliar with thermodynamic notation may find the discussion of Appendix B useful.

From the thermodynamic viewpoint, the persistence of living systems is intrinsically improbable. All the processes cited in the opening paragraph are basically endergonic; that is, they have a positive value of ΔF. The rigid and implacable laws of thermodynamics permit such endergonic processes only upon the condition that the free energy which they consume is compensated for by a continuous supply of free energy from an external source. In biological systems this free energy is made available by the oxidation of foodstuffs, such as sugars, fats, and proteins and supplied in the form of *chemical* energy.

The universal carrier of chemical energy in living systems is ATP, which may be regarded as an agent for the supply of chemical energy *in usable form.* The general nature of the role of ATP has come to be fully appreciated only in the last 20 years.

ATP is often described as a *high energy* compound. Its high energy content is largely electrostatic in origin and arises from the presence of a cluster of four negative charges in the triphosphate portion of the molecule. The splitting off of the terminal phosphate by hydrolysis to form ADP results in an important release of free energy. Such a reaction is said to be *exergonic*.

(13-5) $ATP + H_2O \rightarrow ADP + P$ ($\Delta F^\circ = -8$ KCal)

For the levels of ATP and ADP present *under physiological conditions*, ΔF is close to -12 KCal.* (1)

The only way in which energy can be transferred from one chemical reaction to another is for the two reactions to be consecutive and to have a common intermediate. The chemical energy stored in ATP is utilized by the participation of this compound in *coupled* reactions in which a dephosphorylation of ATP occurs. If the hydrolysis of ATP is coupled with a second reaction which consumes free energy, the negative change in free energy which it contributes may compensate for the positive ΔF of the other half of the process so that the *over-all* value of ΔF becomes negative, permitting the process to occur spontaneously. The dephosphorylation of ATP is often said to *drive* the reaction (1).

A classical example is the biosynthesis of sucrose from glucose and fructose (1). The reaction

(13-6)   glucose + fructose $\leftrightarrows$ sucrose + H_2O

is strongly endergonic ($\Delta F = 7$ KCal). It could never occur in biological systems were it not coupled with the hydrolysis of ATP. The consecutive steps are the phosphorylation of glucose

(13-7)   ATP + glucose $\leftrightarrows$ glucose-1-phosphate + ADP

followed by the combination of glucose-1-phosphate with fructose

(13-8)   glucose-1-phosphate + fructose $\leftrightarrows$ sucrose + P

Reactions (13-7) and (13-8) are *formally* equivalent to the sum of (13-5) and (13-6), and the over-all free energy change may be computed from the known figures for the latter two processes (1). Under *physiological conditions* ΔF is -12 KCal for (13-5) and $+7$ KCal for (13-6). Thus ΔF is -5 KCal for the sum of (13-5) and (13-6), or for the sum of (13-7) and (13-8), permitting the formation of sucrose to proceed, *provided that ATP and the appropriate enzymes are present*.

A more important example of an ATP-dependent biosynthesis is of course the transfer RNA-mediated incorporation of amino acids into

---

* This figure should not be confused with that for the *standard* free energy change under physiological conditions, which is close to -8 KCal (1).

polypeptides, which requires a preliminary activation of each amino acid by formation of an acyl adenylate with ATP (Chapter 11). The list could be extended indefinitely. Indeed it is no exaggeration to state that ATP is involved, directly or indirectly, in virtually all biological processes (2, 3).

Energy sources of living cells. As we have seen, the maintenance of biological functions is dependent upon a continuous supply of energy, which is furnished in the form of ATP. The synthesis of the energy carrier, ATP, must itself overcome considerable thermodynamic barriers and can only be accomplished by coupling with exergonic reactions. In the cells of organisms which lack the capacity for photosynthesis, this is accomplished by the processes of oxidative metabolism.

The most important single source of metabolic energy is the stepwise oxidation of carbohydrates to CO_2 and H_2O. The free energy change accompanying the complete oxidation of glucose is close to -690 KCal per mole, and this figure is of course independent of the particular mechanism, or pathway, of the process.

The cellular oxidation of glucose, unlike the uncontrolled combustion occurring in a flame, proceeds under rigorous control. The utilization of the oxidation of glucose for the synthesis of ATP depends upon a highly intricate series of stepwise coupled reactions, involving many dozens of oxidative enzymes.

The ultimate source of the energy stored in carbohydrates and hence of the energy supply, directly or indirectly, of almost all living organisms is the radiant energy of the sun, as trapped by the photosynthetic process. The carbohydrates thereby produced account for a major portion of the diet of animals, which are thus indirectly dependent upon photosynthesis for their energy requirements.

The stepwise oxidation of carbohydrates. The oxidative metabolism of carbohydrates proceeds in three main phases. The latter two of these represent the common terminal stages of the oxidation of all nutritional materials, including fats and proteins (Fig. 13-1).

Prior to their absorption and entry into the mainstream of oxidative metabolism, carbohydrates are converted into monosaccharides by the digestive system. An essential preliminary to the processing of monosaccharides by the cellular metabolic apparatus is their conversion to phosphorylated derivatives, as glucose-6-phosphate.

The reactions of the initial phase of carbohydrate metabolism are referred to collectively as *glycolysis*, or *anaerobic glycolysis*. This was the first of the major metabolic stages to be well understood at a molecular

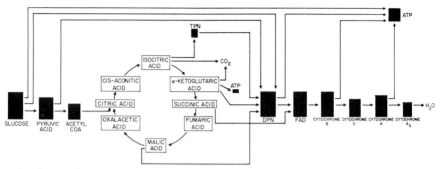

Fig. 13-1a. Schematic version of the oxidative metabolism of glucose (1). The solid blocks indicate the relative free energy remaining at each stage.

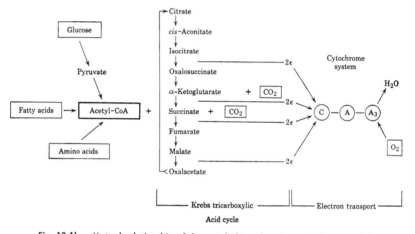

Fig. 13-1b. Mutual relationship of the metabolism of various nutritive materials.

level. The over-all result of glycolysis is the splitting of the hexoses to two molecules each of the three-carbon compound *pyruvic acid*. This seemingly simple process requires the sequential action of a dozen enzymes, many of which have been isolated in purified form.

Unlike the later phases of oxidative metabolism, the reactions of glycolysis involve no net uptake of oxygen. Again in contrast to the terminal stages, the enzyme system involved does not appear to be localized in a particular preformed structural element of the cell, nor does the over-all process depend upon any special geometrical organization.

Although only a minor fraction of the total energy supplied by the oxidative metabolism of glucose is made available *directly* by glycolysis, this set of reactions is necessary to convert glucose to a compound which may be further processed by the later stages.

The pyruvic acid formed by glycolysis is the basic substrate for the second phase of oxidative metabolism. This is the well-known *tricarboxylic acid cycle* (TCA cycle), sometimes known as the *Krebs cycle*. Prior to entry into the TCA cycle, pyruvate is converted into acetate and supplied to the cycle as the acetyl derivative of *coenzyme A*. It is at this point that fats and proteins, broken down to acetate by other enzyme systems, join the common terminal stages of oxidation.

After an initial condensation with oxalacetic acid, a four-carbon acid, to form citric acid, a six-carbon compound, the cycle proceeds by a series of dehydrogenations and decarboxylations to convert citric acid to oxalacetic acid. The two carbon atoms lost appear as CO_2.

The third phase of oxidative metabolism is much less well understood than the initial two. It is during this phase that most of the energy released by oxidative metabolism is converted to useful form. Basically, this final stage, which is known as *oxidative phosphorylation*, is concerned with the coupled phosphorylation of ADP. This occurs simultaneously with certain oxidation-reduction reactions.

The processes of oxidative phosphorylation operate in conjunction with the TCA cycle. However, this cooperation does not take the form of a direct supply of particular substrates by the latter, but rather of a conversion of certain coenzymes to their reduced, hydrogenated forms.

The subsequent reoxidation of these provides the energetic basis for the coupled conversion of ADP to ATP. For each oxygen taken up by the TCA cycle, three moles of ATP are formed. The over-all equations for the combined operations of the TCA cycle and oxidative metabolism may be written:

(13-9)
$$CH_3COCOOH + 5O \rightarrow 3CO_2 + 2H_2O$$
$$15P + 15ADP \rightarrow 15ATP$$

The labile and evanescent nature of the oxidative phosphorylation system has long been recognized. It is dependent upon the presence of relatively "native" extracts or homogenates of tissues and is easily inactivated by application of many of the usual preparative procedures for the isolation of enzymes, in contrast to the relative stability of the enzymes involved in the TCA cycle. The great fragility of the coupling mechanism has been a major technical obstacle to a detailed understanding of this process on a molecular level.

Oxidative phosphorylation, together with the reactions of the TCA cycle, has been found to be localized in highly organized subcellular structures, the *mitochondria* (Fig. 13-2). These are self-contained units, in which all the enzymes and coenzymes essential to the operation of the

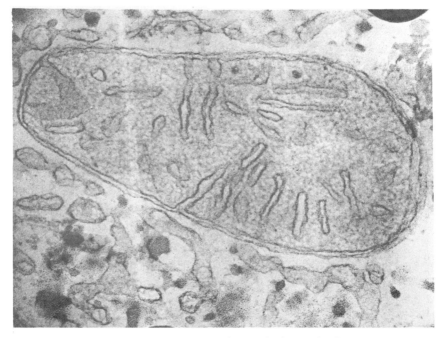

Fig. 13-2. Electron microscope photograph of a mitochondrion.

TCA cycle and oxidative phosphorylation are present. No prominent bottlenecks appear to be present, indicating the remarkably efficient organization of the mitochondria.

13-2 THE COENZYMES OF OXIDATIVE METABOLISM

A recurring theme in the complicated conversions of oxidative metabolism is the transfer of a particular atom or group between two intermediates by a set of coupled reactions. This generally occurs by way of the attachment of the group to one or more carriers, or *coenzymes*. There are many instances of a whole family of enzymes which share a requirement for a particular coenzyme.

No generalization can be made as to the strength of attachment of coenzymes to their enzyme partners. All gradations are known, from systems in which the two have only a transient association to those in which the coenzyme is so tightly bound as to be essentially a prosthetic group.

The coenzymes which will figure in the discussion of this chapter are listed below. No attempt will be made to discuss their chemistry in detail here.

(1) Coenzyme A: This compound (Fig. 13-3) has a very widespread

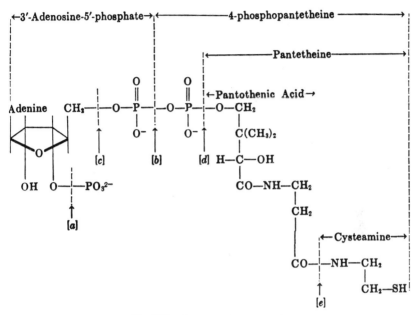

Fig. 13-3. Structure of coenzyme A.

occurrence in biological systems and serves as a cofactor for many enzymes. It contains one residue each of adenosine-3'-phosphate, pantothenic acid, and cysteamine (Fig. 13-3).

The active end of the molecule is that containing the sulfhydryl group, which is readily acylated to yield acyl coenzyme A. Coenzyme A (abbreviated as Co A or Co A.SH) functions biologically as a carrier of acyl groups, especially acetyl. As such, it frequently acts as a link between two enzyme systems which are engaged in the transfer of an acyl group between two intermediates.

(13-10)   acyl donor ⟶ HS.Co A ⟵ ⟶ acylated acceptor
          donor   ⟵ acyl Co A ⟶   acceptor

or

$$R-CO-X + HS.Co\ A \rightarrow R-CO.Co\ A + X$$
$$R-CO.Co\ A + Y \rightarrow R-CO-Y + HS.Co\ A$$

where X and Y are the donor and acceptor, respectively.

In such a transfer process, Co A is acylated in the initial step and deacylated in the second. The pair of reactions, with acyl Co A as a common intermediate, thus forms a coupled reaction sequence.

Acetyl Co A, sometimes known as *active acetate,* is of particular interest as the common end product of the initial phases of the metabolism of carbohydrates, fats, and proteins and the substrate for the common terminal phases of oxidative metabolism.

(2) Diphosphopyridine nucleotide (DPN): DPN, formerly known as coenzyme I,* contains one residue each of adenine and nicotinic acid amide, and two each of ribose and phosphoric acid (Fig. 13-4).

This coenzyme can exist in oxidized or reduced forms (Fig. 13-4). Nomenclature is somewhat confused, the oxidized form being variously designated as DPN or DPN^+ and the reduced form as DPNH, $DPNH_2$, or $DPN.H_2$. In this book DPN and DPNH will be used. The reduction of DPN results in the loss of the positive charge of a base nitrogen (Fig. 13-4).

The reduced and oxidized forms of DPN differ strikingly in several physical properties. The principal absorption band shifts from 260 mμ to 340 mμ upon reduction, which is also accompanied by the appearance of fluorescence.

DPN functions biologically as a *hydrogen carrier.* In particular, it is often associated with *dehydrogenases* and serves as a hydrogen acceptor, being converted to the reduced form DPNH (Fig. 13-4). The hydrogen is then passed on to a second acceptor, generally a flavoprotein, with the regeneration of DPN.

DPN may thus be regarded as a kind of "shuttle" which transports hydrogen from a metabolite to a flavoprotein acceptor (see [4] below).

(13-11) $\quad\quad\quad X.2H + DPN \rightarrow DPNH + X$

$DPNH + \text{flavoprotein} \rightarrow \text{flavoprotein}.2H + DPN$

(3) Triphosphopyridine nucleotide (TPN): TPN, or coenzyme II, is structurally similar to DPN, except for the presence of a third phosphate (Fig. 13-5). Its biological function is similar to that of DPN, and like the latter it serves as a hydrogen carrier. The reduced form of TPN will be designated as TPNH.

(4) The flavin coenzymes: The flavin coenzymes are usually rather tightly bound by their enzyme partners and may be regarded as *prosthetic*

---

* The confusion prevailing as to the nomenclature of these compounds has recently worsened with the introduction of the following new abbreviations: NAD (for nicotinamide adenine dinucleotide) in place of DPN; $NADH_2$ in place of DPNH; NADP in place of TPN.

It is likely that the older terms will be gradually replaced by these in future publications.

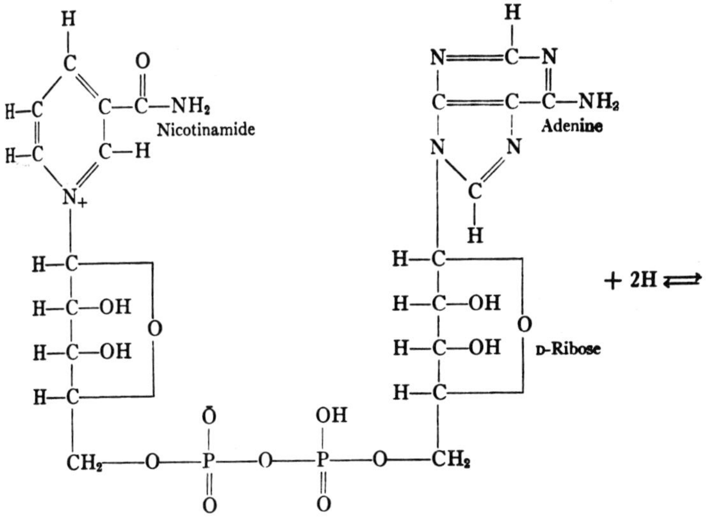

(a)

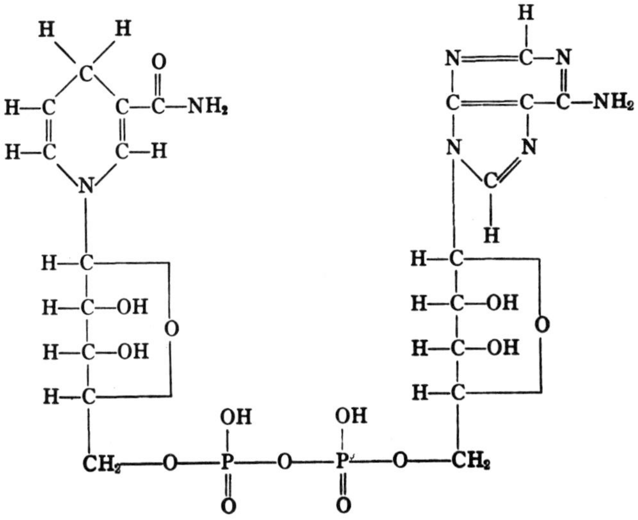

(b)

Fig. 13-4. Structure of oxidized (a) and reduced (b) forms of DPN.

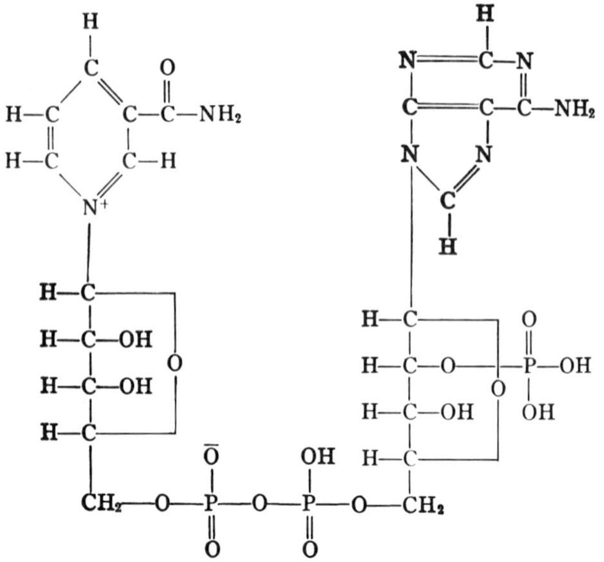

Fig. 13-5. Structure of oxidized form of TPN.

groups. Enzymes of this type are referred to as *flavoproteins*. The three most important flavin coenzymes are *riboflavin, flavin mononucleotide* (riboflavin-5′-phosphate), and *flavin adenine dinucleotide* (Fig. 13-6). Riboflavin, whose structure resembles that of the nucleosides, consists of a nitrogenous base, *isoalloxazine,* joined to a pentose derivative. Flavin mononucleotide (FMN) is the 5′-phosphate derivative of riboflavin. *Flavin adenine dinucleotide* (FAD) contains one residue each of FMN and adenosine-5′-phosphate united by a pyrophosphate linkage (Fig. 13-6).

The biological function of the flavin coenzymes is centered about their oxidation-reduction properties. During reduction, hydrogen is added to the isoalloxazine ring (Fig. 13-7) to produce the *leuco* form, which has an altered absorption spectrum.

The flavin coenzymes serve as hydrogen carriers and frequently act as acceptors for hydrogen from the reduced forms of the pyridine coenzyme, thereby regenerating DPN or TPN. In this manner, the coupled DPN-flavoprotein system serves as a link between the TCA cycle and the oxidative phosphorylation system. The reduced flavoprotein (F) is subsequently reoxidized by a cytochrome.

(13-12) $\qquad$ DPNH + F $\leftrightarrows$ DPN + F·2H

(5) The cytochromes: The cytochromes are a group of hemoproteins

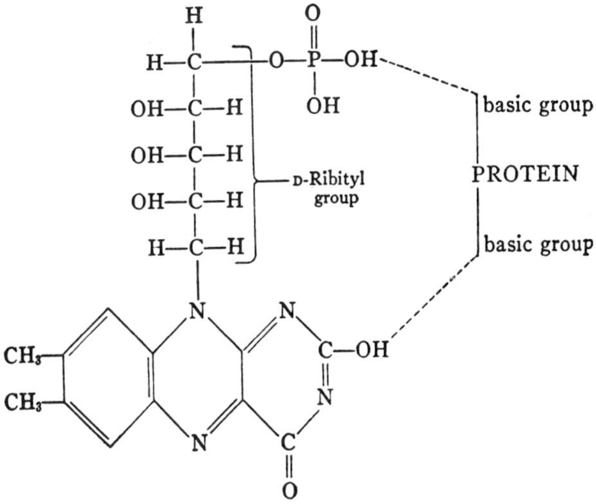

Isoalloxazine group

(a)

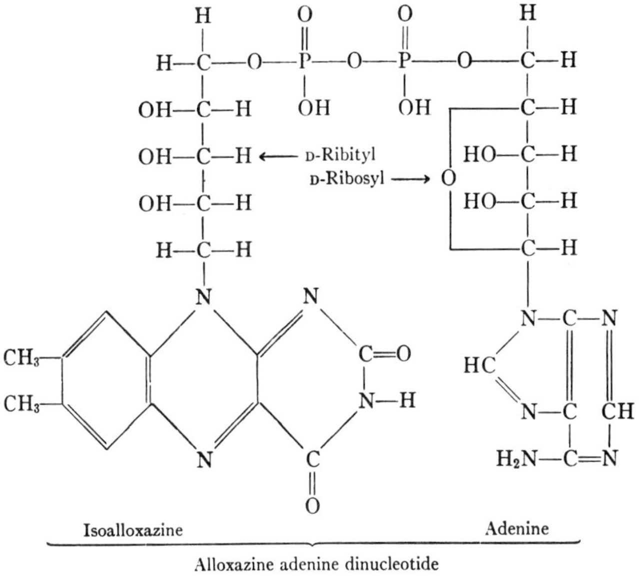

(b)

Fig. 13-6. Structures of FMN (a) and FAD (b).

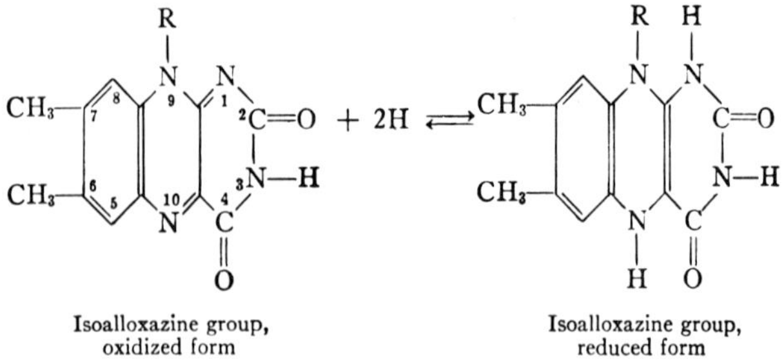

Fig. 13-7. Reduction of a flavin coenzyme.

which have a central collective role in the terminal stages of oxidative metabolism. At least four different cytochromes—a, a_3, b, and c—occur in the mitochondria of mammalian cells. Their prosthetic groups are similar but not identical (Fig. 13-8).

Each of the cytochromes contains an atom of iron as a complex with the prosthetic group, which can exist in either oxidized (Fe^{+++}) or reduced (Fe^{++}) form (Fig. 13-8).

The cytochromes of mammalian mitochondria form the final link in the chain of oxidation-reduction reactions by which the pyridine coenzymes reduced by the operation of the TCA cycle are reoxidized by molecular oxygen. They act in series with a flavoprotein (*DPNH dehydrogenase*) which is reduced by DPNH. The electrons transferred from the reduced flavoprotein are passed along the chain of cytochromes

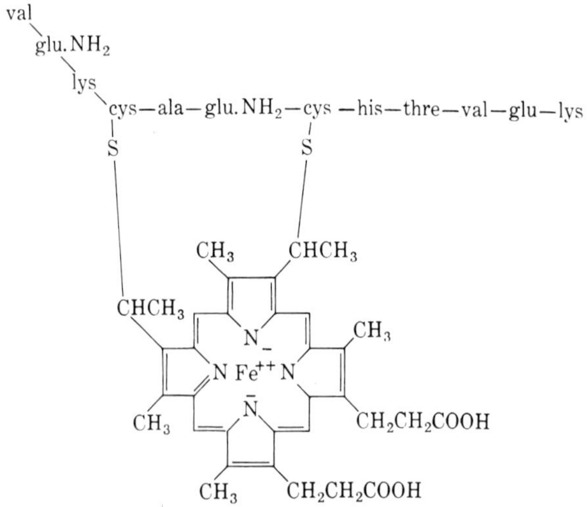

Fig. 13-8. Cytochrome c.

ENERGY TRANSFORMATIONS BY BIOLOGICAL SYSTEMS 425

in a series of coupled oxidations and reductions, the final step of which is the reoxidation of the terminal cytochrome by molecular oxygen.

The tentative sequence of *primary* reactions occurring in the cytochrome system of mitochondria is as follows:

(13-13)
$$F \cdot 2H \quad 2\,Cyt\,b-Fe^{+++} \quad 2\,Cyt\,c-Fe^{++} \quad 2\,Cyt\,a-Fe^{+++} \quad 2\,Cyt\,a_3-Fe^{++} \quad \tfrac{1}{2}O_2$$
$$\qquad\qquad\qquad\qquad\qquad\qquad\qquad\qquad\qquad\qquad\qquad 2H^+$$
$$F \quad 2\,Cyt\,b-Fe^{++} \quad 2\,Cyt\,c-Fe^{+++} \quad 2\,Cyt\,a-Fe^{++} \quad 2\,Cyt\,a_3-Fe^{+++} \quad H_2O$$
$$2H^+$$

where F represents the flavoprotein.

The above reactions are actually of a complex nature, being coupled with ancillary reactions which result in the phosphorylation of ADP (section 13-6).

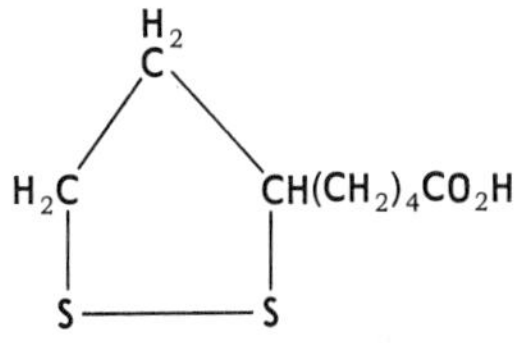

(6) α-Lipoic acid: This compound (Fig. 13-9) consists of a five-membered, disulfide-containing ring plus an aliphatic hydrocarbon chain terminating in a carboxyl group. The biological role of lipoic acid appears to arise from the presence of an intramolecular disulfide linkage and the reactions which it makes possible. In particular, lipoic acid can undergo a reduction to the sulfhydryl form, followed by acylation of the latter. Thus this coenzyme can serve both as an oxidant and an acyl transfer agent.

Lipoic acid (thioctic acid)
1,2—Dithiolane-3-valeric acid

Fig. 13-9. Structure of α-lipoic acid.

(7) Thiamine diphosphate (TPP): This coenzyme (Fig. 13-10) participates in the enzymatic catalysis of reactions by which C-C bonds are cleaved immediately *adjacent* to carbonyl groups. These include the de-

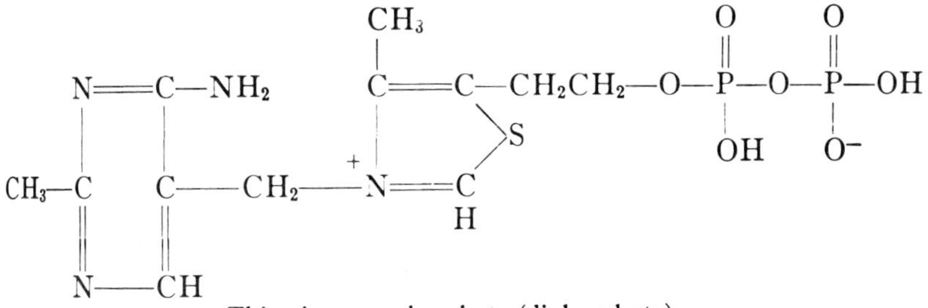

Thiamine pyrophosphate (diphosphate)
cocarboxylase

Fig. 13-10a. Structure of thiamine diphosphate.

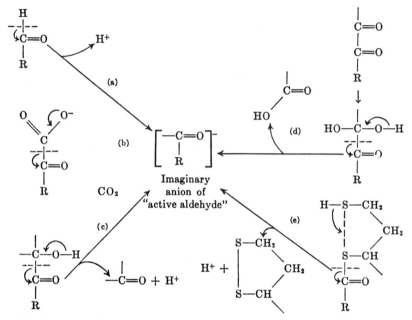

Fig. 13-10b. Some of the types of reaction in which TPP figures as a coenzyme. The common intermediate is believed to form a complex of uncertain structure with TPP.

carboxylation of α-keto acids. In each case, the over-all reaction is a removal of a group X and replacement by a group Y:

$$\text{(13-14)} \quad -\overset{\text{O}}{\underset{\|}{\text{C}}}-\text{XH} + \text{Y} \rightarrow -\overset{\text{O}}{\underset{\|}{\text{C}}}-\text{YH} + \text{X}$$

The X groups are invariably of such a structure that they can leave as stable compounds provided that the bond ruptured is cleaved by withdrawal of the bonding electron pair *toward* the carbonyl and into the thiamine-substrate-enzyme complex. Thus a hydrogen atom is removed as a proton, a carboxylate group as CO_2, and an alcoholic group as an aldehyde plus a proton. The detailed mechanism remains uncertain, as does the role of TPP in the process.

13-3 THE INITIAL PHASE OF THE OXIDATIVE METABOLISM OF CARBOHYDRATES

Preliminary conversions. Carbohydrates, which account for a major fraction of the mammalian diet, are for the most part ingested as polysaccharides or disaccharides, including amylose and amylopectin, glycogen, sucrose, and lactose. These are hydrolyzed to monosaccharides, primarily glucose, by the enzymes of the digestive tract and, as such, are absorbed through the intestinal wall and processed by the enzymes of oxidative metabolism.

Glucose itself has a rather limited participation in metabolic processes. Its entry into the reactions of glycolysis requires its preliminary conversion to a phosphorylated derivative, glucose-6-phosphate (4). The reaction is catalyzed by *hexokinase* and involves the transfer of the terminal phosphate of ATP to the 6-position of glucose (Fig. 13-11).

(13-15)   glucose + ATP $\xrightarrow{\text{hexokinase}}$ glucose-6-phosphate + ADP

The reaction is so highly exergonic that it is virtually irreversible.

Glucose-6-phosphate is reversibly isomerized to glucose-1-phosphate by the enzyme *phosphoglucomutase* (5) (Fig. 13-11). Glucose-1-phosphate is in turn reversibly polymerized to glycogen by the active form of *phosphorylase, phosphorylase* a, with the splitting off of inorganic phosphate (4, 7, 8).

Glycogen is the reserve carbohydrate of animal tissues. As such, its reversible conversion to glucose-1-phosphate appears to be an essential factor in the regulatory mechanism whereby the rate of oxidative metabolism is kept in pace with energy requirements. It is currently believed that phosphorylase may be a key regulatory enzyme.

This enzyme can exist in two well-defined forms; an inactive monomer, phosphorylase b; and an active dimer, phosphorylase a. The interconversion of the two, as catalyzed by *phosphorylase b kinase*, involves the addition of four phosphate groups to the dimer (4, 7, 8).

(13-15)   2 phosphorylase b + 4ATP → phosphorylase a + 4ADP

The rate at which glucose-1-phosphate is withdrawn from the glycogen reservoir and becomes available for glycolysis is governed by the level of phosphorylase a. This is in turn governed by a delicate balance between the synthetic reaction (13-15) and various inactivation processes (4). According to this model, the action of various regulatory hormones including *epinephrine* and *glucagon* consists in stimulating the production of phosphorylase a.

The 6-phosphates of glucose, fructose, and mannose are reversibly interconverted by specific *phosphohexoisomerases* (Fig. 13-11) and thus form a common metabolic pool with glucose-1-phosphate and glycogen. Fructose-6-phosphate is the point of departure for the reactions of glycolysis proper (4).

In the discussion of the following sections, only the *principal* metabolic pathway will be traced. The numerous side reactions, metabolic

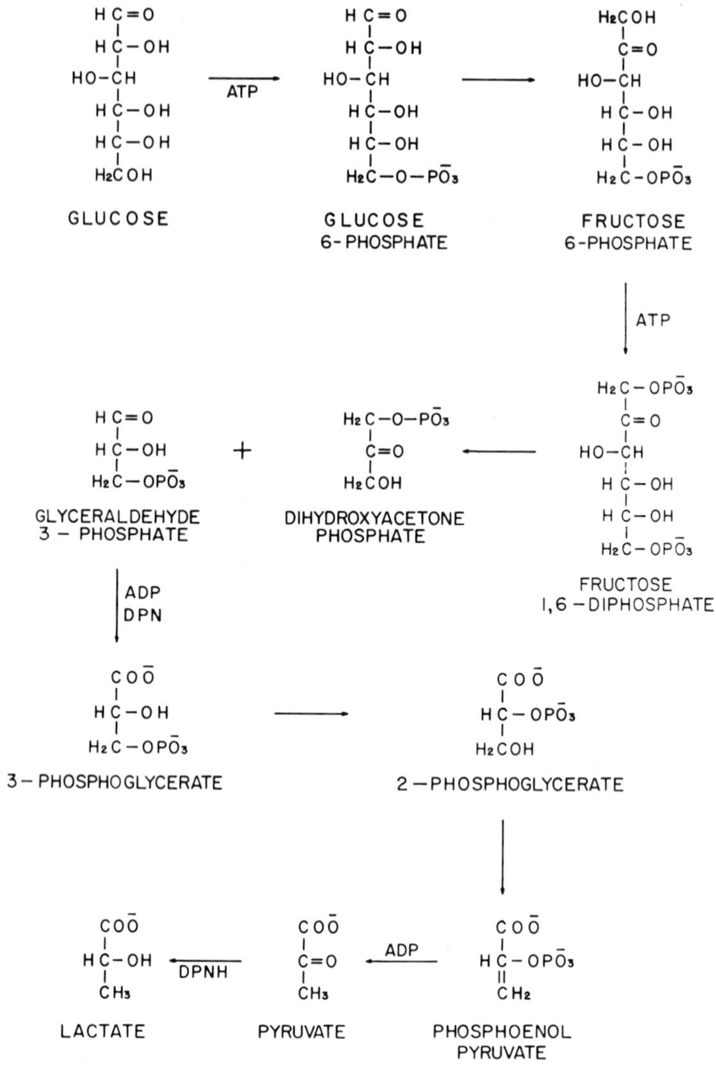

Fig. 13-11a. Conversions of the glycolytic process.

$$\text{GLUCOSE} \xrightarrow{\text{ATP}} \text{GLUCOSE 6-PHOSPHATE} \longrightarrow \text{FRUCTOSE 6-PHOSPHATE}$$

Glucose:
HC=O
HC—OH
HO—CH
HC—OH
HC—OH
H$_2$COH

Glucose 6-phosphate:
HC=O
HC—OH
HO—CH
HC—OH
HC—OH
H$_2$C—O—$\overline{\text{PO}_3}$

Fructose 6-phosphate:
H$_2$COH
C=O
HO—CH
HC—OH
HC—OH
H$_2$C—O$\overline{\text{PO}_3}$

↓ ATP

Fructose 1,6-diphosphate:
H$_2$C—O$\overline{\text{PO}_3}$
C=O
HO—CH
HC—OH
HC—OH
H$_2$C—O$\overline{\text{PO}_3}$

Glyceraldehyde 3-phosphate:
HC=O
HC—OH
H$_2$C—O$\overline{\text{PO}_3}$

+ Dihydroxyacetone phosphate:
H$_2$C—O—$\overline{\text{PO}_3}$
C=O
H$_2$COH

↓ ADP, DPN

3-Phosphoglycerate:
CO$\overline{\text{O}}$
HC—OH
H$_2$C—O$\overline{\text{PO}_3}$

⟶ 2-Phosphoglycerate:
CO$\overline{\text{O}}$
HC—O$\overline{\text{PO}_3}$
H$_2$COH

↓

Phosphoenol pyruvate:
CO$\overline{\text{O}}$
HC—O$\overline{\text{PO}_3}$
‖
CH$_2$

⟵ ADP ⟶ Pyruvate:
CO$\overline{\text{O}}$
C=O
CH$_3$

⟵ DPNH ⟶ Lactate:
CO$\overline{\text{O}}$
HC—OH
CH$_3$

"shunts" and so on, will be omitted from the discussion. The interested reader may find a more complete account in the reviews cited at the end of this chapter (3, 4, 11).

The reactions of glycolysis. Glycolysis proper may be considered to begin with the phosphorylation of fructose-6-phosphate (Fig. 13-11). By virtue of the set of reactions which interconverts glycogen and the sugar

ENERGY TRANSFORMATIONS BY BIOLOGICAL SYSTEMS 429

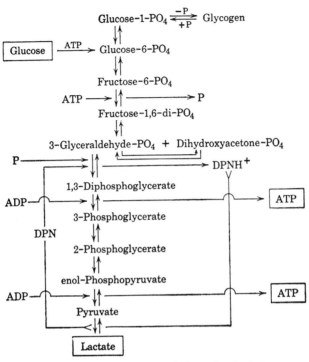

Fig. 13-11b. Transformation of glucose by glycolysis.

phosphates, the glycolytic process is enabled to draw upon the glycogen reservoir.

The reactions of glycolysis split one molecule of hexose to two molecules of pyruvic acid. In contrast to the TCA cycle, most of the reactions involve phosphorylated intermediates. The consecutive steps may be summarized as follows:

(1) Fructose-6-phosphate is further phosphorylated to form fructose-1, 6-diphosphate. The reaction is catalyzed by *phosphohexokinase*.

(13-16) fructose-6-phosphate + ATP → fructose-1,6-diphosphate + ADP

(2) Fructose-1,6-diphosphate is cleaved by *aldolase* to one molecule each of glyceraldehyde-3-phosphate and dihydroxyacetone phosphate (4, 6). The two products are reversibly interconverted by *triosephosphate isomerase*.

(13-17) fructose-1,6-diphosphate ⟨ glyceraldehyde-3-phosphate ⇅ dihydroxyacetone phosphate

(3) The next step is of particular interest as an example of an oxidative

reaction coupled with a phosphorylation of ADP to ATP. The over-all reaction, by which glyceraldehyde-3-phosphate is converted to 3-phosphoglycerate, occurs in two stages.

The first of these, which is catalyzed by *glyceraldehyde phosphate dehydrogenase* (9, 10), consists in the conversion of glyceraldehyde-3-phosphate to 1,3-diphosphoglycerate, with a concomitant reduction of DPN.

(13-18)   glyceraldehyde-3-phosphate + P + DPN ⇌ 1,3-diphosphoglycerate + DPNH + H^+

The product, a high energy mixed anhydride of a carboxylic acid and phosphoric acid, next serves to phosphorylate ADP in a reaction catalyzed by *phosphoglyceryl kinase*.

(13-19)   1,3-diphosphoglyceric acid + ADP ⇌ 3-phosphoglyceric acid + ATP

The over-all reaction may be regarded *formally* as equivalent to the sum of two partial reactions. The first of these, which is strongly exergonic ($\Delta F = -16$ KCal), is the oxidation of an aldehyde to a carboxylic acid. (Here and elsewhere the values of ΔF refer to physiological conditions.)

(13-20) $RCHO + DPN \leftrightarrows RCO\bar{O} + DPNH + H^+$

The second which is endergonic ($\Delta F = +12$ Kcal) is the phosphorylation of ADP

(13-21) $ADP + P \leftrightarrows ATP$

The sum of (13-20) and (13-21), which is equivalent to the sum of (13-18) and (13-19), is exergonic ($\Delta F = -4$ Kcal). The energy involved in oxidation of the aldehyde is recovered as ATP with some to spare. Thus two molecules of ATP are synthesized at this stage for each molecule of glucose which enters glycolysis.

One molecule of DPN is reduced during the above process. If there were no mechanism for the reoxidation of DPNH, the available DPN would soon be converted to this form, with a resultant blockage of glycolysis. In the presence of an adequate oxygen supply, the reoxidation of DPNH is accomplished by the cytochrome system. If the oxygen supply is insufficient, DPN is regenerated during the conversion of pyruvate to lactate, which becomes important under these conditions.

(4) The next step is an isomerization of 3-phosphoglyceric acid to 2-phosphoglyceric acid, catalyzed by *phosphoglyceryl mutase* (4). This

enzyme is believed to utilize 2,3-diphosphoglyceric acid as a coenzyme and to transfer a phosphate from the 3-position of the latter to the 2-position of the former (Fig. 13-12).

(13-22) 3-phosphoglycerate + 2,3-phosphoglycerate ⇌
2-phosphoglycerate + 2,3-phosphoglycerate

Thus the supply of 2,3-phosphoglyceric acid is not depleted.

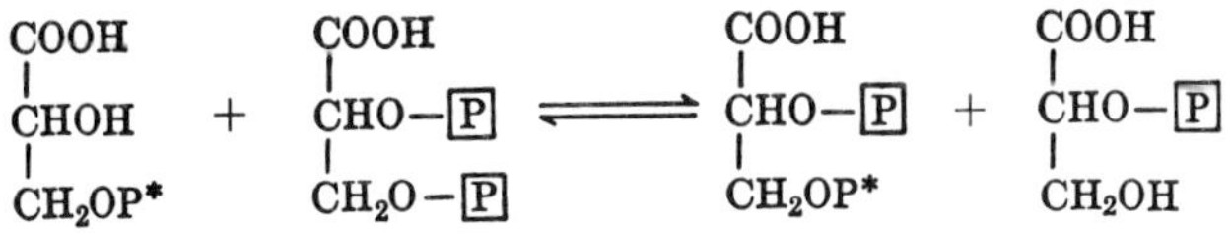

$$\begin{array}{c}COOH\\|\\CHOH\\|\\CH_2OP^*\end{array} + \begin{array}{c}COOH\\|\\CHO-\boxed{P}\\|\\CH_2O-\boxed{P}\end{array} \rightleftharpoons \begin{array}{c}COOH\\|\\CHO-\boxed{P}\\|\\CH_2OP^*\end{array} + \begin{array}{c}COOH\\|\\CHO-\boxed{P}\\|\\CH_2OH\end{array}$$

Fig. 13-12. Interconversion of 2- and 3-phosphoglycerate.

(5) 2-Phosphoglyceric acid is then converted to phosphoenol pyruvic acid by *enolase*.

(13-23) 2-phosphoglycerate ⇌ 2-phosphoenol pyruvate + H_2O

The removal of a molecule of water considerably enhances the ΔF for hydrolysis of the phosphate.

(6) Finally, the phosphate group of 2-phosphopyruvic acid is transferred to ADP in a reaction catalyzed by *pyruvate kinase*, thereby forming pyruvic acid. This is the second of the two glycolytic reactions in which ATP is synthesized.

(13-24) 2-phosphopyruvate ⇌ pyruvate
         + ADP   + ATP

Pyruvic acid then enters the second, oxidative phase of metabolism, being converted to CO_2 and H_2O by the TCA cycle. If the oxygen supply is inadequate, a fraction of the pyruvic acid produced is diverted to the formation of lactic acid by the action of *lactic dehydrogenase*. This is a DPN-requiring enzyme, and the process is accompanied by the reoxidation of one mole of DPNH.

(13-25) pyruvate + DPNH ⇌ lactate + DPN

This reaction serves to restore the DPN used up in step (3). If the oxygen supply is adequate, DPNH is reoxidized primarily by the cytochrome system, and the lactic acid step is unnecessary. Hence lactic acid does not accumulate under these conditions.

Over-all result of glycolysis. As we have seen, the preliminary conversion of glucose to glucose-6-phosphate and the phosphorylation of fructose-6-

phosphate in step (1) consume one mole of ATP each. This combined loss of two moles of ATP is made good by the formation of two moles (for each mole of glucose) in step (3) and in step (5), so that a net gain of two moles of ATP accompanies the conversion of one mole of glucose to two moles of pyruvate or lactate.

The free energy change for the reaction

(13-26)   glucose + 2ADP + 2P → 2 lactate + 2ATP

is about -26 Kcal (1). This represents a rather minor fraction of the total energy yield for the complete oxidation of glucose to CO_2 and H_2O ($\Delta F = -686$ Kcal). About half of the energy released by glycolysis is stored in the form of ATP.

Linkage of glycolysis and TCA cycle. Pyruvic acid is supplied to the enzymes of the TCA cycle in "activated" form as acetyl Co A. This compound represents the common end product of the initial metabolic processing of carbohydrates, fats, and amino acids. Although still incompletely understood, the over-all reaction appears to be analogous to the conversion of α-ketoglutarate to succinyl Co A, which will be described in section 13-5, and likewise involves the formation of a complex with TPP, followed by a transfer of the acetyl group first to lipoic acid and then to Co A.

13-4 THE MITOCHONDRIA

Structure and function. The mitochondria are cigar-shaped bodies, 0.5 to 3μ long and 0.1 to 0.5μ wide, that occur in the cells of all living organisms which require molecular oxygen (16). A conspicuous feature of their structure is the presence of external and internal double membranes. The latter are known as *cristae* (Fig. 13-3). It is still uncertain whether the cristae are attached directly to the external membrane.

The mitochondria have thus a large internal surface area. The structural elements within the mitochondrion are in contact with a continuous fluid medium which is sealed off from the surrounding cytoplasmic fluid. This contains essential enzymes and cofactors. A rupture of the external wall of the mitochondrion results in a leakage out of these, with a consequent loss of enzymatic functions.

Mitochondria from different sources vary greatly in their capacity to withstand physical deformation without loss in activity. Two generalizations can be extracted from the mass of available information:

(1) The capacity to carry out the TCA cycle is lost if the mitochondrial form is destroyed.

(2) The capacity for oxidative phosphorylation is eliminated by the obliteration of the double membrane structure.

Thus mitochondria which have been fragmented to smaller particles no longer mediate the complete TCA cycle but may retain the ability to carry out the processes of oxidative phosphorylation. The enzymes involved in the TCA cycle occur in the region exterior to the cristae (or in the intracristal space). Those which participate in oxidative phosphorylation are present entirely in the membranes in organized "solid-state" systems.

Indeed, the complete oxidative phosphorylation system appears to recur numerous times in each mitochondrion. The mitochondrial membrane is not merely an inert barrier, but a highly complex and organized structure, which is the site of intense metabolic activity. The mutual position of the molecules participating in oxidative phosphorylation is crucial for their combined function. Any major dislocation of the membrane structure results in the elimination of their over-all function.

13-5 THE TCA CYCLE

General remarks. The TCA cycle occupies a central position in intermediary metabolism. It represents the terminal oxidative phase for the biological processing of all the nutritional elements, including carbohydrates. It is primarily at this stage that the CO_2 and H_2O produced by oxidative metabolism arise.

The successive reactions of the cycle may be visualized as commencing with the condensation of acetyl Co A with oxalacetate to form citrate. The following steps consist of a series of oxidations and decarboxylations in which oxaloacetic acid is ultimately regenerated. For each revolution of the cycle, one mole of acetyl Co A is consumed, and two moles of CO_2 are evolved. The process can operate in a more or less continuous manner as long as acetyl Co A is fed into the cycle, and its products are removed.

The enzymes of the TCA cycle, like those involved in oxidative metabolism, occur in the mitochondria. However, unlike the latter, they do not appear to be integrated into the substructure, with one exception. Thus it has been possible to isolate the various enzymes and reproduce the consecutive steps separately *in vitro*.

Steps of the TCA cycle. The reaction sequence of the TCA cycle has been well established (Figs. 13-1 and 13-14). The consecutive steps are as follows:

(1) The initial reaction is the condensation of oxalacetate with acetyl Co A to form citrate and regenerate Co A.

(13-27)   acetyl Co A + oxalacetate $\leftrightarrows$ citrate + Co A + H^+

The reaction is catalyzed by *condensing enzyme* (11, 13). The energetics

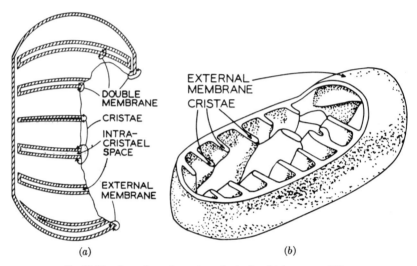

Fig. 13-13. Two schematic versions of mitochondrial structure (16).

are such as to favor the forward reaction strongly. It is probable that citroyl Co A is an intermediate, but its formation has yet to be demonstrated.

(2) Citrate is next reversibly isomerized to isocitrate by way of the intermediate cis-aconitic acid. The process is catalyzed by *aconitase*.

(13–28)   citrate $\leftrightarrows$ cis-aconitate $\leftrightarrows$ isocitrate

There is some evidence that aconitate may not be an obligatory intermediate and that the enzyme may also catalyze the direct conversion of citrate to isocitrate.

Aconitase is unusual in that it catalyzes the dehydration of both the tertiary hydroxyl of citrate and the secondary hydroxyl of isocitrate. This has led to suggestions that two different enzymes might be involved, but efforts to resolve them have been unsuccessful.

(3) The next step is a complex one whose over-all result is the oxidation and decarboxylation of isocitrate to α-ketoglutarate (11, 14). The enzyme involved, *isocitric dehydrogenase*, is TPN specific, and the complete process may be represented by:

(13-29)   isocitrate + TPN $\leftrightarrows$ α-ketoglutarate + CO_2 + TPNH

At least two major steps are involved. The first is dehydrogenation which presumably leads to oxalosuccinate.

(13-30)   isocitrate + TPN ⇌ oxalosuccinate + TPNH + H$^+$

The second step is a decarboxylation of oxalosuccinate (Fig. 13-14).

(13-31)   oxalosuccinate + H$^+$ ⇌ α-ketoglutarate + CO$_2$

(4) α-Ketoglutarate is next converted to succinate by a complicated mechanism catalyzed by the *α-ketoglutaric dehydrogenase* system. The

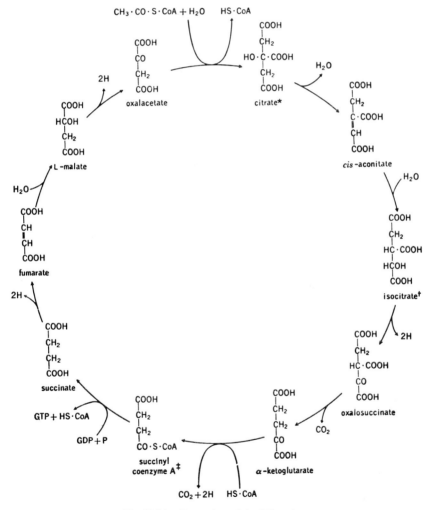

Fig. 13-14. Conversions of the TCA cycle.

complete system includes TPP, α-lipoic acid, Co A, DPN, and $\overset{++}{\text{Mg}}$ (11, 15).

The initial step is believed to be a reaction between α-ketoglutarate and diphosphothiamine (TPP), in which a succinic semialdehyde-TPP complex is formed and CO_2 is liberated.

$$(13\text{-}32) \quad R-\overset{O}{\underset{\|}{C}}-COOH + TPP \rightleftharpoons R-\overset{O}{\underset{\|}{C}}-H\,[TPP] + CO_2$$

where $\quad R = COOHCH_2CH_2-$

The next step is probably the reaction of the complex with the disulfide form of α-lipoic acid in such a way that the aldehyde group of succinic semialdehyde is oxidized to the corresponding carboxyl, while the disulfide is simultaneously reduced to the sulfhydryl form, with concomitant formation of an acylthiol. TPP is regenerated by this reaction.

(13-33)

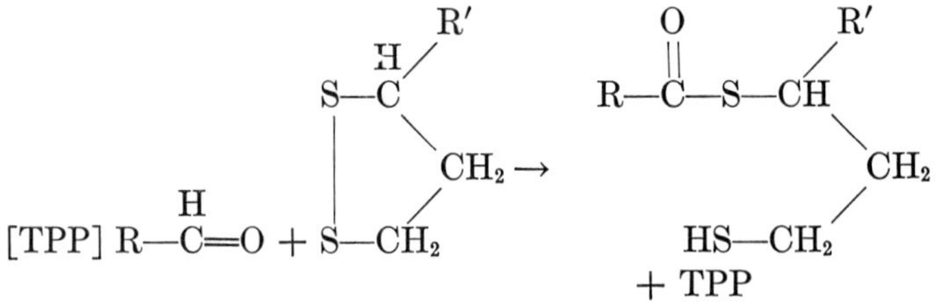

$$[TPP]\,R-\overset{H}{\underset{}{C}}\!=\!O + \underset{S-CH_2}{\overset{S-C\overset{H}{\diagup}\diagdown_{CH_2}}{|}}\overset{R'}{\diagup} \rightarrow \underset{HS-CH_2}{\overset{R-\overset{O}{\underset{\|}{C}}-S-CH\diagdown_{CH_2}}{}}\overset{R'}{\diagup} + TPP$$

In the next step, the succinyl group is transferred to the thiol group of Co A, forming reduced α-lipoic acid and succinyl Co A:

(13-34)

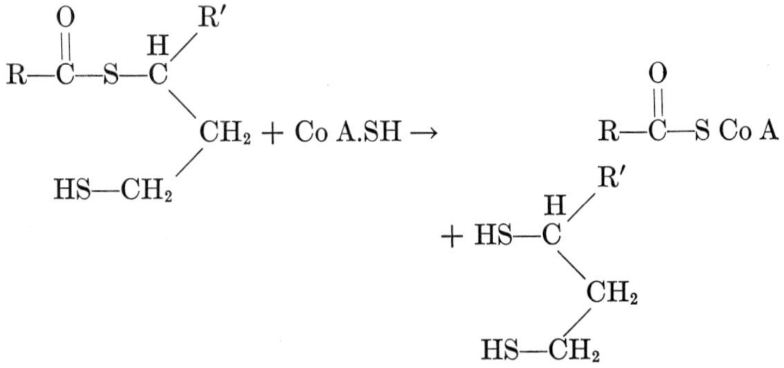

$$\underset{HS-CH_2}{\overset{R-\overset{O}{\underset{\|}{C}}-S-C\overset{H}{\diagup}\diagdown_{CH_2}}{}}\overset{R'}{\diagup} + Co\,A.SH \rightarrow R-\overset{O}{\underset{\|}{C}}-S\,Co\,A + \underset{HS-CH_2}{\overset{HS-C\overset{H}{\diagup}\diagdown_{CH_2}}{}}\overset{R'}{\diagup}$$

The reduced form of α-lipoic acid is reoxidized to the disulfide by *lipoic acid dehydrogenase*, a DPN-specific enzyme, and thereby made available for further transformations.

(13-35)
$$\text{HS–C}\begin{array}{c}\text{H}\\\diagup\\\diagdown\end{array}\begin{array}{c}\text{R}'\\\\\text{CH}_2\end{array} + \text{DPN} \rightarrow \begin{array}{c}\text{H}\\\text{S–C}\\|\\\text{S–CH}_2\end{array}\begin{array}{c}\text{R}'\\\\\text{CH}_2\end{array} + \text{DPNH}$$
$$\text{HS–CH}_2$$

The final stage of this process is the reaction of succinyl Co A with GDP and inorganic phosphate to regenerate Co A and form GTP and succinate.

(13-36)   succinyl—S.Co A + GDP + P $\leftrightarrows$ Co A.SH + GTP + succinate

This reaction is probably itself complex, consisting of several as yet unresolved steps. The GTP produced phosphorylates ADP to ATP in a reaction catalyzed by *nucleoside diphosphokinase*.

(13-37)   GTP + ADP $\leftrightarrows$ GDP + ATP

(5) Succinate is next dehydrogenated by succinic dehydrogenase to form fumaric acid. This reaction represents a link between the TCA cycle and the oxidative phosphorylation system. Unlike the other enzymes of the TCA cycle, succinic dehydrogenase is tightly bound to the structure of the mitochondrion, where it has a close spatial association with the cytochrome system. The dehydrogenation of succinate is coupled with the reduction of the oxidized form of cytochrome b, perhaps indirectly.

(13-38)   succinate + 2 cyt b—Fe^{+++} → fumarate + 2 cyt b—Fe^{++} + 2H$^+$

The process may be regarded as the transfer of a pair of electrons from succinate to cytochrome b. They are then passed along the respiratory chain (section 13-6).

Succinic dehydrogenase is a flavoprotein containing one mole of FAD. The nature of its interaction with cytochrome b suggests that their precise mutual location must be important for the process.

(6) The next step is the addition of water to fumerate to form malate. It is catalyzed by fumarase.

(13-38)   fumarate + H$_2$O $\leftrightarrows$ malate

(7) The final step is the dehydrogenation of malate to oxalacetate, as catalyzed by *malic dehydrogenase*, which is DPN-specific.

(13-39)   malate + DPN $\leftrightarrows$ oxalacetate + DPNH + H$^+$

With the regeneration of oxalacetate the TCA cycle is completed.

Over-all results of the TCA cycle. The TCA cycle is, to a large extent, self-perpetuating. The intermediates consumed in each step are regenerated by a second turn of the cycle. In the passage of a single molecule of pyruvic acid through the cycle, the following net changes occur (1, 2, 3, 11):

(1) Three molecules of CO_2 are produced. These are products of the formation of acetyl Co A from pyruvate, the decarboxylation of oxalosuccinate, and the decarboxylation of α-ketoglutarate.

(2) Three molecules of DPN are reduced and one of TPN.

(3) One molecule of cytochrome b is reduced.

Despite the lack of phosphorylated intermediates in the TCA cycle, it is known that the oxidation of pyruvate by this series of reactions is accompanied by a large number of phosphorylations of ADP. For every molecule of pyruvic acid which is oxidized in the TCA cycle, 15 molecules of ADP are phosphorylated. This is not accomplished directly by the cycle but by the oxidative phosphorylation system which is linked to and driven by it. If these ancillary processes are included, the over-all equation for the complete oxidation of one mole of pyruvic acid becomes:

(13-40) $\quad CH_3COCOOH + 5O \rightarrow 3CO_2 + 15ATP + 2H_2O + 15P + 15ADP$

This reaction is formally equivalent to the sum of the reactions

(13-41) $\quad CH_3COCOOH + 5O \rightarrow 3CO_2 + 2H_2O \quad (\Delta F = -273 \ Kcal)$

and

(13-42) $\quad 15P + 15ADP \rightarrow 15ATP \quad (\Delta F = 180 \ Kcal)$

Thus about two-thirds (180/273) of the energy released by the oxidation of pyruvate is harnessed for the synthesis of ATP.

13-6 OXIDATIVE PHOSPHORYLATION

General remarks. This final, and least understood, phase of oxidative metabolism is responsible for the greater part of ATP synthesis (17, 18, 19). However, the energy which drives this process is supplied by the TCA cycle, which is linked with the oxidative phosphorylation system at four points. One of these is the reduction of cytochrome b by succinic dehydrogenase. The others are the reduction of DPN (or TPN) at steps 3, 4, and 6 (section 13-5).

In the latter case, the over-all primary process of aerobic phosphorylation is the reoxidation of pyridine nucleotide by molecular oxygen.

(13-43) $\quad DPNH + O \rightarrow DPN + H_2O \quad (\Delta F - 55 \ Kcal)$

The highly exergonic nature of this reaction enables it to provide the energetic driving force for a series of phosphorylations coupled to it.

The reoxidation of DPNH is mediated by a series of linked oxidation-reduction reactions involving a flavoprotein and four cytochromes, all of which are integrated into the structure of the mitochondrial wall. This system is known as the *respiratory chain* and may be tentatively represented as (1, 18, 19):

(13-44)   DPN → flavoprotein → cyt b → cyt c → cyt a → cyta$_3$ → O$_2$

The process culminates in the reoxidation of cytochrome a$_3$ by molecular oxygen (equation 13-13). The mechanism cited in equation (13-44) should not be regarded as either complete or final. It is quite possible that additional components, such as coenzyme Q (20), may figure in the complete scheme.

In the case of the conversion of succinate to fumarate by succinic dehydrogenase, DPN is not involved directly. Instead, the reduction of cytochrome b is coupled somehow with the reaction:

(13-45)
$$\begin{array}{l} \text{succinate} \\ \text{fumarate} \end{array} \Bigg] \quad \Bigg[\begin{array}{l} \text{cyt b—Fe}^{+++} \\ \text{cyt b—Fe}^{++} \end{array}$$

The remaining steps are the same as in the case of the reoxidation of DPNH. Equation (13-45) should not be taken literally. It is possible that intermediate steps occur.

No physical transfer of intermediates by diffusion occurs. The geometrical organization of the respiratory chain is so efficient that the process takes the form of a transfer of electrons down the chain to molecular oxygen.

Coupled phosphorylations. The passage of a pair of electrons down the respiratory chain is accompanied by the phosphorylation of three molecules of ADP by coupled reactions. The over-all reaction for the reoxidation of DPNH is

(13-46)   DPNH + O + 3P + 3ADP → DPN + H$_2$O + 3ATP

The detailed mechanism of the phosphorylation steps is still uncertain. However, it has been possible to locate them along the respiratory chain. The prevailing evidence places the first phosphorylation between DPN and the flavoprotein, the second between cytochrome b and cytochrome c, and the third between cytochromes a and a$_3$ (Fig. 13-15).

A number of mechanisms for the phosphorylation reactions have been proposed and are being actively investigated. Basically, they all postu-

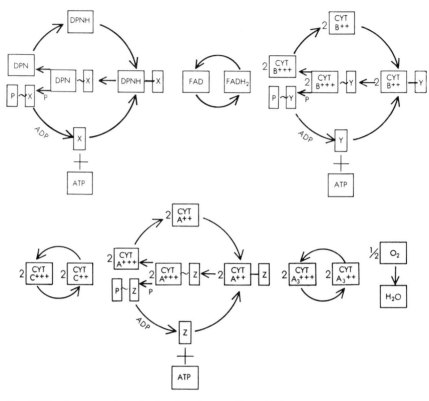

Fig. 13-15. A proposed mechanism for the coupling of electron transport and oxidative phosphorylation. This figure should not be taken too literally and may require considerable revision.

late the existence of a carrier C which couples with a molecule X to form a high energy intermediate, which in the presence of P and ADP produces ATP.

(13-47)
$$C + X \xrightarrow[\text{transfer}]{\text{electron}} C \sim X$$

$$C \sim X + P + ADP \rightarrow C + ATP + X$$

Lehninger has proposed a more explicit scheme involving the coupling of the carrier in a reduced state (1, 18, 19):

(13-48)
$$C_{red} + X + \text{oxidant} \rightarrow C_{ox} \sim X + \text{reductant}$$

$$C_{ox} \sim X + P \rightarrow C + P \sim X$$

$$P \sim X + ADP \rightarrow ATP + X$$

One form of the Lehninger mechanism is illustrated in Fig. 13-15. Here the carriers are identified with known components of the respiratory chain. However, they may figure in intermediate reactions which have not as yet been recognized.

Mitochondrial subunits. Heart mitochondria can be broken down into subunits, which retain in varying degree the functions of the intact particle. By a proper choice of technique it is possible to convert mitochondria almost quantitatively into small fragments in which the double membrane structure is preserved. Such particles retain the capacity to reoxidize DPNH and synthesize ATP, although they can no longer carry out the TCA cycle.

By a more drastic process of disintegration, particles may be obtained which have lost all double membrane structure. Such particles may catalyze the reoxidation of DPNH by molecular oxygen, but do not phosphorylate ADP. They are referred to as *electron transport particles* (ETP).

Each ETP may be regarded as a mitochondrial subunit from which the enzymes required for the TCA cycle and for oxidative phosphorylation have been stripped away, but which retain the components of the respiratory chain. These thus appear to form definite tightly organized subunits within the mitochondrial structure, which may be dissociated from the phosphorylating enzymes.

GENERAL REFERENCES

1. A. Lehninger, *Rev. Mod. Phys.*, **31**, 136 (1959).
2. A. Lehninger, *Sci. Amer.*, **202**, 102 (1960).
3. *General Biochemistry*, J. Fruton and S. Simmonds, p. 284-386 and 520-524, Wiley, New York (1958).

SPECIFIC REFERENCES

Glycolysis

4. B. Axelrod in *Metabolic Pathways*, D. Greenberg (ed.), vol. 1, p. 97, Academic Press, New York (1960).
5. C. T. Cori, S. Colowick, and C. F. Cori, *J. Biol. Chem.*, **124**, 543 (1938).
6. J. Taylor, A. Green, and C F. Cori, *J. Biol. Chem.*, **73**, 591 (1948).
7. E. Krebs, A. Kent, and E. Fischer, *J. Biol. Chem.*, **231**, 73 (1958).
8. E. Fischer and E. Krebs, *J. Biol. Chem.*, **216**, 121 (1955).
9. G. T. Cori, M. Slein, and C. F. Cori, *J. Biol. Chem*, **173**, 605 (1948).

10. J. Taylor, S. Velick, G. T. Cori, C. F. Cori, and M. Slein, *J. Biol. Chem.*, **173,** 619 (1948).

TCA cycle

11. H. Krebs and J. Lowenstein in *Metabolic Pathways*, D. Greenberg (ed.), vol. 1, p. 129, Academic Press, New York (1960).
12. V. Massey, *Biochem. J.*, **51,** 490 (1952).
13. S. Ochoa, *J. Biol. Chem.*, **174,** 133 (1948).
14. S. Ochoa and E. Weisz-Tabori, *J. Biol. Chem.*, **159,** 245 (1945).
15. I. Gunsalus in *Symposium on the Mechanism of Enzyme Action*, W. McElroy and B. Glass (eds.), p. 545, Johns Hopkins, Baltimore (1954).

Mitochondria

16. D. Green and S. Fleischer in *Metabolic Pathways*. D. Greenberg (ed.), vol. 1, p. 41, Academic Press, New York (1960).

Oxidative phosphorylation

17. *Recent Advances in Biochemistry*, T. Goodwin, p. 1, Churchill, London (1960).
18. A. Lehninger, C. Wadkins, C. Cooper, T. Devlin, and J. Gamble, *Science*, **128,** 450 (1958).
19. A. Lehninger, *J. Biol. Chem.*, **234,** 2187, 2467 (1959).
20. A. Lehninger and C. Wadkins, *Ann. Rev. Biochem.*, **31,** 47 (1962).

Appendix A

The Primary Structure of the B-Chain of Insulin

Use of the DNP labeling technique demonstrated that the monomer unit of insulin consisted of two polypeptide chains whose NH_2-terminal residues were phenylalanine and glycine (section 3-5). These are usually denoted as the B- and A-chain, respectively.

The A- and B-chains were separated by performic acid oxidation of the cystine cross-links to cysteic acid and obtained in purified form. They were found to contain 20 and 30 amino acids, respectively. The compositions are cited in Table 3-3.

The amino acid sequence of the B-chain of oxidized insulin was the first to be determined and will be considered in detail here. Inspection of Table 3-3 reveals that five residues occur but once in the B-chain—aspartic acid, threonine, proline, serine, and arginine. Any two peptides which share any one of these must overlap and originate from the same portion of the B-chain.

Conversion of the B-fraction of oxidized insulin to its DNP derivative, followed by partial acid hydrolysis, yielded a mixture of DNP-phenylalanyl peptides. These were fractionated chromatographically, hydrolyzed, and their amino acids identified. The peptides DNP-Phe, DNP-Phe-Val, DNP-Phe-Val-Asp, and DNP-Phe-Val-Asp-Glu were isolated and identified, fixing the NH_2-terminal sequence of the B-chain as Phe-Val-Asp-Glu.

Unlabeled fraction B, after partial hydrolysis with acid, was separated into groups of peptides by a combination of ionophoresis and adsorption chromatography. The fractions thus obtained were further separated into peptides by two-dimensional paper chromatography. The peptide spots were cut out, eluted, hydrolyzed, and their amino acids identified. NH_2-

terminal groups were determined by the DNP method. In this manner, the structure of the dipeptides was completely fixed, and the NH_2-terminal residues of the higher peptides were determined. In writing down sequences, the first three letters of the name of each amino acid will be used as its abbreviation (Table 3-4).

Through knowledge of the NH_2-terminal groups, it was possible to bring the bewildering variety of peptides isolated into partial order. The tetrapeptide Phe-Val-Asp-Glu (1a) has already been located at the NH_2-terminal end of the chain (Table A-1). Since aspartic acid is a singly occurring residue, any peptide containing it must be attached to this NH_2-terminal sequence. Thus the isolation of the pentapeptide containing the above four residues and His permitted enlargement of the NH_2-terminal sequence to Phe-Val-Asp-Glu-His (1b).

There are two cysteic acid residues in the B fragment of oxidized insulin, which must be included in two different sequences. Two tripeptides containing cysteic acid were identified: Leu-($CySO_3H$, Gly) and Val-($CySO_3H$, Gly). Identification of the dipeptides Leu-$CySO_3H$, Val-$CySO_3H$, and $CySO_3H$-Gly permitted the two cysteic acid tripeptides to be written as Leu-$CySO_3H$-Gly (1c) and Val-$CySO_3H$-Gly (1d).

Isolation of the tripeptide Leu-Val-$CySO_3H$ and the tetrapeptide Tyr-Leu-Val-$CySO_3H$ established the sequence containing the second cysteic acid group as Tyr-Leu-Val-$CySO_3H$-Gly (1e).

Similarly, the identification of His-Leu-$CySO_3H$ and Glu-His-Leu-$CySO_3H$ permitted enlargement of the first cysteic acid sequence (1c) to Glu-His-Leu-$CySO_3H$-Gly (1f). The sequence Glu-His is shared with the NH_2-terminal pentapeptide Phe-Val-Asp-Glu-His. This raised the possibility that the two are overlapping parts of the same sequence.

However, the B-chain contains three Glu residues which must be accounted for before the above hypothesis can be accepted. One of these was found in a peptide with NH_2-terminal serine: Ser-(Ala, Glu, His, Leu, Val). Since serine occurs but once, the residue order in this peptide was readily determined. Identification of the peptide Ser-(Glu, His, Leu, Val) placed Ala at the COOH-terminal end of the original serine peptide. The isolation of Ser-His, Ser-His-Leu, and Ser-His-Leu-Val established the order of the first four residues. Hence the serine peptide had the structure Ser-His-Leu-Val-Glu-Ala (1g).

The remaining Glu group occurred in a tripeptide Gly-(Arg, Glu), where Arg is singly occurring. Isolation of Arg-Gly placed Arg at the COOH-terminal end of the Gly-Glu-Arg tripeptide and established the Arg sequence as Gly-Glu-Arg-Gly (1h).

Thus, two of the three Glu groups have been accounted for unambiguously. It follows that the remaining glutamic of the NH_2-terminal phenylalanine sequence and that of the first cysteic acid sequence must be the same and that the two sequences can be combined as Phe-Val-Asp-Glu-His-Leu-CySO$_3$H-Gly (1i).

One more sequence could be established from the acid digest. The tetrapeptide Thr-(Ala, Lys, Pro) was isolated, of which Thr and Pro occur singly. Identification of Thr-Pro and Pro-(Ala, Lys) fixed the sequence as Thr-Pro (Ala, Lys). Identification of Lys-Ala completed the sequence of the tetrapeptide as Thr-Pro-Lys-Ala (1j) since the only other Ala group has been located after Glu in the serine peptide (1g).

While pausing for breath, it is useful to summarize what has been established up to this point. As Table A-1 shows, all the amino acid residues of the B-chain, except for one leucine, one tyrosine, and two phenylalanines have been located in five extended sequences (1e, 1g, 1h, 1i, 1j).

Further progress required the use of enzymes. The extreme lability to acid of the bond involving the amino group of serine precludes its occurrence except as an NH$_2$-terminal residue, rendering it impossible to establish the linkage of the serine peptide by acid hydrolysis. Moreover, the remaining amino acids are concentrated in a nonpolar zone, which gives rise on acid hydrolysis to a mixture of similar nonpolar peptides very difficult to resolve. For these reasons, hydrolysis by enzymes was resorted to.

From the pepsin digest of the B-chain a peptide was isolated which contained the NH$_2$-terminal phenylalanine, plus the amino acids (Asp, Glu, Gly, His, Leu, CySO$_3$H, Ser, Val). All these residues occur in sequences 1i and 1g of Fig. 5. Since they include the singly occurring residues aspartic acid and serine, the peptide must include all of sequence 1i and part of sequence 1g. Thus these two sequences must be connected, and the known NH$_2$-terminal sequence can be expanded to sequence 2a of Table A-2.

By the consecutive action of chymotrypsin and pepsin, a peptide of composition (Ala, Glu, Leu, Tyr, Val) was produced and isolated. This can only arise from sequence 2a, if the latter is extended to include a tyrosine (sequence 2b).

Chain-B contains three glycines, one of which is contained in 2b. Others are found in 1e and 1h, which must therefore be parts of the same sequence (2c). Since a dipeptide Gly-Phe was earlier observed in the acid hydrolysate, a phenylalanine residue must be joined to the COOH-terminal end of sequence 2c to form 2d. This follows from the fact that the residues joined to the carboxyls of all other glycines have already been identified.

From a digest obtained from the consecutive action of chymotrypsin and trypsin, a peptide of structure Gly-Phe-Phe was identified. Thus a second phenylalanine can be added to sequence 2d, forming 2e.

A link between sequences 1j and 2e could also be established. From the chymotrypsin digest a peptide was isolated containing the four residues of 1j, plus an NH_2-terminal tyrosine. This permitted enlargement of 1j to 2f, thereby completing the assignment of all the amino acids in the B-chain. The isolation from the trypsin digest of a peptide containing the singly occurring residues proline and threonine from 2f and also phenylalanine proved that 2f and 2e must be connected. This followed because the only phenylalanine apart from those in the COOH-terminal sequence of 2e is the NH_2-terminal phenylalanine of the B-chain. Thus 2f and 2e can be combined to form 2g.

There are only two tyrosines in the B-chain. One of these is accounted for in the interior of 2g. The other is present both as the COOH-terminal residue of 2b and as the NH_2-terminal residue of 2g. Therefore, 2b and 2g must be joined.

Thus the sequence determination of the B-chain of insulin was at last completed. The complete sequence has the form shown in Fig. 3-6, including amide groups.

TABLE A-1. SEQUENCES DETERMINED FROM ACID HYDROLYSATE OF B-CHAIN

1a	Phe-Val-Asp-Glu
1b	Phe-Val-Asp-Glu-His
1c	Leu-CySO$_3$H-Gly
1d	Val-CySO$_3$H-Gly
1e	Tyr-Leu-Val-CySO$_3$H-Gly
1f	Glu-His-Leu-CySO$_3$H-Gly
1g	Ser-His-Leu-Val-Glu-Ala
1h	Gly-Glu-Arg-Gly
1i	Phe-Val-Asp-Glu-His-Leu-CySO$_3$H-Gly
1j	Thr-Pro-Lys-Ala

TABLE A-2. SEQUENCES OF CHAIN B DETERMINED BY ENZYMATIC HYDROLYSIS

2a	Phe-Val-Asp-Glu-His-Leu-CySO$_3$H-Gly-Ser-His-Leu-Val-Glu-Ala
2b	Phe-Val-Asp-Glu-His-Leu-CySO$_3$H-Gly-Ser-His-Leu-Val-Glu-Ala-Tyr
2c	Tyr-Leu-Val-CySO$_3$H-Gly-Glu-Arg-Gly
2d	Tyr-Leu-Val-CySO$_3$H-Gly-Glu-Arg-Gly-Phe
2e	Tyr-Leu-Val-CySO$_3$H-Gly-Glu-Arg-Gly-Phe-Phe
2f	Tyr-Thr-Pro-Lys-Ala
2g	Tyr-Leu-Val-CySO$_3$H-Gly-Glu-Arg-Gly-Phe-Phe-Tyr-Thr-Pro-Lys-Ala

Appendix B
Basic Thermodynamic Concepts

Classical thermodynamics may be said to stem largely from three fundamental laws. These laws may be regarded as concise summations of a large amount of experience, rather than as theorems which can be derived from more fundamental principles.

It is desirable to begin with several definitions. A *system*, in the thermodynamic sense, is any material which can be differentiated from, and be considered independently of, its surroundings. By the *surroundings* is meant the rest of the universe, and in particular that part which is in immediate contact with the system and can exchange energy with it. A thermodynamic system may, for example, be a solution of chemicals in a beaker, a solid crystal, or a living organism.

A system is said to be at *equilibrium* with its surroundings when its composition and physical parameters (such as temperature and pressure) are invariant with respect to time and when no net transport of mass or energy occurs between system and surroundings.

Changes in the state or condition of a system arise from the action of external factors, such as temperature or pressure. Such changes, as the crystallization of a liquid by cooling, may generally be accomplished by a number of paths. If the transition is brought about by infinitesimally small displacements of the system from the original state so that its displacement from equilibrium at any moment is very small, the path is said to be *reversible*.

For example, the slow crystallization of a liquid at its melting point (T_m) approximates a reversible process. If the liquid is supercooled to a temperature below T_m, then a temperature (T_c) will be ultimately reached at which a spontaneous crystallization occurs. If the crystal is rewarmed liquid is not reformed at T_c. The forward path is not retraced, and the initial process is said to be irreversible.

The First Law of Thermodynamics is concerned with the relationship between changes in total energy (E), increments in thermal energy or heat (Q), and external work performed (W). All three quantities have the dimensions of energy and may be expressed as ergs, calories, and so on. The external work may take a number of forms, such as the expansion of a gas which drives a piston or the flow of electricity through an external circuit.

In differential form:

(B-1) $$dE = DQ - DW$$

or, for large changes: $\Delta E = Q - W$

Here DQ and DW are used to designate *inexact* differentials, the values of whose integrals depend upon the particular path of a transition, as well as the initial and final points.

In other words, the energy increment of a system which has undergone a transformation is equal to the difference between the increment of heat and the work done upon the surroundings. It should be particularly noted that dE, although it represents the difference between two inexact differentials, is itself an exact differential whose integral is independent of path. The energy change (ΔE) of a system occurring in the course of a transition is a function *only* of the initial and final states.

(B-2) $$\Delta E = \int_{\text{state 1}}^{\text{state 2}} dE$$

In general terms, the heat content of a system may be regarded as that fraction of its total energy which takes the form of thermal motion of its constituent atoms. An increase in the heat content of a system is always reflected by a rise in temperature. The relationship between the two quantities may be formalized in terms of the *heat capacities* at constant volume (C_v) and at constant pressure (C_p).

(B-3) $$C_v = \left(\frac{\partial Q}{\partial T}\right)_v$$

$$C_p = \left(\frac{\partial Q}{\partial T}\right)_p$$

The First Law is often stated in the alternative and equivalent form:

The total energy of a system plus its surroundings is constant and independent of any transformations which they may undergo. Energy can neither be created nor destroyed by chemical means.

It is often convenient to introduce a new quantity which is a measure of the total heat content of a system. The *enthalpy* (H) is defined in terms of E, pressure (P), and volume (V) by

(B-4) $$H = E + PV$$

Enthalpy, like energy, is a function only of the state of a system, and ΔH for a transformation depends only on the initial and final states, being independent of the path. The energy and enthalpy per unit mass of a chemical compound are well-defined quantities.

At constant pressure it may be shown that ΔH and Q are equivalent quantities.

(B-5) $$\Delta H = Q_p$$
and $$dH = C_p\, dT$$

At this point it is necessary to introduce a new basic parameter, *entropy* (S). For a *reversible* process

(B-6) $$dS = \frac{DQ}{T}$$

where T is the *absolute* temperature (°K), which is related to the centigrade scale by

(B-7) $$T_{\text{absolute}} = T_{\text{centigrade}} + 273.2$$

Entropy, like energy and enthalpy, is a function only of the final state of a system. The entropy change of a system undergoing a transition between two states is given by:

(B-8) $$\Delta S = \int_{\text{state 2}}^{\text{state 1}} \frac{dQ}{T}$$

where it is specified that the path between the two states consists of reversible steps only.

The significance of entropy lies in its close relation to the degree of order characteristic of the system. In general, the greater the extent to which the positions of the constituent atoms of a system can be predicted, the lower its entropy. A process which is accompanied by a loss of order, such as the melting of a crystal, whose atoms are arranged in a regular three-dimensional array, to form a liquid in which they are in a relatively random arrangement, results in an increase of entropy. The following general statements can be made as to the entropy change accompanying the transition of a system between two states.

If the transformation occurs by one or more reversible steps only, then the net entropy change of the system plus its surroundings is zero.

(B-9) $$\Delta S_{\text{system}} = \Delta S_{\text{surroundings}}$$

In this case the entropy change of the system is given by equation B-8.

If the transformation occurs by an irreversible process, then the net entropy change of the system plus its surroundings is greater than zero.

(B-10)
$$\Delta S_{system} \neq \Delta S_{surroundings}$$
$$\Delta S_{system} + \Delta S_{surroundings} > 0$$

For an irreversible process the entropy change of the system is no longer given by equation B-8. The above two statements may be regarded as one form of the Second Law of Thermodynamics.

The absolute entropy of a pure crystalline material at a given temperature may be computed from thermal data, by making use of the Third Law of Thermodynamics, which assigns a value of zero for the entropy of such a material at absolute zero. At constant pressure,

(B-11)
$$dH = C_p \, dT$$

and
$$dS = \frac{dH}{T} = \frac{C_p}{T} dT$$

The entropy of a pure crystalline material at temperature T is given by:

(B-12)
$$S = \int_0^{T_1} \frac{C_p}{T} dT + \frac{Q_1}{T_1}$$
$$+ \int_{T_1}^{T_2} \frac{C_p}{T} dT + \frac{Q_2}{T_2} + \cdots$$
$$+ \int_{T_{n-1}}^{T_n} \frac{C_p}{T} dT + \frac{Q_n}{T_n}$$
$$+ \int_{T_n}^{T} \frac{C_p}{T} dT$$

Here $T_1 \cdots, T_n$ represent the temperatures of the various phase transitions which the material undergoes and $Q_1 \cdots, Q_n$ are the heat changes accompanying each transition. Thus if the appropriate data are available, the entropy of a pure substance may be tabulated for standard conditions (25°C, atmospheric pressure). The entropy change to be expected for a hypothetical chemical reaction is equal to the algebraic difference of the summed entropies per mole of reactants and products. Thus for the reaction

(B-13)
$$aA + bB \rightarrow cC + dD$$
$$\Delta S = c\, S_c + d\, S_D - a\, S_A - b\, S_B$$

where S_c, etc., represent entropies per mole.

A central objective of thermodynamics is the prediction of the capability of a particular process to occur spontaneously, without a supply of external energy. The introduction of the concept of *free energy* is very useful in this connection. Two types of free energy have been defined. These are Helmholtz free energy (A) and Gibbs free energy (F).

(B-14)
$$A = E - TS$$
$$F = H - TS = A + PV$$

The significance of the Gibbs free energy may be concisely stated. *The Gibbs free energy represents that fraction of the total energy of a system which is available for external work.* For a particular transformation, the value of ΔF represents the maximum possible work which is obtainable by the process.

Like the enthalpy and the entropy, the free energy is a well-defined property, changes in which depend only upon the initial and final states of a system. It may be shown that, for a *spontaneously* occurring process, ΔF is always less than zero. Processes with a net positive value of ΔF do not occur spontaneously.

The free energy change to be expected for a hypothetical chemical reaction is equal to the algebraic difference between the summed free energies of reactants and products. From the sign of ΔF, as computed from entropic and enthalpic data, a prediction may be made as to the possibility of the reaction occurring spontaneously.

For a chemical reaction of the type of equation B-13, we have (at constant pressure):

(B-15)
$$\Delta F = \Delta F° + RT \ln \frac{[C]^c[D]^d}{[A]^a[B]^b}$$

Here $[C]$, $[D]$, etc., are the concentrations of products and reactants in moles per liter. ΔF represents the change in free energy accompanying the conversion of one mole of each reactant into products, *when the concentrations of all species have the specified values*. $\Delta F°$ is the free energy change when all species are in their *standard state* (1 mole/liter, atmospheric pressure).

When the system is at chemical equilibrium, no net interconversion of reactants and products occurs. In this case, $\Delta F = 0$, and no net reaction occurs in either direction. Under these conditions

(B-16)
$$\Delta F = 0$$
$$\Delta F° = -RT \ln \left(\frac{[C]^c[D]^d}{[A]^a[B]^b}\right)_{\text{equilibrium}}$$

or, from the definition of the equilibrium constant (K),

(B-17)
$$\Delta F° = -RT \ln K$$

REFERENCES

1. *Chemical Thermodynamics*, I. Klotz, Prentice-Hall, Englewood Cliffs (1950).
2. *Theoretical Chemistry*, S. Glasstone, Van Nostrand, Princeton (1944).
3. *The Nature of Thermodynamics*, P. Bridgman, Harvard University Press, Cambridge, Mass. (1941).

Appendix C

Synthesis of Polypeptides

In recent years synthetic methods have been developed which permit the conversion of α-amino acids into polymers of high molecular weight (10^3 to 10^6). Polymers of almost all the natural amino acids have been prepared, as well as a number of copolymers. While polymers composed of a few α-amino acids in a definite repeating sequence have been synthesized, it is not at present possible to prepare a polypeptide of high molecular weight which has a predetermined, nonrepeating sequence, such as occurs in natural proteins. Nevertheless, synthetic polypeptides have proven to be invaluable models for the fibrous class of protein. As is discussed in Chapter 5, physical studies upon the synthetic polymers were crucial in providing insight into the spatial organization of natural proteins.

The simplest method of initiating polymerization is undoubtedly the thermal process studied by Fox and co-workers. By heating amino acids in dry form at 200° for several hours, heterodisperse tars are formed, which appear to contain true polypeptides which can be hydrolyzed to amino acids. The wealth of side reactions makes this process unsuitable as a preparative technique, but it retains some paleobiological interest as a possible mechanism for the evolution of proteins under primal conditions on earth.

A number of procedures are available for converting amino acids to an active form which is capable of polymerization. To be satisfactory, any process must preserve the steric configuration of the amino acid and avoid any chemical alteration of the side chain. The most widely used method which satisfies these criteria involves the conversion of the α-amino acid to a cyclic anhydride, the *N-carboxyanhydride* (NCA) derivative.

A large number of procedures are available for preparing the NCA derivatives of the α-amino acids. One of the more commonly employed processes involves the conversion of the α-amino group to its N-carbobenzoxy derivative according to the following scheme:

(C-1)

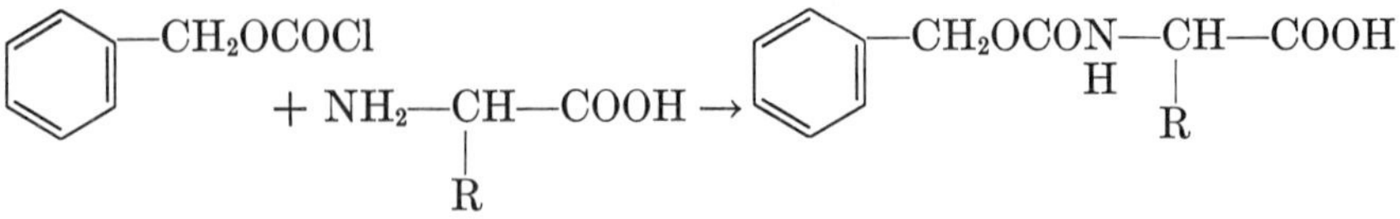

$$\text{C}_6\text{H}_5\text{—CH}_2\text{OCOCl} + \text{NH}_2\text{—CH(R)—COOH} \rightarrow \text{C}_6\text{H}_5\text{—CH}_2\text{OCON(H)—CH(R)—COOH}$$

The carboxyl group is converted to an acyl chloride by treatment with phosphorus pentachloride (PCl_5).

(C-2)

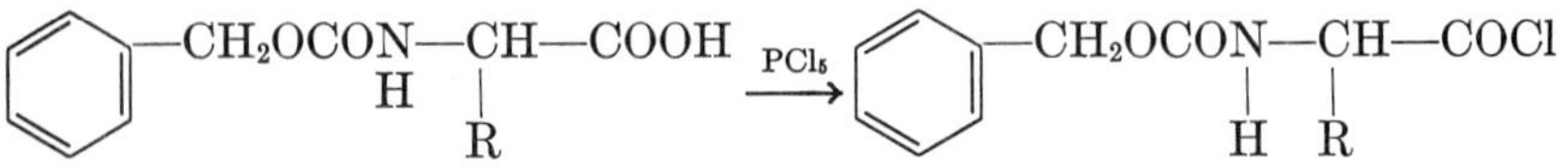

$$\text{C}_6\text{H}_5\text{—CH}_2\text{OCON(H)—CH(R)—COOH} \xrightarrow{PCl_5} \text{C}_6\text{H}_5\text{—CH}_2\text{OCON(H)—CH(R)—COCl}$$

Upon heating to 60° *in vacuo*, an intramolecular cyclization occurs to form the NCA, with the splitting off of benzyl chloride.

(C-3)

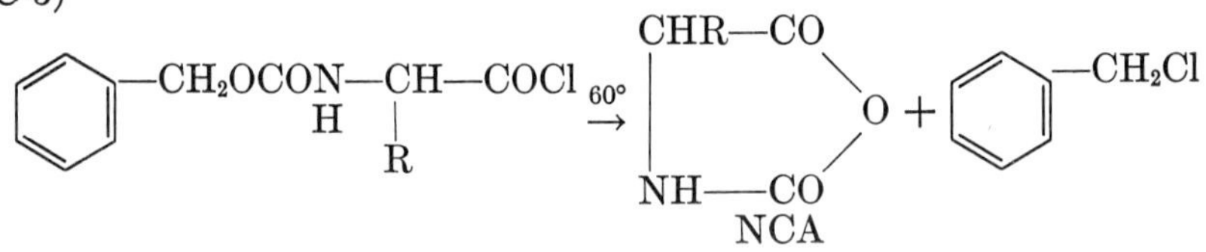

$$\text{C}_6\text{H}_5\text{—CH}_2\text{OCON(H)—CH(R)—COCl} \xrightarrow{60°} \underset{\text{NCA}}{\underset{|}{\overset{\text{CHR—CO}}{\overset{|}{}}}\!\!\!\!\!\!\!\!\!\!\!\!\!\!\!\!\!\text{>O}} + \text{C}_6\text{H}_5\text{—CH}_2\text{Cl}$$

(NCA: CHR—CO\O/NH—CO cyclic anhydride)

Polymerization of the NCA derivatives may be initiated either in bulk or in solution. The polymerization in bulk requires high temperatures and usually yields polymers of relatively low molecular weight. The reaction in solution is more readily controlled and is capable of producing polymers of high molecular weight.

A chemically inert solvent, such as dioxane or dimethylformamide, is normally used. The addition of a suitable initiator is necessary. The chemical nature and concentration of the initiator are important in determining the mechanism of polymerization and the yield and molecular size of the product. Among the most commonly used initiators are primary, secondary, and tertiary amines.

The polymerization of the cyclic anhydride is accompanied by the evolution of carbon dioxide. This provides a means for monitoring the reaction.

(C-4)

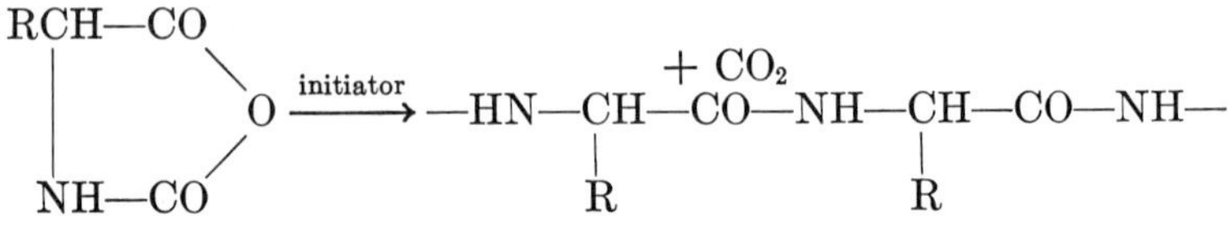

$$\underset{\text{NH—CO}}{\overset{\text{RCH—CO}}{\underset{|}{}\!\!\!\!\!\!\!\!\!\!\!\!\!\!\text{>O}}} \xrightarrow{\text{initiator}} \text{—HN—CH(R)—CO—NH—CH(R)—CO—NH—} + CO_2$$

APPENDIX C—SYNTHESIS OF POLYPEPTIDES

In the case of amino acids with an amino group in the side-chain, such as lysine, both amino groups are converted to their carbobenzoxy derivatives. Cyclization of the carboxyl and α-amino groups is carried out as before to form the NCA. After polymerization, the carbobenzoxy groups may be removed by treatment with phosphonium iodide (PH_4I).

(C-5)

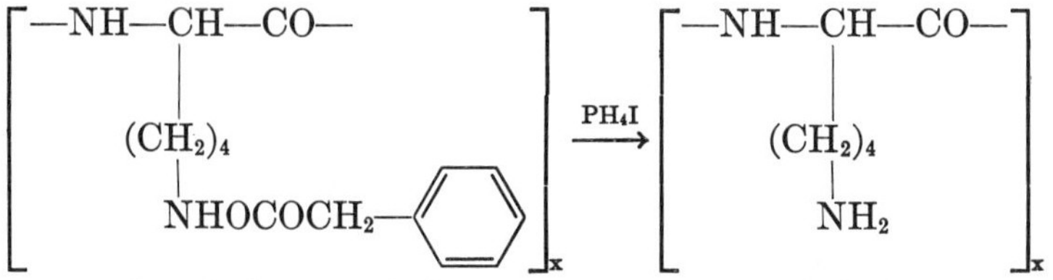

$$\begin{bmatrix} -NH-CH-CO- \\ | \\ (CH_2)_4 \\ | \\ NHOCOCH_2-\phi \end{bmatrix}_x \xrightarrow{PH_4I} \begin{bmatrix} -NH-CH-CO- \\ | \\ (CH_2)_4 \\ | \\ NH_2 \end{bmatrix}_x$$

polycarbobenzoxylysine   polylysine

If the side chain is acidic, as in the case of glutamic acid, both the α- and the side-chain carboxyl groups are capable of participating in peptide bond formation. If polymers involving only the α-carboxyl are desired, then it is necessary to protect the side-chain carboxyl by converting it to an ester.

In the case of glutamic acid, the side-chain may be selectively esterified by treatment with benzyl alcohol (ϕ-CH_2OH) in the presence of hydriodic acid (HI).

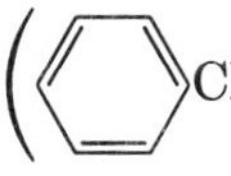

(C-6)

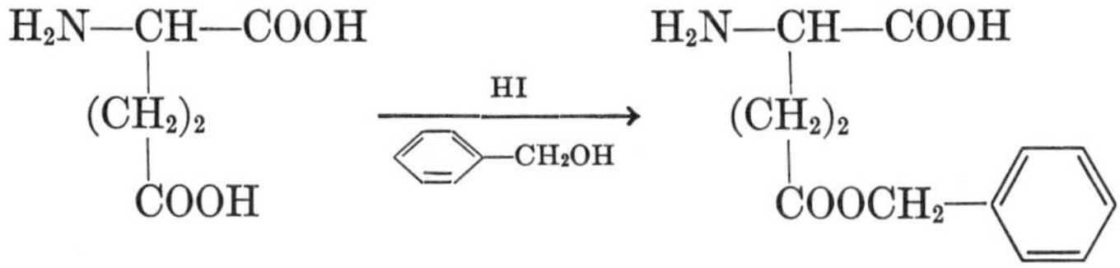

$$\begin{array}{c} H_2N-CH-COOH \\ | \\ (CH_2)_2 \\ | \\ COOH \end{array} \xrightarrow[\phi-CH_2OH]{HI} \begin{array}{c} H_2N-CH-COOH \\ | \\ (CH_2)_2 \\ | \\ COOCH_2-\phi \end{array}$$

Cyclization to the NCA derivative and polymerization of the anhydride may be carried out as before.

(C-7)

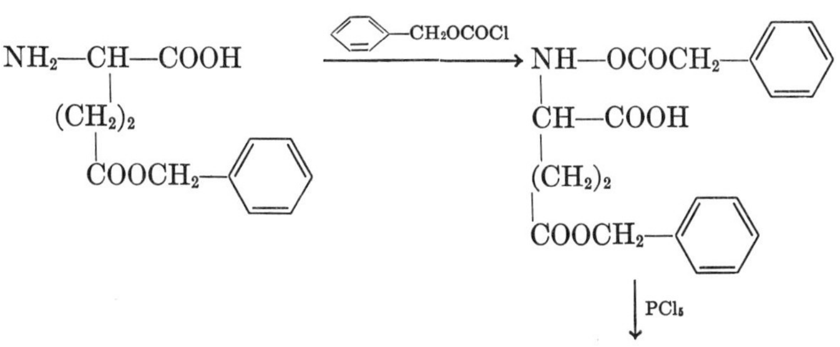

$$\begin{array}{c} NH_2-CH-COOH \\ | \\ (CH_2)_2 \\ | \\ COOCH_2-\phi \end{array} \xrightarrow{\phi-CH_2OCOCl} \begin{array}{c} NH-OCOCH_2-\phi \\ | \\ CH-COOH \\ | \\ (CH_2)_2 \\ | \\ COOCH_2-\phi \end{array}$$

$\downarrow PCl_5$

$$\left[\begin{array}{c}-\text{HN}-\text{CH}-\text{CO}-\\|\\(\text{CH}_2)_2\\|\\\text{COOCH}_2-\!\!\bigcirc\end{array}\right]_x \leftarrow \begin{array}{c}\text{OC}\text{O}\\||\\\text{HN}-\text{CH}-\text{CO}\\|\\(\text{CH}_2)_2\\|\\\text{COOCH}_2-\!\!\bigcirc\end{array}$$

poly-γ-benzylglutamic

After polymerization, the benzyl groups may be removed by treatment with phosphonium iodide.

(C-8) $$\left[\begin{array}{c}-\text{HN}-\text{CH}-\text{CO}-\\|\\(\text{CH}_2)_2\\|\\\text{COOCH}_2-\!\!\bigcirc\end{array}\right]_x \xrightarrow{\text{PH}_4\text{I}} \left[\begin{array}{c}-\text{HN}-\text{CH}-\text{CO}-\\|\\(\text{CH}_2)_2\\|\\\text{COOH}\end{array}\right]_x$$

poly-γ-benzylglutamic polyglutamic

Appendix D

Biological Oxidation and Reduction

In the broad sense, *oxidation* of an atom by a chemical process is said to occur when the atom loses electrons, and *reduction* is said to take place when it gains electrons. A simple example is the reduction of cupric ion in solution by metallic zinc.

(D-1) $$Zn + Cu^{++} \rightarrow Zn^{++} + Cu$$

or
$$\begin{cases} Zn \rightarrow Zn^{++} + 2\epsilon \\ Cu^{++} + 2\epsilon \rightarrow Cu \end{cases}$$

In this case, Zn is oxidized, while Cu^{++} is reduced.

In the oxidation of organic compounds, H^+ ions are often involved. For example, the oxidation of hydroquinone to quinone by ferric ion may be written:

(D-2) $$\underset{\substack{|\\OH}}{\overset{\substack{OH\\|}}{C_6H_4}} + 2Fe^{+++} \rightleftarrows \underset{\substack{\|\\O}}{\overset{\substack{O\\\|}}{C_6H_4}} + 2H^+ + 2Fe^{++}$$

In this process hydroquinone loses electrons to Fe^{+++} and is oxidized to quinone, with a concomitant loss of two H^+ ions. In the reverse process quinone acquires two electrons from two Fe^{++} ions and combines with two H^+ ions to reform hydroquinone.

If all the reactants of equation (2) are in the same vessel, the transport of electrons occurs by direct interchange between the interacting species.

If the $\overset{++}{\text{Fe}}\text{-}\overset{+++}{\text{Fe}}$ and quinone-hydroquinone systems are in different vessels, then reaction can still occur, provided that some arrangement exists for the flow of electrons from one system to the other. This may be accomplished by having each solution in contact with an inert electrode (Pt or Au) and the two electrodes joined by an external conductor (Fig. D-1).

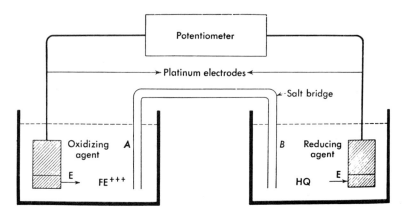

Fig. D-1. Schematic picture of an oxidation-reduction cell.

Ionic conductance between the solutions is achieved by a salt bridge. If a potentiometer is included in a series arrangement with the external connector, the electrical potential between the solutions may be measured.

If $\overset{+++}{\text{Fe}}$ ion is placed in vessel A and hydroquinone in vessel B (Fig. D-1), then electrons will be withdrawn from the A electrode according to the half-reaction.

(D-3) $\qquad 2\overset{+++}{\text{Fe}} + 2\epsilon \leftrightarrows 2\overset{++}{\text{Fe}}$

Simultaneously, the B electrode will gain electrons from hydroquinone (HQ) with the release of H^+ ions and the formation of quinone (Q).

(D-4) $\qquad HQ \leftrightarrows Q + 2H^+ + 2\epsilon$

Transport of electrons through the external circuit occurs until the reaction attains equilibrium, when the potential difference disappears.

The electrical potential developed by an oxidation-reduction system, such as the $\overset{+++}{\text{Fe}}\text{-}\overset{++}{\text{Fe}}$ mixture in contact with a Pt electrode, is generally expressed with reference to a standard *hydrogen electrode*. The latter consists of a platinized platinum electrode immersed in 1 M H^+ and under one

APPENDIX D—BIOLOGICAL OXIDATION AND REDUCTION

atmosphere of hydrogen pressure. The half reaction occurring at the hydrogen electrode is

(D-5) $$H_2 \leftrightarrows 2H^+ + 2\epsilon$$

The potential of an oxidation-reduction system is normally defined as the potential which would be developed if it were connected in series with a reference hydrogen electrode in an arrangement like that of Fig. D-1. This of course amounts to assigning arbitrarily a potential of zero to the hydrogen electrode.

In the most general case, if the half-reaction of such a system is of the type,

(D-6) $$ox + n\epsilon + xH^+ \leftrightarrows red$$

then the *redox* potential (E_h) of the system with respect to the hydrogen electrode is given by:

(D-7) $$E_h = E_o + \frac{RT}{nf} \ln \frac{[ox][H^+]^x}{[red]}$$

$$= E_o + \frac{RT}{nf} \ln \frac{[ox]}{[red]} + x \frac{RT}{nf} \ln [H^+]$$

If H^+ ion is not involved in the process ($x = 0$), then equation (7) is replaced by

(D-8) $$E_h = E_o + \frac{RT}{nf} \ln \frac{[ox]}{[red]}$$

Here R is the gas constant; T is the absolute temperature; f is the faraday (96,500 coulombs); and [ox], [red] represent the molar concentrations of the oxidized and reduced forms. The parameter E_o is the potential for standard conditions ([ox] = [red], [H^+] = 1).

In the case of the quinone-hydroquinone systems, $x = 2$ and $n = 2$, so that the expression for the potential becomes:

(D-9) $$E_h = E_o + \frac{RT}{2f} \ln \frac{[quinone]}{[hydroquinone]} + \frac{RT}{f} \ln [H^+]$$

For the Fe^{+++}-Fe^{++} system ($n = 1$, $x = 0$), we have:

(D-10) $$E_h = E_o + \frac{RT}{f} \ln \left[\frac{Fe^{+++}}{Fe^{++}} \right]$$

$$= 0.748 + \frac{8.32T}{96,500} \ln \left[\frac{Fe^{+++}}{Fe^{++}} \right]$$

at 25°C,

$$E_h = 0.748 + \frac{8.32 \times 298}{96{,}500} \ln \frac{[\text{Fe}^{+++}]}{[\text{Fe}^{++}]}$$

$$= 0.748 + 0.059 \log_{10} \frac{[\text{Fe}^{+++}]}{[\text{Fe}^{++}]}$$

When, as in the quinone-hydroquinone case, the process involves H^+ ions, it is customary to incorporate the term expressing the pH-dependence into E_o to yield a standard potential which is specified for a particular pH and temperature. Equation (7) is therefore replaced by:

(D-11) $$E_h = E_m + \frac{RT}{nf} \ln \frac{[\text{ox}]}{[\text{red}]}$$

where $$E_m = E_o + x \frac{RT}{nf} \ln [H^+]$$

The standard redox potentials of many organic compounds of biological interest have been determined (Table D-1). The numerical value of the potential is a direct measure of the oxidizing power of the system. The more positive the value, the greater is the oxidizing power. A system of more positive E_m will oxidize (or be reduced by) a system of less positive E_m.

More precisely, in order to predict which of two such systems will undergo a net oxidation, their respective values of E_h may be computed from equation (D-11). The system of more positive E_h will be reduced and the system of less positive E_h will undergo oxidation until equilibrium is attained.

If all species (A, A^+, B, B^+) are present, for the reaction

(D-12) $$A + B^+ \to A^+ + B$$

to proceed to a finite extent, the value of $\Delta E_h (= E_B - E_A)$ must be positive. If ΔE_h is negative, the reaction will proceed in the reverse direction, that is

(D-13) $$A^+ + B \to A + B^+$$

If ΔE_h is zero, then the system is at equilibrium, and no net reaction occurs.

The potential difference is directly related to the free energy change (Appendix B) for the process.

(D-14) $$\Delta F = -nf \, \Delta E_h$$

Indeed, the two approaches are basically equivalent.

Table D-1. Standard Redox Potentials

System	E_m	pH
Cytochrome c	+0.27	7.0
Ascorbic acid	+0.136	4.6
Flavoprotein	−0.059	7.0
DPN	−0.325	7.4
Riboflavin	−0.208	7.0

Index

A and B chains (insulin), 64, 65, 66, 443, 446
acetyl coenzyme A, 434
acetylcholinesterase, 210, 238, 239, 240, 241, 242
acid hydrolysis, 51, 52
acidic amino acids, 39
aconitase, 434
acrylamide, 54
actin, 171, 172, 173, 174, 175, 176, 177, 178
activators, 214
actomyosin, 175
acyl adenylate, 333
adaptor, 308
adenine, 247, 250, 251
adenosine-5′-diphosphate, 260, 438, 439
adenosine-5′-triphosphate, 173, 174, 175, 260, 398, 399, 400, 401, 402, 413, 414, 415, 438, 439
adenovirus, 350
adrenocorticotropin, 72
adsorption chromatography, 55
aldolase, 429
aldose, 382
aliphatic side-chains, 33
amino acids, 26-42
α-amino group, 26
α-amylase, 210
amylopectin, 391, 392
amylose, 391
antibodies, 178, 179, 180, 181, 182, 183, 184, 185, 186, 187, 188
antigens, 178, 179, 180, 181, 182, 183, 184, 185, 186, 187
Archibald, 114
Archimedes, 112
Arnon, 400, 402
aromatic side-chains, 37
Arrhenius, 228, 229
asymmetric center, 31
Avery, 299

bacteria, 17
bacteriophage, 347, 361-381
basic amino acids, 39
Benzer, 378
biopolymers, 2
Bragg equation, 145
Brownian motion, 85, 106

Calvin, 403
Canfield, 69
capsid, 349
capsomere, 349
carbohydrates, 383-411
α-carboxyl group, 26
carboxylase, 210
carboxypeptidase, 53, 61
cardiac muscle, 171
cell, 3, 4, 5
cellobiose, 389
cellulose, 2, 394
centromeres, 10
centrosomes, 4
cesium chloride, 287, 288
Champe, 378
chiasmata, 13
Chlorella pyrenoidosa, 403
chlorophyll, 397, 398, 399, 400, 401, 402, 403
chloroplasts, 4, 397, 398, 399
chromatids, 10
chromatin, 5
chromatography, 48, 55, 56, 57, 58, 59
chromosomes, 5, 9, 10, 11, 12, 266
chymotrypsin, 52, 232, 233, 234, 235, 236, 237
chymotrypsinogen, 233, 234, 235
cis-trans effect, 22
cistron, 23, 374, 375, 376, 377, 378
citrate, 434
coenzyme A, 419, 420
collagen, 163-170
concentration gradient, 95

condensing enzyme, 434
β-conformation, 141, 142
conjugated proteins, 44
conjugation, 18
COOH-terminal group, 49, 50, 61
Corey, 131, 141
corticotropin, 53, 67
coulombic interaction, 133, 134
cyclic photophosphorylation, 400, 401
cysteine, 34, 35
cystine, 34, 35, 43, 46
cytochrome a, 439
cytochrome b, 437, 439
cytochrome c, 69, 439
cytochromes, 422, 424
cytoplasm, 4
cytosine, 247, 253, 254

Debye-Huckel constant, 122
denaturation, 157, 170, 280, 281, 282, 283, 284, 285, 286, 321
density gradient ultracentrifugation, 287, 288
deoxyribofuranose, 257
deoxyribonuclease, 210, 268, 271
deoxyribonucleic acid, 3, 266-305
deoxyribose, 247
deuterium exchange, 151
dextro-rotation, 30
diethylaminoethyl cellulose, 57
diffusion, 105
dihydroxyacetone, 383
dihydroxyacetone phosphate, 429
diisopropylfluorophosphate, 231, 235
dimethylaminoethyl acetate, 239
1-dimethylaminonaphthalene-5-sulfonyl chloride, 105
dinitrophenyl reagent, 49, 59
Dintzis, 341, 342, 343, 344
2,3-diphosphoglyceric acid, 431
diphosphopyridine nucleotide, 420, 436, 437, 438, 439
diphosphothiamine, 436
diploid cells, 10
dipolar ion, 26
dipole moment, 134, 135, 136
dispersion forces, 136
Dixon, 232
DNA-polymerase, 261, 271, 292
dominant gene, 6
Doty, 147, 166, 291

Drosophila melanogaster, 14
Drude, 150

eclipse period, 363
Einstein, 101
electron microscopy, 78
electron transport particles, 441
electrophoresis, 77, 120-127
ellipsoid of revolution, 92
enantiomorphs, 30
endoplasmic reticulum, 5
energy, 447, 448, 449
enolase, 431
enthalpy, 448, 449
entropy, 155, 449, 450, 451
enzyme induction, 338, 339, 340
enzyme repression, 338
enzymes, 207-246
erythrocytes, 197
Escherichia coli, 18, 73
excited lifetime, 103

F agent, 19
fibrin, 188
fibrinogen, 188, 189, 190, 191, 192
fibrinopeptide, 192
fibrous long spacing, 165
fibrous proteins, 148
Fick's laws, 106, 107
fingerprinting, 61, 73
flavin adenine dinucleotide, 422
flavin coenzymes, 420
flavin mononucleotide, 422
Flory-Mandelkern, 102
fluorescence polarization, 102, 103, 104, 105
Formvar, 78
Fraenkel-Conrat, 67
free energy, 412, 413, 414, 451, 452
frictional coefficient, 92, 93, 107
fructose, 414
fructose-1,6-diphosphate, 429
fructose-6-phosphate, 428, 429
fumarase, 437
fumaric acid, 437
furanose, 387

gametes, 10
genes, 7, 8, 9, 19, 70
genetic code, 308, 336, 337
genetic map, 15

genetics, 5
globin, 198
globular proteins, 133, 148
γ-globulin, 178-188
glucose, 432
glucose-1-phosphate, 414, 427
glucose-6-phosphate, 427
glucoside, 387
glyceraldehyde, 383
glyceraldehyde-3-phosphate, 429, 430
glyceraldehyde phosphate dehydrogenase, 430
glycogen, 2, 393, 394, 409, 410
glycolysis, 415, 416, 427-433
glycosidase, 210
glycosidic bond, 247, 258
glycyl-leucine dipeptidase, 214
Goldberg, 187
Golgi bodies, 5
grana, 399
guanine, 247, 250, 251

Hanson, 176, 177
haploid cells, 10, 17
hapten group, 180
α-helix, 132, 138, 139, 140, 141
heme, 198
hemoglobin, 3, 44, 73, 197, 198, 199, 200, 201, 202, 341, 342, 343, 344
Henry equation, 125
herpes virus, 350
heterozygote, 6, 21
hexokinase, 427
Hfr, 19
histidine, 39
homozygote, 6
Hotchkiss, 301
Huxley, 176, 177
hyaluronic acid, 394
hybridization, 290, 291, 292
hydration, 94
hydrodynamic methods, 90
hydrogen bond, 46, 137, 138
hydrogen ion titration curve, 121, 122
hydrophobic bonds, 158
5-hydroxymethyl cytosine, 269, 364, 365
hydroxyproline, 37
hypochromism, 278, 279, 317
hypoxanthine, 258

ichthyocol, 166

icosahedron, 347
imidazole, 39, 237
α-imino acids, 37
induction, 381
influenza virus, 350
inhibitors, 214, 221-225
inosine, 258
insulin, 62, 63, 64, 65, 66, 71, 72
intrinsic dissociation constant, 40
intrinsic pK, 121
intrinsic viscosity, 108
ion-exchange, 55
ionophoresis, 53
isocitrate, 434
isocitric dehydrogenase, 435
isoelectric point, 54

α-ketoglutarate, 435, 436
ketose, 383
Kornberg, 271, 292

lactic acid, 431, 432
lactic dehydrogenase, 431
lactose, 389
laevo-rotation, 30
Landsteiner, 180
Lane, 303
Lathyrus odoratus, 13
Lederberg, 18
Lehninger, 440, 441
light scattering, 83-90
α-lipoic acid, 425, 436
lipoic acid dehydrogenase, 436
lysogenic state, 379
lysogenization, 379, 380
lysosome, 4
lysozyme, 69

MacLeod, 299
malic acid, 437
malic dehydrogenase, 437
maltose, 389
Marmur, 291, 303
Matthaei, 334, 336
McCarty, 299
meiosis, 10
β-melanocyte stimulating hormone, 67
Mendel, 5
Meselson, 297, 298, 299, 371
messenger RNA, 308, 331
methylamine, 240

Michaelis constant, 219
Michaelis-Menten, 216-221
minimum molecular weight, 49
missense mutation, 377, 378
mitochondria, 4, 417, 432, 433
mitosis, 10
mobility, 124
Monod, 339, 340
monosaccharides, 2, 383-389
Moore, 55
Morgan, 14
mutagenic agents, 17
mutants, 358, 359
mutation, 15, 71, 74
muton, 24
myoglobin, 204, 205
myosin, 171-178

N-carboxyanhydride, 453, 454, 455
negative staining, 80
neostigmine, 238
Neurath, 232
NH_2-terminal groups, 49, 50, 60, 61
ninhydrin, 38, 56, 59
Nirenberg, 334, 336
noncyclic photophosphorylation, 402, 403
nonsense mutations, 376
nucleolus, 5
nucleoside diphosphokinase, 437
nucleosides, 256, 257, 258, 259
nucleotides, 2, 247-265
nucleus, 4

Ochoa, 334, 336
operator, 339, 340
operon, 339, 340
optical activity, 27
optical rotation, 148-151, 317
orf virus, 350
osmotic pressure, 81
ova, 10
oxalacetate, 434, 437
oxalosuccinate, 435
oxidation, 457-461
oxidative phosphorylation, 417, 418, 438-441

paper chromatography, 58, 59
paper electrophoresis, 54
parent gelatin, 171

partial specific volume, 95, 112
Pauling, 131, 141, 184
pepsin, 52
peptidase, 210
peptide bond, 43, 44, 45
performic acid, 34
phenotype, 6
phenylisothiocyanate, 60, 61
phenylthiocarbamyl derivative, 60, 61
phenylthiohydantoin derivative, 61
phosphodiester bond, 250
2-phosphoenol pyruvate, 431
phosphoesterase, 210
phosphoglucomutase, 427
2-phosphoglyceric acid, 430, 431
3-phosphoglyceric acid, 406, 407, 430
phosphoglyceryl kinase, 430
phosphohexoisomerase, 427
phosphorolysis, 323
phosphorylase, 410, 427
phosphotungstic acid, 79
photosynthesis, 396-408
physostigmine, 239
Pisum sativum, 5
platelets, 188
pleated sheet, 141
pneumococcus, 300
poly-γ-benzyl glutamate, 148, 154
polyelectrolyte, 315
polyglutamic acid, 153
polynucleotide phosphorylase, 213, 214, 261, 322, 323
polyoma virus, 350
polypeptides, 43
polyribonucleotides, 322-326
polysaccharides, 2, 389-396, 409, 410
polysomes, 332
porphyrin, 198
Porter, 187, 188
precipitin reaction, 180
primary structure, 44
primers, 215, 322, 323
proline, 37
prophage, 379
protein synthesis, 329-344
proteins, 3
prothrombin, 188
purine, 247
pyranose, 386
pyrimidine, 247
pyrophosphatase, 210

pyruvate kinase, 431
pyruvic acid, 431, 432

radius of gyration, 86, 87
random coils, 90, 91, 132
Rayleigh, Lord, 84
recessive gene, 7
recombination, 15, 370-379
recon, 24
reconstitution of TMV, 357
reduced intensity, 85
reduction, 457-461
regulator gene, 339, 340
renaturation, 286, 287, 288, 289, 290, 291, 292
replication, 267, 296, 297, 298, 299
repressor, 339, 340
respiratory chain, 439
reticulocytes, 341, 342, 343, 344
reticulo-endothelial system, 179
R_f, 58
riboflavin, 422
ribofuranose, 257
ribonuclease, 67, 210, 242, 243, 244, 245, 311, 312
ribonucleic acid, 3, 307-346
ribose, 247
ribosomal RNA, 310, 314
ribosomes, 5, 309, 330
ribulose-1,5-diphosphate, 406, 407, 408
ribulose-5-phosphate, 407, 408
Richards, 243
rII mutants, 369-379
RNA-polymerase, 307, 326, 327, 328, 329
rotational relaxation time, 94, 103

Sanger, 49, 62, 72
sarcomeres, 176
schlieren technique, 96, 98, 99, 115
Schwann, 3
secondary structure, 131
sedimentation, 77, 109-120
sedimentation coefficient, 111
sedimentation equilibrium, 111, 114, 115
sedimentation velocity, 109, 111, 112, 113
segregation, 6
selenomethionine, 41
Sephadex, 57
serum albumin, 194, 195, 196, 197

sickle-cell anemia, 73
side-chain, 26
Simha, 101
Sinsheimer, 279
skeletal muscle, 171
smooth muscle, 171
snake venom phosphodiesterase, 268, 311
soluble RNA, 309
specific activity, 214
specific rotation, 27, 149
specific viscosity, 100
spermatozoa, 10
spleen phosphodiesterase, 259, 311
Stahl, 297, 298, 299
starch, 2, 389
Stein, 55
Stent, 367, 368
Stokes' law, 93
structural gene, 339, 340
Sturtevant, 236, 237
subtilisin, 52
succinate, 435, 436, 437
succinic dehydrogenase, 437
sucrose, 389, 414
sucrose gradient sedimentation, 118, 119, 120
synthesis of polypeptides, 453-456

Tatum, 18
temperate bacteriophages, 379, 380, 381
template, 308
tertiary structure, 131, 133, 157
thermodynamics, 447-452
thiamine diphosphate, 425, 426
thrombin, 188
thymine, 247
Tiselius, 108
tobacco mosaic virus, 67, 80, 81, 352-360
transcription, 267
transduction, 381
transfer RNA, 308, 321, 332, 333, 334
transforming principle, 299-304
transmutation, 22
tricarboxylic acid cycle, 417, 418, 433-438
triphosphopyridine nucleotide, 398, 399, 400, 402, 420, 435
tropocollagen, 166
trypsin, 52, 229, 230, 231, 232
trypsinogen, 229, 230

tryptophan, 51, 52
tryptophan synthetase, 73, 338
turnover number, 214

ultracentrifugation, 109-120
uracil, 247, 254
urea, 210
uridine diphosphate D-glucose, 408, 409, 410

vaccinia virus, 351
Van't Hoff, 228
vegetative state, 363
Virchow, 3

virial coefficient, 82
viruses, 347-382
viscosity, 99, 100, 101, 102

Watson-Crick model, 272-276
Weigle, 371

xanthine, 258
X-ray diffraction, 76, 143, 144, 145, 146, 319

Yanofsky, 73

Zimm grid, 87, 166